Sustainable Plant Nutrition in a Changing World

This book series aims to provide new insight and improve understanding of plant nutrition, a crucial component in plant production and food security. Plant nutrition encompasses the study of plant growth and metabolism, external supply of necessary chemical compounds and the biochemistry of various nutrients in plants. This field of study has grown in light of recent scientific advances and climate change and has expanded to include the impact of climate change, pollution, nanomaterials, biofortification and genomics on the nutrition of plants.

The component volumes are written by leading scholars from around the globe and contain the latest research and applications. Each volume is fully illustrated with in-depth references for further reading. This series will be of interest to all scientists, academics, students and professionals in the plant sciences and agriculture, as well as biotechnology, ecology and biochemistry.

More information about this series at http://link.springer.com/series/16351

Qaisar Mahmood
Editor

Sustainable Plant Nutrition under Contaminated Environments

Springer

Editor
Qaisar Mahmood
Department of Environmental Sciences
COMSATS University Islamabad
Abbottabad, Pakistan

Department of Biology, College of Science
University of Bahrain
Sakhir, Bahrain

ISSN 2662-2394 ISSN 2662-2408 (electronic)
Sustainable Plant Nutrition in a Changing World
ISBN 978-3-030-91498-1 ISBN 978-3-030-91499-8 (eBook)
https://doi.org/10.1007/978-3-030-91499-8

This Springer imprint is published by the registered company Springer Nature Switzerland AG
The registered company address is: Gewerbestrasse 11, 6330 Cham, Switzerland

To all disease-affected poor humans of the world.

Preface

With the growth of various industries, many pollutants have been introduced in the environment. Likewise, decreasing water availability for agricultural activity in many parts of developing countries has compelled farmers to use wastewater irrigation. In advanced agricultural system, farmers are adapting various strategies to achieve higher yield to sustain crop productivity. Consequent to introduction of contaminants in the environment, soil pollutants have become a burning issue. Selection of disease-resistant, high-yielding crop varieties and extensive fertilizer applications are quite common among the farming community. Unplanned and uncontrolled discharge of atmospheric contaminants is threatening crop health and yield. There are certain qualitative and quantitative threats to agricultural crops due to atmospheric primary and secondary pollutants such as particulates, sulfur dioxide (SO_2), oxides of nitrogen (NOx), and ozone (O_3). The detrimental effects of these pollutants depend on their concentration and dose thus resulting in acute or chronic toxic effects on quality and yield of a certain crop. A plant may undergo biochemical and physiological modifications at the cellular level as well as plant level as a response to unwanted absorption of gases and chemicals. The physiological processes, most importantly photosynthetic CO_2 fixation and energy metabolism, suffer negatively by air pollutants. In brief, food security is threatened by air pollution.

Knowledge of the sources, availability, effects of heavy metals in contaminated soils is essentially important to reduce the associated risks and enhance the safe supply of food utilizing available management options. Chapter 2 reported the mitigation processes of heavy metals in soils using organic manures. The mechanism of heavy metals remediation in manure amended soils has been discussed. There is a general consensus on the effects of manure amendments for the improvement of physicochemical and biological properties of soils. Adding composted manures to soil particularly changes the bioavailability and mobility of heavy metals, and reduces the harmful effects of polluted soils via several processes. Inorganic fertilizers are efficient source of nutrients for crop plants because the addition of inorganic fertilizers into soil makes nutrients to release more quickly and be readily absorbed by plants. In addition, the influence of inorganic fertilizers in the modulation of soil components such as microbial biomass carbon (MBC), soil community structures, and enzymatic functions is substantial. Due to these properties, inorganic fertilizers are a better alternative to traditional fertilizers, which pose the risk of groundwater contamination via nitrate leaching. Furthermore, the addition of inorganic nutrients maintains soil nutrients profile, water status, nutrients mobility in soil, permeability of soil, and physical and chemical properties of soil, which change with a massive addition of traditional fertilizers.

Soil health relates directly to plant and human health. Soil pollution is mainly caused by the presence of excessive quantities of essential and/or other toxic elements/chemicals, either by natural or anthropogenic activities. Plants cultivated on contaminated soils show a reduction in growth, nutrients uptake, and yield. Concentrations exceeding optimal levels in plants have both direct and indirect adverse effects on plants. Contamination by toxic elements not only reduces plant growth and development but also affects other activities such as photosynthesis and plant mineral nutrition, and decreases the activities of certain enzymes. A common practice in water-stressed regions since early times is irrigation of crops with untreated wastewater. The production of huge volumes and nutrients availability, such as phosphorus and nitrogen, are the other two reasons to use untreated wastewater for irrigation purposes.

Injudicious use of fertilizer may cause accumulation of various substances like nitrate in edible parts of farm produce, which may reach toxic level, thereby raising environmental issues and increasing concerns about supply of hygienic food items. Cultural practices, farming policies, and extensive application of fertilizers may contribute to environmental risks. To combat, Zn deficiency in the human diet and for the proper function of the plants, various strategies have been devised. There is soil/foliar application of Zn using chemical fertilizers of which solubility in soil and its bioavailability is the main issue. To combat this problem, an economical and eco-friendly strategy is required to enhance Zn availability to crop plants.

In the past few decades, consequent contaminations of environment and soil have become a burning issue. Among wild and cultivated flora, medicinal plants have well-known pharmaceutical applications. They attribute their effectiveness against life-threatening ailments to their biologically active organic compounds and inorganic elemental composition. Among elements present in medicinal plants, macro, micro, beneficial, trace, and rare earth elements are found to be influenced by environmental abiotic (temperature, drought, salinity, and heavy metal stress) and biotic (herbivores and pathogens) stress and physicochemical properties of soil. Climate smart agriculture (CSA) may help to resolve these challenges. Climate smart agriculture consists of spreading and updating awareness among local farmers to adapt farming practices to meet the challenges of agriculture, including efforts where adoption of modified farming practices generates positive outcomes through risk-management strategies and support resource sustainability/productivity. The main significant benefits of CSA may be to increase agricultural productivity, to adapt the resilience mechanisms for climate change, and to mitigate greenhouse emissions from agricultural sector completely.

The universal existence of PCBs can be observed in every aspect of the natural habitat involving flora, fauna, soil, air, and water, and its definite origin in the ecosystem is yet to be revealed. Due to their lipophilic nature, PCBs can easily be part of the food chain. In addition to their lipophilic nature, PCBs are environmentally stable and can resist very high temperatures; therefore, they can survive for a long time in the environment. A lot of focus has been put on different uptake and driving mechanisms behind organic pollutants, and efforts are being put to discover more effects at a molecular level. These are commonly found organic pollutants that are responsible for disrupting various plant parts and functions, such as membrane, nuclear functions, uptake functions, and other anomalies. Under abiotic stress, production of reactive oxygen species (ROS) such as hydrogen peroxide or superoxide causes harmful effects on the survival of rhizobacteria, which play an important role in the growth and yield of various crop plants. To cope with ROS stress, rhizobacteria activate certain regulons that are controlled by the OxyR, PerR, or PerR-like homolog and SoxR transcription factors. All these transcription factors sense peroxides during the oxidation of iron, manganese, zinc, nickel, etc. moieties and stimulate overlapping sets of proteins which defend their weak metalloenzymes.

Salinity is a major threat to global agricultural production. Salt stress causes accumulation of higher concentrations of Na^+ in the cytoplasm, which has deleterious effects on cell metabolism. For genetic manipulation of plant salt tolerance, it is necessary to understand the mechanism of Na^+ sensing, influx, efflux, and compartmentalization. Na^+ influx through the root cell plasma membrane mainly takes place via nonselective cation channels while its compartmentalization into vacuoles occurs through antiport proteins. Similarly, the efflux of Na^+ from the plant roots is catalyzed by the salt overly sensitive 1 (SOS1) protein. Bacteria in rhizosphere along with plant roots create an entirely unique interaction between each other and make the soil alive. Rhizobacteria through multiple metabolic reactions produce different chemicals in different scenarios, which help plants with efficient nutrient acquisition. Rhizobacteria make N, P, and Fe available to plant via nitrogen fixation, phosphate solubilization, and siderophore-Fe chelation. Rhizobacteria communicate with each other via chemical signals (Acyle homoserine lactone) and assist plants to not only improve their vigor but also cope with pollution stress. These chemical signals regulate bacterial assistance in nutrient acquisition and phytoremediation. The current edited book will provide deep insights into metabolism of various pollutants in the environment with special reference to interference with plant nutrition; the physiological aspects of plant nutrition will be specially discussed in this book which will enhance the current knowledge on effects of pollutants on plant growth and physiology.

Being editor of this book, I am thankful to all contributors for their interests, significant contributions, and cooperation that eventually made the present work possible. Many thanks are due to all the well-wishers, teachers, seniors, research students, and affectionate family members. Without their unending support, motivation, and encouragements, the present grueling task would have never been accomplished. I would like to offer my sincere thanks to Mrs. Sowmya Thodar, Professor Hassan Elramady, Eric Stannard, Kate Lazaro, and Nicholas DiBenedetto and their team at Springer for their continuous support which made our efforts successful.

Abbottabad, Pakistan Qaisar Mahmood

Contents

Chapter 1
Effects of Air Contamination on Agriculture

Romana Khan, Alireza Noorpoor, and Abdol Ghaffar Ebadi

1.1 Introduction

The importance of plants for existence of life on earth is undeniable. They maintain the atmosphere; they are source of oxygen, fix CO_2, and absorb other pollutants besides providing food, shelter, medicine, and many other benefits. Plants are crucial for environment, yet plants need a certain environment to live. Agriculture crops are plants grown to feed the rapidly expanding population; hence, the success of cultivated crops lies in their decent yields. Crops grow well under certain favorable conditions. Their life cycle from sowing to harvesting depends on environmental conditions such as sunlight, temperature, precipitation, soil characteristics, biotic parameters, and even the shape of the landscape of the area they are adapted to. Any change in these parameters causes stress to the crop; for instance, stress of crops due to rainfall variation from one year to the next year is a well-observed phenomenon. Same is utterly true for the air live in, which is a mixture of gases, few chemicals, and particulate matter. The composition of air is variable yet principally consists of abundant components such as oxygen and compounds of nitrogen and less-abundant components such as carbon dioxide, hydrocarbons, and oxides of sulfur. As all these components originate from both natural and anthropogenic sources, it can be concluded that the air was never free from contaminants and major air pollutants always existed in the natural atmosphere. The natural sources of air pollutants vary from

R. Khan (✉)
Department of Environmental Sciences, COMSATS University Islamabad, Abbottabad, Pakistan
e-mail: romanakhan@cuiatd.edu.pk

A. Noorpoor
School of Environment, College Engineering, University of Tehran, Tehran, Iran

A. G. Ebadi
Department of Agriculture, Jouybar Branch, Islamic Azad University, Jouybar, Iran

Q. Mahmood (ed.), *Sustainable Plant Nutrition under Contaminated Environments*, Sustainable Plant Nutrition in a Changing World,
https://doi.org/10.1007/978-3-030-91499-8_1

volcanic eruptions, strong winds or storms, drought to volatile organic compounds, and pollen emission from plants. As crops evolution took place in the presence of these background concentrations, they have some adaptability. Man-made pollutants are generated from stationary sources such as industries, power plants, and burning of fossil fuels and from mobile sources such as automobile emissions. There exists a threshold concentration of atmospheric components to which crops gets adapted and beyond which its tolerance is affected leading to stress, competition, and death. Economic growth, industrialization, and associated energy demands are contaminating atmosphere, with pollutants concentrations beyond its capacity to tackle. Agricultural losses due to air pollution are measured by various methods such as dose–response equation, field comparison test, regional comparison, and models. The dose–response equation is taken from field experimental data to relate losses and amount of pollutants. Field comparison test is performed to examine series of crop yield change when exposed to specific pollutant dose. Regional comparison is accomplished by comparing the same crop yields of two regions, one near a pollution source and one that is not exposed to pollutants having similar ecological conditions. Lastly, the same can be estimated by using models. According to the figures published by the World Health Organization, seven million people died due to air pollution exposure in 2012. As the direct human health effect of air pollution is so intense, less attention is given to its impacts on staple crops yields and.

1.2 General Facts About Air Pollution

Air pollution is the introduction of chemicals, particulate matter, or biological materials into the atmosphere in concentrations that can cause harm or discomfort to humans or other living beings or damage the natural or built materials. The term air pollution is also implied to describe a situation in which new substances or energy forms which were not present before reach the atmosphere. The term may also include a condition when some substances that are normally present in upper atmosphere (stratosphere) are produced in the troposphere, for instance, ozone. Similarly, air pollutant can be an otherwise normal air component whose concentration is above the normal range. More than 3000 substances can be considered air pollutants. Following is the list of some common air pollutants that exist in air either in gas form or as suspended particles.

Air pollutants include compounds of carbon (CO_2, CO, CH_4), nitrogen (NO_2, NO, N_2O_4, NH_3, NH_4^+), sulfur (SO_2, H_2S, COS, CS_2), O_3, C_6H_6 vapors, Hg, volatile phenols, Cl_2, PM_{10} and $PM_{2.5}$ particulate matters, heavy metals with toxic effect (Pb, Ni, Cd, As), and polycyclic aromatic hydrocarbons (PAHs). These pollutants can be classified into acidifying agents – sulfur dioxide (SO_2), nitrogen oxides (NO_x), ammonia (NH_3) fluoride and Cl_2, hydrogen chloride (HCl) – and oxidizing agents – carbon monoxide (CO), PAN (peroxyacetylnitrate-$CH_3CO{\bullet}O_2{\bullet}NO_2$), and ozone ($O_3$). Table 1.1 enlists the response of crops to air pollutants.

Table 1.1 Summary of agriculture crops behavior toward air pollutants

Pollutant	Sensitive	Tolerant
Sulfur dioxide (SO_2)	Alfalfa, barley, buckwheat, cotton, Douglas fir, soybeans, wheat, clover, oats, pumpkin, radish, rhubarb, spinach, squash, Swiss chard, and tobacco	Asparagus, cabbage, celery, corn, onion, and potato
Nitrogen dioxide	Apple, barley, beans, clover, radish, raspberry, and soybean	Alfalfa, beet, carrot, corn, cucumber, eggplant, onion, peach, rhubarb, and tomato
Fluorides	Apricot, barley (young), blueberry, peach (fruit), gladiolus, grape, plum, prune, sweet corn, and tulip	Alfalfa, asparagus, bean (snap), cabbage, carrot, cauliflower, celery, cucumber, eggplant, pea, pear, pepper, potato, squash, tobacco and wheat
Peroxyacetyl nitrate (PAN)	Bean, celery, lettuce, pepper, spinach, and tomato	Broccoli, cabbage, cauliflower, cucumber, onion, radish, squash
Ozone	White bean, peanut, sweet potato, turnip, wheat, soybean, cucumber, grape, green bean, lettuce, onion, potato, radish, rutabagas, spinach, sweet corn, tobacco and tomato, rice, barley, oilseed, mustard, alfalfa, olive	Endive, oat, broccoli, pear, and apricot

The air pollutants can be classified as primary or secondary depending upon their sources. Primary pollutants are directly emitted from their source, such as burning of fossil fuels, vehicular emissions, power plants, and industries. The above shared list comprises primary pollutants. Secondary pollutants, on the other hand, are certain photo-oxidants generated by chemical reaction of primary pollutants among themselves or with the other atmospheric components under certain atmospheric conditions (sunlight, humidity, presence of particulate matter, etc.). Two examples whose ambient concentrations can be high enough to cause problems are PAN (peroxyacetyl nitrate) and ozone.

1.3 Effect of Primary Pollutants on Vegetation

1.3.1 Sulfur Dioxide (SO_2)

The main culprit of sulfur dioxide (SO_2) production is burning of fossil fuels for industrial or domestic applications. Smelting of sulfide ores is also known to produce SO_2. The extent of injury that sulfur dioxide causes to plants depends on the concentration of SO_2 and the duration of exposure. Sensitive species are susceptible to SO_2 concentration range above 0.3–0.5 ppm persistent for a period of 2–3 h. Exposure to lower concentrations for weeks is likely to cause serious metabolic effects, particularly growth inhibition, even if there are no visible symptoms. The

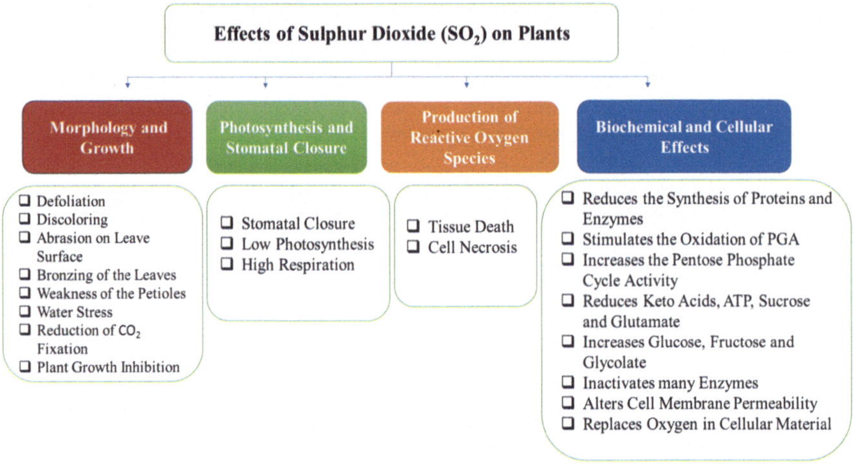

Fig. 1.1 Summarized effects of SO_2 on vegetation

adverse effects of SO_2 on plants such as defoliation and discoloring are visible within a five-mile radius from the source. Some plants being more sensitive suffer from damage even if they are 30 miles away from the source. On exposure to SO_2, plants immediately respond by closing their stomata, which inhibits further entry of the pollutant and results in low photosynthesis, high respiration and general water stress. Upon entering cell, SO_2 dissolves into bisulfite and sulfite ions.

It has been scientifically known that the main toxic effect of SO_2 on plant system is exerted either by oxidative stress or by generation of sulfite ions. Significant adverse effect can be observed on the overall plant growth, photosynthetic efficiency, and turnover. Plants detoxify the excess sulfur using enzymes like superoxide dismutase, peroxidase, and polyphenol oxidase and form S-containing sulfur compounds (Brahmachar 2017). It is well known that sulfur is an important micronutrient of plants that is utilized in the synthesis of amino acids (cysteine and methionine), vitamins (biotin), hormones (ethylene and polyamines); in the production of photosynthetic oxygen; in electron transport; and in the synthesis of several other defense-related enzymes and compounds (Capaldi et al. 2015). The overall damage to vegetation on SO_2 exposure is summarized in Fig. 1.1.

1.4 Effect on Overall Morphology and Growth of Plants

The injury to leaves caused by this obnoxious gas depends on its concentration and acute or chronic exposure patterns. SO_2 first interferes with stomata to remain if humidity is high. Within the intercellular spaces of the leaf, SO_2 interacts with plasmalemma that surrounds the cell, thus disrupting it and the ion flow and nutrient balance which is controlled by it. On entering into the cell, SO_2 interacts with mitochondria and chloroplast. It interferes with the structure and permeability of chloroplast membrane. The chloroplast then swells and degenerates. Exposure to SO_2 at low concentrations causes discoloration of leaf tissues and upward curling of leaf lamina, whereas exposure to high concentration causes lamina curling associated with drying and brittleness. In *Prosopis juliflora,* petiole length is reported to become reduced in response to SO_2 decontamination, and at the same time, dry weight of leaf increased (Seyyednejad and Koochak 2011). It was also observed that less growth and expansion of leaf surfaces resulted in increased stomatal frequency. In *Cajanus cajan and Amaranthus paniculatus*, SO_2 exposure increased the stomatal frequency in both lower and upper surfaces of leaves (Sujatha et al. 2016). Negative impact of SO_2 on nitrogen content of leaf was also observed in *Alnus sieboldiana* (Choi et al. 2014).

Acute injury is visible as abrasion on leave surface that is prominent between the veins and sometimes on margins of the leaves. The colors of the damaged area can vary from a tan or white to an orange-red or brown depending upon the plant species, weather conditions, and time period.

Chronic injury symptoms in general appear as leaf discoloration (chlorosis) and under-surface bronzing of the leaves. However, sensitivity to SO_2 differs considerably among different plant species and varieties and even members of the same species. This observation is explained by the fact that plant species differs in geographical location, weather, stage of growth, and mellowing. The following crop plants are commonly considered vulnerable to SO_2: alfalfa, barley, buckwheat, cotton, Douglas fir, soybeans, wheat, clover, oats, pumpkin, radish, rhubarb, spinach, squash, Swiss chard, and tobacco. Resistant crop contains asparagus, cabbage, celery, corn, onion and potato.

Long-term exposure to SO_2 fumigation is found to have an adverse effect upon plant growth. It is observed that the length of root and shoot get stunted after chronic SO_2 exposure. As a response of SO_2 exposure beyond permissible limits, up to 50% reduction of annual height and 70% reduction of diameter increase rate have been reported (Rai et al. 2011). Besides this, dry weight of seedlings and weakness of the petioles were also observed to get reduced by SO_2 contact (Choi et al. 2014).

Reduction of relative water content (RWC) in flannel weed and periwinkle has been reported due to SO_2 stress. Higher RWC helps in maintaining water balance and provides resistance during osmotic stress and drought stress. Thus, plants sensitive to SO_2 fumigation tend to lose more water content (Verma and Chandra 2014).

Reduction of CO_2 fixation and increased respiration rate that result in breakdown of stored carbohydrate products have been reported as an effect of SO_2 exposure in *Prosopis sp.* (Seyyednejad and Koochak 2011).

1.5 Effect of SO_2 on Photosynthesis and Stomatal Closure

The process of photosynthesis in plants is very sensitive to SO_2 fumigation. SO_2 apparently has an antagonistic effect on plant photosynthesis which, consequently, adversely correlates with the height and girth of plant axis. The quantum yields drop by 1.5-fold and the photochemical efficiency of photosystem II drops severely (Sha et al. 2010). Inhibition of essential Calvin cycle enzymes like fructose bisphosphatase and ribulose bisphosphate carboxylase along with reduction of total chlorophyll content was also reported (Ling et al. 2014). In SO_2 stress conditions, an increase in pheophytin is observed, which may indicate the degradation of chlorophyll because pheophytin is considered the by-product of chlorophyll degradation (Chung et al. 2011).

One effect of SO_2 fumigation is acidification in cell resulting from SO_2 dissolution into cellular water content consequently producing sulfuric acid. It is one of the major factors in SO_2-induced stomatal closure (Hu et al. 2014).

$$SO_2 + H_2O \rightarrow [SO_2 \cdot H_2O] \rightarrow HSO_3^- + H^+ \rightarrow SO_3^{2-} + 2H^+$$

This low pH affects the H^+ channel and membrane polarity which ultimately results in inhibition of K^+ pump and stomatal closure. This SO_2 stress also increases the production of a hormone abscisic acid (AbA). Another inhibitory factor responsible for stomatal closure thus causing inhibition in gaseous exchange and physiological processes of photosynthesis and respiration (Hu et al. 2014).

1.6 Production of Reactive Oxygen Species

The main toxic effect of SO_2 is attributed to the production of sulfite (SO_3^{2-}) and bisulfite (HSO_3^-) radicals after dissolution of SO_2 into cellular water. The plant detoxifies this condition by converting sulfite to less harmful sulfate radicals, but reactive oxygen species (ROS) like peroxide (H_2O_2), superoxide radicals ($O_2^-\bullet$), and hydroxyl radical ($OH^\cdot$) are generated as by-products of the reaction. Production of excess ROS is one of the key indications of SO_2 stress (Li and Yi 2012). The damage of PS-II is also a prominent source of ROS production as studied in *Fragaria sp.* The hyper-accumulation of ROS is detrimental as this oxidative stress exerts negative impact on nucleic acids and proteins of plants. The most destructive effect of ROS is exerted on cellular and membrane lipid which gets rapidly peroxidized (Apel and Hirt 2004). The excessive amount of peroxidized lipid has a cytotoxic effect that causes tissue death additionally cell necrosis by hypersensitive reaction. Thus, tissue necrosis in SO_2 stress can be explained by the formed ROS and its consequences. Evidence was found in *Arabidopsis* (rockcress) – this plant on exposure to higher concentration of SO_2 showed hyper-accumulation of ROS after 72 h of exposure. Whereas peroxide accumulation was increased by about 90% in the stressed plant compared to untreated plants, thus elucidating the inter-relation between SO_2 and ROS. It should be noted that ROS exacts damage in plant system still its role in initiating stress response and defense pathways cannot be overlooked (Foyer and Noctor 2005).

1.7 Biochemical and Cellular Effects

SO_2 enters plant in gaseous or sulfite form via stomata or root hairs, then, first to the guard cells and later towards a chain of epidermal cells. Plants tackle it through normal sulfur metabolism till it is in sufficient amount that causes injuries to plant (Mazid et al. 2011). At low concentrations, sulfite is metabolized to sulfate by chloroplast that acts as a sulfur source in protein synthesis of plants. At high concentration, it accumulates as sulfhydryl (RSH) decreasing the sulfides (S^{2-}) content of plants. It interferes in electron transport chain, replaces oxygen in cellular material, and alters the cell membrane permeability by affecting its structural protein. SO_2 interferes with amino acid metabolism and reduces the synthesis of proteins and enzymes. It stimulates the oxidation of PGA (phosphoglyceric acid) and increases the activity of pentose phosphate cycle. It reduces the level of keto acids, ATP (adenosine 5′-triphosphate), sucrose, and glutamate in plants and increases the level of glucose, fructose, and glycolate. It inactivates many enzymes either by breaking their S-S bonds or by changing their stereo structure.

1.7.1 Nitrogen Dioxide

Atmospheric nitric oxide (NO) and nitrogen dioxide (NO_2) are both considered as harmful and useful for plant development. It was established that NO is a phytohormone that impacts diverse physiological processes in plants. Exogenous NO_2 positively normalizes the vegetative and reproductive growth of plants (Takahashi and Morikawa 2014). The use of ecological methods such as plant absorption and catabolism of atmospheric NO_2 is very important to control its concentrations. In the nitrate assimilation pathway, NO_2 can form organic nitrogenous compounds in plants after metabolization incorporated in leaves without causing harm. Thus, only a small amount of this NO_2 can be assimilated into organic compounds.

The oxides of nitrogen are rising in the atmosphere of the urban environment due to industrialization and automobile exhaust emissions. The oxides of nitrogen have a negative effect on plants in terms of poorer growth and loss of productivity and thus crop value. They affect plants either directly or indirectly. High levels of NOx directly cause leaf damage and reduced growth. They make vegetation more susceptible to disease and frost damage. Indirectly, they act as primary substances in the production of atmospheric ozone and acid rain, as well as they act as a center for the formation of fine particles (PM).

NO_2 exposure induces complex physiological responses in plants that comprises changes in antioxidant enzyme activity, N metabolic enzyme activity, and both the components and distribution of nitrogenous metabolic products in plant tissues. Low concentration of NO_2 in leaves form nitrate and nitrite that are used by plants in nitrate metabolism. High concentration of NO_2 causes leaf necrosis as it can result in unnecessary accumulations of nitrite (NO_2^-) and cell acidification because of the reaction of NO_2 with water. It can also result in the generation of reactive oxygen species (ROS) and interfere with N assimilation as well as plant growth. The overall effect is acute damage to leaves, whole-plant chlorosis, or death.

An interesting study illustrating the effects of nitrogen dioxide (NO_2) was conducted by Sheng et al. on plants that are divided into 13 functional groups according to the Angiosperm Phylogeny Group classification system. The results showed that nitrogen oxide causes significant damage to vegetation. The results also indicated that in most of the functional groups, NO_2 exposure affected leaf chlorophyll. Plants signal self-protection by showing an increase in both peroxidase activity and soluble protein and malondialdehyde concentrations (Sheng and Zhu 2019). As the study was based on the functional groups of plants, the results can be generalized that provided an indirect evidence of NO_2 ill effects and plant response.

The major sources of nitrogen oxides are power plant and vehicular emissions; however, agriculture itself is contributing its share in its stock in the form of agri-ammonia. It vaporizes into the air and is deposited dry or in rainfall where in the ground bacteria breaks it into nitrogen and nitric acid. Exposure of ammonia results in irregular, bleached, bifacial, necrotic lesions. Discoloration on grasses dark upper surface or reddish, interveinal necrotic streaking may also occur. Generally, flowers, fruits, and woody tissues are not affected, and even if effected, plants recover by producing new leaves. Sensitive species include apple, barley, beans, clover, radish, raspberry, and soybean. Resistant species include alfalfa, beet, carrot, corn, cucumber, eggplant, onion, peach, rhubarb, and tomato. With increasing population, the demand of nitrogen fertilizer increases. All these sources of NOx give rise to acid rain, which was previously linked to SO_2 concentration in air emitted from power plants and due to fossil combustion. Sulfur dioxide emissions reduced almost 70% from 1990 to 2008, as it was recognized as a basic component of acid rain. On contrary, the emissions of nitrogen dioxide (NO_2) decreased only 35% during that same period. The scientific community is concerned over NOx emissions as it may cause acid rain, which is capable of leaching critical soil nutrients and injuring crops (Tennesen 2010).

1.7.2 Fluorides

Fluorides occur as gaseous fluorides, particulate fluorides, or hydrogen fluorides. The effect of fluoride on foliar parts and its accumulative nature in plants have long been known (Moraes et al. 2002).

Fluorides enter plants through two main routes: via air through stomata and via soil and water. Through stomata, they enter the leaves, pass through the cell wall, and travel to the leaf tips and margins. Similarly roots are the primary source of absorption of fluorides from the soil (Kamaluddin and Zwiazek 2004).

A study on fluoride toxicity in two varieties of *Solanum lycopersicum* (tomato) concluded that fluoride stress has a negative effect on plant growth and development, especially on seed germination, leaf area, and net assimilation rate (Ahmad et al. 2018). Recently, a study was conducted on the damage caused to *Triticum aestivum* (wheat) cultivated near brick kilns (a fluoride source). The results revealed deposition of fluorides on the leaves that caused foliar injury and concluded that wheat plants cultivated within 100 m of the brick kilns were seriously damaged (Urooj et al. 2019).

Studies on the susceptibility of plant species to fluorides showed that apricot, barley (young), blueberry, peach (fruit), gladiolus, grape, plum, prune, sweet corn, and tulip are the most sensitive pants. Resistant plants include alfalfa, asparagus, bean (snap), cabbage, carrot, cauliflower, celery, cucumber, eggplant, pea, pear, pepper, potato, squash, tobacco, and wheat (Heather 2003).

Fluorides interfere with the metal components of proteins, thereby hindering the activity of many enzymes. This affects several metabolic activities of plants: cell wall composition; photosynthesis; respiration; synthesis of carbohydrate, nucleic acids, and nucleotides; and energy balance of the cell. In the form of HF, fluorides reduce the leaves endoplasmic reticulum (ER) which causes reduction in the number of ribosomes, detached from ER and mitochondria become swollen. Synthesis of chlorophyll and cellulose are repressed and the activities of UDP-glucose-fructose transglucosylase, phosphoglucomutase, enolase, and polyphenol oxidase are reduced. Instead, activities of catalase, peroxidase, pyruvate kinase, PEP-carboxylase, glucose-6-phosphate dehydrogenase, cytochrome oxidase, and pentose phosphate pathway are initiated (Kumar et al. 2017).

1.7.3 Ethylene

Among the organic gaseous pollutants, ethylene is the most common. It plays a significant role in regulating several processes of plant growth and development and even death of plants; hence, they are known as "death" or "ripening hormone." Propane, gasoline, and natural gas contain a substantial amount of ethylene; ethylene is produced when these substances are combusted. Fruits, vegetables, and flowers contain receptors that absorb free ethylene molecules from the atmosphere (Dhall 2013). Ethylene is a natural plant growth substance, so its damaging effects on plants are very similar to the symptoms of abnormal growth (Gheorghe and Ion 2011). Ethylene is one of the main hormones in the regulation of leaf senescence. The response of plant to ethylene differs greatly from one species to another. Exposure to ethylene causes early senescence effects, such as yellowing of the stems, abscission, or necrosis. Chlorophyll loss and leaf abscission are the visible signs of leaf senescence. At the molecular level, ethylene is involved in organized cell dismantling and the activation of recycling of nutrients from senescing leaves to the other parts. Leaf cells undergo sequential and systematic dismantling that encompasses nucleic acid depletion, protein and lipid degradation, membrane disruption, peroxidation, and breakdown of leaf pigment. For broad-leaf plants, damage includes curling of the leaves and shoots downward (epinasty) and stunted growth. Ethylene-exposed tomato plants may form twisting, defoliating, and bloom drop (Kumar et al. 2020). Chlorophyll reduction has also been observed in rocket salad leaves exposed to 1 μL L^{-1} of ethylene during storage, a condition that reduces the shelf life for about 2 days. Abscission or initiation of necrosis is another consequence that is inducted by ethylene on leaf senescence. Leaf abscission is a coordinated process involving numerous structural modifications of the cell in the abscission region zone (Iqbal et al. 2017). For the first time, it has been found that ethylene affects the plant life of cotton over large field areas. Acceleration of ripening is a valuable use of ethylene, but it can also be undesirable, for example, when cucumbers turn yellow prematurely in its presence. When the concentration of ethylene in the cold storage is above 20 ppb, the firmness of kiwi fruit is significantly reduced. The damage level depends on the ethylene concentration, period of exposure, and temperature. In leafy vegetables such as spinach, broccoli, parsley, and cucumbers, ethylene accelerates senescence, as specified by loss of chlorophyll, thus decreasing the market quality of leafy, floral, and immature fruit vegetables (Dhall 2013).

1.7.4 Particulate Pollutants

Solid and liquid particles suspended in air are known as particulate matter (PM). They are sulfates, nitrates, ammonium, inorganic ions (Ca, Cl, K, Mg, and Na), organic and elemental carbon, metals (including Cd, Cu, Ni, V, and Zn), polycyclic aromatic hydrocarbons (PAH), biological components (allergens, microbial compounds), and particle-bound water (Shrivastava et al. 2018; Weerakkody et al. 2017, 2018). The physical and chemical characteristic of PM is representative of its location (Turkyilmaz et al. 2018). Plants are known to reduce PM from air, as these are deposited on leaf surface and stick to leaf wax layer. However, during this process, plant growth gets disrupted as dust deposited on leaves blocks the stomata. This blockage reduces sunlight absorption and also hinders the transport of CO_2, thereby slowing down photosystem II (Sett 2017).

The threats related to dust deposition on leaf surface are defined by its size and type. PM interferes with the leaf gas exchange and results in the reduction of plant growth, flowering, reproduction, number of leaves, and leaf area. Reduction in leaf area and leaf number may be due to decreased leaf production rate and enhanced senescence (Seyyednejad et al. 2011). The inhibition of net photosynthesis would restrain the translocation, and ultimately leaf area would decrease. It has been demonstrated that cement dust-treated plants of *Brassica campestris* (mustard) showed a steady reduction in growth, photosynthetic pigments, yield, and oil content compared to control plants, and these factors together decreased the biomass of this plant (Shukla et al. 1990; Prasad et al. 1991). The structure and morphology of epicuticular wax are reliable indicators of plant health, which act as a barrier between the plant and the environment. Deposition of road dust as well as cement dust containing high levels of MgO and PM causes degradation of epicuticular wax by increasing the erosion rate of wax structure and by inducing changes in leaf wettability. Thus, inhibiting transpiration which may result in extensive physiological costs, such as prevention of gas exchange and photosynthesis, and loss of solutes from leaf cells. PM obtained from stone dust due to mining activities cause foliar anomalies and injuries such as tissue necrosis, brown and yellow patches, black spots, and, in extreme cases, death of leaves (Saha and Padhy 2011). Once deposited on the leaf surface, some elements may be taken up into the leaf via the stomata, thereby affecting the overall development of plant and reducing the resistance of plants to drought, frost, insect, and fungi (Shanker et al. 2005; Reimann et al. 2001).

Presence of heavy metals greater than their threshold level could be harmful for local vegetation. Increased levels of heavy metals may cause oxidative stress by inducing the generation of reactive oxygen species (ROS) within sub-cellular compartments. Likewise, it may decrease enzymatic and non-enzymatic antioxidants due to an affinity with sulfur-containing group (–SH) (Gupta et al. 2012; Benavides et al. 2005). Heavy metals like zinc, copper, and iron are essential for biosynthesis of enzymes, auxin, and some proteins that are crucial for normal growth and development of plants (Onder and Dursun 2006). Alteration in their concentration can result in noteworthy change of biochemical processes in plants leading to loss of production, lower yield, and poor quality of agricultural crops (Bucher and Schenk 2000). High zinc concentration in the leaves of *Corylus avellana* (hazel) resulted in loss of hazel nut production. Excessive copper may destroy subcellular structure of plants (Sresty and Madhava 1999). Photosynthetic efficiency of most of the plants is affected by heavy metals (Krupa and Baszyński 1995; Burzynski and Klobus 2004). Excess levels of Cu, Cd, or Pb inhibit directly the photosynthetic electron transport (Krupa and Baszyński 1995; Myśliwa-Kurdziel et al. 2002) and activities of Calvin-Benson cycle enzymes or net assimilation of CO_2 (Prasad and Strzałka 1999; Burzynski and Klobus 2004). Cerium (Ce) in conjunction with UV-B radiation decreased the chlorophyll content, net photosynthetic rate, Hill reaction activity, photophosphorylation rate, and Mg2þ-ATPase activity in soybean (Glycine max L.) (Liang et al. 2010). Cu was more toxic than Cd and Pb and showed decreased net photosynthetic rate and stomatal conductance in leaves of cucumber (Burzynski and Klobus 2004).

1.8 Effect of Secondary Pollutants on Vegetation

1.8.1 Peroxyacetyl Nitrate (PAN)

PAN is the second main source of pollution, causing damage to crops. It induces tissue collapse on the lower surface of most plants, resulting in leaves that form glazed, tanned, or silver areas with bands. The leaves that are affected get premature senescence. The leaves' entire width can be damaged in some plants like tomatoes and potatoes. PAN is most toxic to young and smaller plants. Matured leaves are less likely to be damaged (Kumar et al. 2020).

Young plants and rapidly expanding leaves are more susceptible to PAN. It interacts complexly with SO_2 and O_3, resulting in variable impact conditions (Gheorghe and Ion 2011). The collapse of the tissue completely appears as a diffuse band in the leaf that helps to distinguish PAN damage from other damages. Peroxyacetyl nitrate also causes silver-leaf effect (Bhushan 2018). Species sensitive to PAN are bean, celery, lettuce, pepper, spinach, and tomato, whereas those tolerant are broccoli, cabbage, cauliflower, cucumber, onion, radish, and squash.

Chlorosis and necrosis in the leaves are distinctive visible symptoms of PAN exposure. Photosynthesis, respiration, and carbohydrates and protein synthesis are often intervened. It halts photorespiration, reduces nicotinamide adenine dinucleotide phosphate (NADP), and fixes carbon dioxide, cellulose formulations, and photosynthesis-related enzymes (Gheorghe and Ion 2011).

1.8.2 *Ozone (O_3)*

Tropospheric ozone (O_3) is recognized as the most hazardous secondary air pollutant affecting crop productivity in most parts of the world. Ozone injury to crops has been observed on large scale in North America and Europe. The growing emissions of hydrocarbons and nitrogen oxide have significantly increased the ground-level O_3 concentrations. It can be carried for long distances, which may be the reason for its high concentrations in rural areas compared to urban areas whereas later is considered as its source (Tiwari et al. 2008; Rai and Agrawal 2008; Rai and Madhoolika 2012).

Plants absorb ozone via stomata, and once it enters the plant, it damages cell walls and membranes by the reactive oxygen species that are formed. Accelerated chlorophyll destruction is reported due to induced metabolic changes within the plant cells. This leads to cell death or serious reductions in photosynthesis. Visible injuries that occur due to chronic exposure to low ozone concentrations are changes in pigmentation or bronzing, chlorosis, and premature senescence. Acute exposures to the high ozone levels result in flecking and stippling (Benjamin et al. 2007). The results of these negative outcomes depend on both the concentration and duration of O_3 exposure. Plants can detoxify low concentrations of O_3 but only to a certain threshold level. Above the detoxification level, a yellowing of leaves and premature leaf loss, decreased seed production, and reduced root growth, resulting in reduced yield quantity and/or quality and reduced resilience to other stress such as drought, occur. Physiological effects of ozone exposure include reduced photosynthesis, increased turnover of antioxidant systems, damage to reproductive processes, increased dark respiration, lowered carbon transport to roots, reduced decomposition of early successional communities, and reduced forage quality of C4 grasses (Benjamin et al. 2007; Calatayud et al. 2002; Chappelka 2002). Staple food crops, most importantly wheat and soybean, are sensitive to ozone and showed 18% reduction of yield on exposure to a 7 h mean O_3 concentration of 60 ppb. Yield reduction is also observed for crops like rice, maize, and potato, with a calculated 10% yield reduction at 60 ppb ozone O_3 concentration. Reich and Amundson (Powell et al. 2003) found a 50% reduction in photosynthesis for clover and wheat, but only a 10% reduction for white pine. It was postulated that an ozone dose of 20 ppm results in a photosynthesis reduction of 7% for conifers, 36% for hardwoods, and 73% for crops (Benjamin et al. 2007). Economic loss is significant in Europe, where eight out of the nine crops in Europe are sensitive or moderately sensitive to O_3 concentration. Although O_3 sensitivity varies between cultivars, modern cultivars of crops such as wheat seem to be more O_3 sensitive than traditional cultivars. It seems that breeding crops to obtain high yield might have resulted in more sensitive species. Extensive damage to plant leaves and reduction in yield are well-recognized negative effects of O_3; however, little information exists about its effects on food quality. So far, negative impacts of O_3 have been found in protein yield of wheat, sugar content of potato, and oil quality of oilseed rape (Reich and Amundson 1985; Gina and Harmens 2011).

1.9 Conclusions and Future Recommendations

The production of high-quality nutritious and fresh agricultural crops is always demanded. There is a substantial evidence from the published research that air contamination restrains the yield and quality of agriculture crops. Yet this field requires further attention from researchers, agriculturalists, and even common man, whose life depends on food.

A recent study showed decrease in air pollution in three states of India due to COVID lockdown situation (Hari et al. 2021). As the air pollution source was identified as crop residual burning, it was observed that the lockdown improved air quality in Delhi. Recently, it has been concluded by many researchers that beyond any doubt food security and agricultural crop yields are at risk due to air pollution (Devrajani et al. 2020). Moreover, a study done by Wei and Wang (2021) explains crop loss from an economic perspective. The authors relate the decrease in crop yield to the output elasticity of production and change in important factors such as labor, chemicals, capital, seeds. Authors also demonstrated that the relationship among these important factors change from substitutable to complementary and vice versa (Wei and Wang 2021). It is documented that the economic loss due to air pollution in United States was found to be 40–50 billion dollar in 1997 (Lee 2000). In Pakistan, rice is the second major cash crop. The yield of rice dropped by 43%, 39%, and 18%, respectively, during 2003–2004 when the seasonal average concentrations of O_3, NO_2, and SO_2 were 70 ppb, 28 ppb, and 15 ppb (Wahid 2006).

It has been concluded that implementation of effective policies to control air pollution will result in increased food production and security. The formation and implementation of such policies are only possible when data about the knowledge of the causes and effects of air pollution exists. It is recommended that the impacts of air pollution on agriculture should be widely disseminated and should be properly incorporated into agricultural and environmental policy development.

References

Ahmad MA, Bibi H, Munir I, Ahmad MN, Zia A, Mustafa G, Ullah I, Khan I (2018) Fluoride toxicity and its effect on two varieties of Solanum Lycopersicum. Research Report. Fluoride 51(3):267–277

Apel K, Hirt H (2004) Reactive oxygen species: metabolism, oxidative stress, and signal transduction. Annu Rev Plant Biol 55:373–399

Benavides MP, Gallego SM, Tomaro ML (2005) Cadmium toxicity in plants. Braz J Plant Physiol 17:21–34

Benjamin S, Felzer TC, John MR, Jerry MM, Xiaodong W (2007) External geophysics, climate and environment (climate) impacts of ozone on trees and crops. C R Geosci 339:784–798

Bhushan A (2018) Impact of air pollution on vegetable crops. Open Access J Sci 2(6):367–368

Brahmachar S (2017) SO_2 stress: its effects on plants, plant defense response and strategies for developing enduring resistance. Int J Adv Res Sci Eng Technol 4(7):303–309

Bucher AS, Schenk MK (2000) Toxicity level for phytoavailable zinc in compost peat substrate. Sci Hortic 83:339–352

Burzynski M, Klobus G (2004) Changes of photosynthetic parameters in cucumber leaves under Cu, Cd, and Pb stress. Photosynthetica 42(4):505–510

Calatayud A, Ramirez JW, Iglesias DJ, Barreno E (2002) Effects of ozone on photosynthetic CO_2 exchange, chlorophyll a fluorescence and antioxidant systems in lettuce leaves. Physiol Plant 116:308–316

Capaldi FR, Gratão PL, Reis AR, Lima LW, Azevedo RA (2015) Sulfur metabolism and stress defense responses in plants. Trop Plant Biol 8:60–73. https://doi.org/10.1007/s12042-015-9152-1

Chappelka AH (2002) Reproductive development of blackberry (Rubus cuneifolius), as influenced by ozone. New Phytol 155:249–255

Choi D, Toda H, Kim Y (2014) Effect of sulfur dioxide (SO_2) on growth and physiological activity in Alnus sieboldiana at Miyakejima Island in Japan. Ecol Res 29:103–110

Chung YC, Chung PL, Liao SW (2011) Carbon fixation efficiency of plants influenced by sulfur dioxide. Environ Monit Assess 173:701–707

Devrajani SK, Qureshi M, Imran U, Nisa TU (2020) Impact of gaseous air pollutants on agricultural crops in developing countries: a review. J Environ Sci Public Health 4(2):71–82

Dhall RK (2013) Ethylene in post-harvest quality management of horticultural crops: a review. Res Rev J Crop Sci Technol 2(2):9–24

Foyer CH, Noctor G (2005) Oxidant and antioxidant signalling in plants: a reevaluation of the concept of oxidative stress in a physiological context. Plant Cell Environ 28:1056–1071

Gheorghe IF, Ion B (2011) The effects of air pollutants on vegetation and the role of vegetation in reducing atmospheric pollution. In: The impact of air pollution on health, economy, environment and agricultural sources, IntechOpen, IntechOpen Limited 5 Princes Gate Court, London, SW7 2QJ, UK. pp 241–280

Gina M, Harmens H (2011) Ozone pollution: a hidden threat to food security; a report. ICP Vegetation Programme Coordination Centre, Centre for Ecology and Hydrology, Environment Centre Wales, Bangor

Gupta S, Nayek S, Bhattacharya P (2012) Effect of airborne heavy metals on the biochemical signature of tree species in an industrial region with an emphasis on anticipated performance index. Chem Ecol 27(4):381–392

Hari M, Sahu RK, Tyagi B, Kaushik R (2021) Reviewing the crop residual burning and aerosol variations during the COVID-19 pandemic hit year 2020 over North India. Pollutants 1:127–140

Heather G (2003) Revision of factsheet. Air Pollut Agric Crops, Order No. 85-002; Printed June

Hu KD, Tang J, Zhao DL, Hu LY, Li YH, Liu YS, Jones R, Zhang H (2014) Stomatal closure in sweet potato leaves induced by sulfur dioxide involves H_2S and NO signaling pathways. Biol Plant 58(4):676–680

Iqbal N, Khan NA, Ferrante A, Trivellini A, Francini A, Khan MIR (2017) Ethylene role in plant growth, development and senescence: interaction with other phytohormones. Front Plant Sci 8:475

Kamaluddin M, Zwiazek JJ (2004) Effects of root medium pH on water transport in paper birch (Betula papyrifera) seedlings in relation to root temperature and abscisic acid treatments. Tree Physiol 24:1173–1180

Krupa Z, Baszyński T (1995) Some aspects of heavy metals toxicity towards photosynthetic apparatus – direct and indirect effects on light and dark reactions. Acta Physiol Plant 17:177–190

Kumar K, Giri A, Vivek P, Kalaiyarasan T, Kumar B (2017) Effects of fluoride on respiration and photosynthesis in plants: an overview. Ann Environ Sci Toxicol 2(1):043–047

Kumar B, Pattnaik P, Mishra D (2020) Increasing air pollutants and its impact on vegetable crops. Agric Food 2(5):13–15

Lee EH (2000) Early detection mechanisms for tolerance and amelioration of ozone stress in crop plants. In: Agrawal SB, Agrawal M (eds) Environmental pollution and plant response. Lewis Publishers, Boca Raton

Li L, Yi H (2012) Effect of sulfur dioxide on ROS production, gene expression and antioxidant enzyme activity in Arabidopsis plants. Int J Plant Physiol Biochem 58:46–53

Liang C, Zhang G, Zhou Q (2010) Effect of cerium on photosynthetic pigments and photochemical reaction activity in soybean seedling under ultraviolet-B radiation stress. Biol Trace Elem. https://doi.org/10.1007/s12011-010-8786-y

Ling W, Lin-ying D, Xi L (2014) Chlorophyll biosynthesis metabolism of golden-leaf plants to SO_2 stress. Chin J Ecol 9:7–9

Mazid M, Khan TM, Mohammad F (2011) Response of crop plants under sulphur stress tolerance. J Stress Physiol Biochem 7:25–57

Moraes RM, Klumpp A, Furlan CM, Klumpp G, Domingos M, Rinaldi MCS, Modesto IF (2002) Tropical fruit trees as bioindicators of industrial air pollution in Southeast Brazil. Environ Int 28:367–374

Myśliwa-Kurdziel B, Prasad MNV, Strzałka K (2002) Heavy metal influence on the light phase of photosynthesis. In: Prasad MNV, Strzałka K (eds) Physiology and biochemistry of metal toxicity and tolerance in plants. Kluwer Academic Publisher, Dordrecht, pp 229–257

Onder S, Dursun S (2006) Air borne heavy metal pollution of Cedrus libani (A. Rich) in the city centre of Konya (Turkey). Atmos Environ 40:1122–1133

Powell MC, Muntifering RB, Lin JC, Chappelka AH (2003) Yield and nutritive quality of sericea lespedeza (Lespedeza cuneata) and little bluestem (Schizachyrium scoparium) exposed to ground-level ozone. Environ Pollut 122:313–322

Prasad MNV, Strzałka K (1999) Impact of heavy metals on photosynthesis. In: Prasad MNV, Hagemayer J (eds) Heavy metal stress in plants. From molecules to ecosystems. Springer-Verlag, Berlin/Heidelberg/New York, pp 117–139

Prasad MNV, Subramanian RB, Inamdar JA (1991) Effect of cement kiln dust on Cajanus cajan (L.) Millsp. Indian J Environ Health 33:11–21

Rai R, Agrawal M (2008) Evaluation of physiological and biochemical responses of two rice (Oryza sativa L.) cultivars to ambient air pollution using open top chambers at a rural site in India. Sci Total Environ 407:679–691

Rai R, Madhoolika A (2012) Impact of tropospheric ozone on crop plants. Proc Natl Acad Sci India, Sect B Biol Sci 82(10):1–16

Rai R, Rajput M, Agrawal M, Agrawal SB (2011) Gaseous air pollutants: a review on current and future of emissions and impact on agriculture. J Sci Res 55:77–102

Reich PB, Amundson RG (1985) Ambient levels of ozone reduce net photosynthesis in tree and crop species. Science 230:566–570

Reimann C, Koller F, Kashuline G, Nilksvaara H, Englmaier E (2001) Influence of extreme pollution on the inorganic chemical composition of some plants. Environ Pollut 115:239–252

Saha DC, Padhy PK (2011) Effects of stone crushing industry on Shorea robusta and Madhuca indica foliage in Lalpahari forest. Atmos Pollut Res 2:463–476

Sett R (2017) Responses in plants exposed to dust pollution. Hortic Int J 1(2):00010
Seyyednejad SM, Koochak H (2011) Some morphological and biochemical responses due to industrial air pollution in Prosopis juliflora (Swartz) DC plant. Afr J Agric Res 8(18):1968–1974
Seyyednejad SM, Niknejad M, Koochak H (2011) A review of some different effects of air pollution on plants. Res J Environ Sci 5:302–309
Sha C, Wang T, Lu J (2010) Relative sensitivity of wetland plants to SO_2 pollution. Wetlands 30:1023–1030
Shanker AK, Cervantes C, Loza-Tavera H, Avudainayagam S (2005) Chromium toxicity in plants. Environ Int 31:739–753
Sheng Q, Zhu Z (2019) Effects of nitrogen dioxide on biochemical responses in 41 garden plants. Plants 8(2):45. https://doi.org/10.3390/plants8020045
Shrivastava B, Dipak NK, Karajagi A, Dosi R (2018) Severe neonatal hypercalcemia secondary to aluminium toxicity. Int J Contemp Pediatrics 5(1):254
Shukla J, Pandey V, Singh SN, Yunus M, Singh M, Ahmad KJ (1990) Effect of cement dust on the growth and yield of Brassica campestris L. Environ Pollut 66:81–88
Sresty TVS, Madhava RKV (1999) Ultrastructural alterations in response to zinc and nickel stress in root cells of pigeon pea. Environ Exp Bot 41(1):3–13
Sujatha B, Priyadarshini B, Umamahesh C, Kumar MVVP, Divya Jyothi LB, Parvathi ASG, Saraswathi J (2016) Morphological and biochemical changes in the leaves of Cajanus cajan and Amaranthus paniculatus under foliar application of aqueous sulphur dioxide. Int J Eng Invention 5(8):36–47
Takahashi M, Morikawa H (2014) Nitrogen dioxide regulates organ growth by controlling cell proliferation and enlargement in Arabidopsis. New Phytol 201:1304
Tennesen M (2010) Acid rain returns – this time it is caused by nitrogen emissions. www.scientificamerican.com
Tiwari S, Rai R, Agrawal M (2008) Annual and seasonal variations in tropospheric ozone concentrations around Varanasi. Int J Remote Sens 29:4499–4514
Turkyilmaz A, Sevik H, Cetin M, Saleh EAA (2018) Changes in heavy metal accumulation depending on traffic density in some landscape plants. Pol J Environ Stud 27(5):2277–2284
Urooj R, Ahmad SS, Ahmadc MN (2019) Assessment of fluoride damage to wheat (Triticum Aestivum) cultivated near brick kilns in the Rawalpindi District. Pakistan Assess 265:265
Verma V, Chandra N (2014) Biochemical and ultrastructural changes in Sida cordifolia L. and Catharanthus roseus L. to auto pollution. Int Sch Res Notices. https://doi.org/10.1155/2014/263092
Wahid A (2006) Influence of atmospheric pollutants on agriculture in developing countries: a case study with three new varieties in Pakistan. Sci Total Environ 371:304–313
Weerakkody U, Dover JW, Mitchell P, Reiling K (2017) Particulate matter pollution capture by leaves of seventeen living wall species with special reference to rail-traffic at a metropolitan station. Urban For Urban Green 27:173–186
Weerakkody U, Dover JW, Mitchell P, Reiling K (2018) Quantification of the traffic-generated particulate matter capture by plant species in a living wall and evaluation of the important leaf characteristics. Sci Total Environ 635:1012–1024
Wei W, Wang Z (2021) Impact of industrial air pollution on agricultural production. Atmos 12:639

Chapter 2
Soil Metal Contamination and Its Mitigation

Bushra Haroon, Muhammad Irshad, Abdol Ghaffar Ebadi, and Ping An

2.1 Introduction

Soil pollution by heavy metals is an emerging environmental and ecological issue throughout the world. In terrestrial ecosystems, soil works as one of the significant sinks for heavy metals (Sun et al. 2018). Heavy metals are obtained through two known sources: natural and anthropogenic. Anthropogenic sources include vehicle emissions, mineral fertilizers, industries and sewage sludge, etc. (Srivastava et al. 2017). Human health is indirectly or directly affected by heavy metal pollution. If these toxic metals cross their limits, they negatively affect the environment and also the growth of plants and animals and fertility of soils (Asirifi 2017). Wastewater reuse has become an important practice in agricultural fields nowadays due to water scarcity. People have started to use raw city effluent for vegetables and crops in agricultural sector to meet their demand because good-quality water is decreasing day by day (Libutti et al. 2018). Wastewater comprises toxic substances and nutrients that are beneficial and sometimes harmful for crop production (Sims et al. 2018). Continuous irrigation with urban and industrial wastewater leads to heavy metal buildup in agricultural soils and plant tissues (Edelstein and Ben-Hur 2018). Heavy metals are not degradable and have a long half-life period; thus, they may cause issues related to the pollution in crops and soils and bioaccumulate in human

B. Haroon · M. Irshad (✉)
Department of Environmental Sciences, COMSATS University Islamabad, Abbottabad, Pakistan
e-mail: mirshad@cuiatd.edu.pk

A. G. Ebadi
Department of Agriculture, Jouybar Branch, Islamic Azad University, Jouybar, Iran

P. An
Arid Land Research Center, Tottori University, Tottori, Japan

Q. Mahmood (ed.), *Sustainable Plant Nutrition under Contaminated Environments*, Sustainable Plant Nutrition in a Changing World,
https://doi.org/10.1007/978-3-030-91499-8_2

Table 2.1 Average values of water quality parameters of wastewater used for irrigation, along with standard limits of drinking water by NSDWQ-Pak and irrigation water by the USA

Parameters	Unit	Present study (mean)	NSDWQ-Pak	US irrigation water quality
pH	–	9.20 ± 0.3	6.5–8.5	6.5–8.4
EC	$\mu S\ cm^{-1}$	1124 ± 26	–	<700
COD	$mg\ L^{-1}$	459 ± 14	–	–
Pb	$mg\ L^{-1}$	2.50 ± 0.31	0.05	5.0
Ni	$mg\ L^{-1}$	1.64 ± 0.23	0.02	0.2
Mn	$mg\ L^{-1}$	0.98 ± 0.14	>0.5	0.2
Fe	$mg\ L^{-1}$	0.72 ± 0.1	–	5.0
Zn	$mg\ L^{-1}$	1.51 ± 0.1	5.0	2.0
Cu	$mg\ L^{-1}$	0.81 ± 0.02	2.0	0.2
Cd	$mg\ L^{-1}$	0.32 ± 0.02	0.01	0.01
K	$mg\ L^{-1}$	0.72 ± 0.10	–	–
Ca	$mg\ L^{-1}$	4.10 ± 0.31	–	–
Mg	$mg\ L^{-1}$	0.91 ± 0.02	–	–
Na	$mg\ L^{-1}$	6.81 ± 0.42	–	–

Radojevic and Bashkin (2010)

body (Obosu-Mensah 2018). Toxic substances namely cadmium (Cd), copper (Cu), zinc (Zn), and lead (Pb) are the main pollutants observed in vegetables (Alves et al. 2018). Eating contaminated vegetables and other crops irrigated with wastewater is harmful for human health. Heavy metal contaminated food is essentially important for the safety and quality of food (Zafarzadeh et al. 2018). Heavy metal pollution of soils has been causing adverse impacts on public health, even if they are present in low concentration in the environmental systems (Tóth et al. 2016). In Pakistan, industrial wastewater is being widely used for irrigation (Zia et al. 2017). Sewage water and industrial effluents contain high amounts of heavy metals that are harmful to freshwater bodies, agriculture, natural environment, and anthropogenic population (Calmon 2018). However, these effluents are also rich sources of organic matter and contain necessary plant nutrients.

Different plants exhibit different properties with regard to the uptake of heavy metals from the soil. Therefore, it is very necessary to understand soil–plant characteristics (Abbas et al. 2017). It has been noticed that heavy metals were present in the crops cultivated in the restricted polluted areas (Shen et al. 2017). Edible crops grown in the agricultural soils of Hattar, Pakistan, showed accumulation of heavy metals beyond allowable limits (Irshad et al. 2015). Through food chain, toxic heavy metals enter into the animal system; therefore, heavy metal monitoring in soil has greatly attracted the attention of researchers (Sarwar et al. 2017). Restoration of heavy metal contaminated soil is very complicated (Liu et al. 2017). To avoid the entry of heavy metals in the surface water, underground water, soil, and ultimately food chain, the metals in soil have to be immobilized (Table 2.1).

Table 2.2 Heavy metal concentrations in soil irrigated with contaminated water (mg kg^{-1}) and the maximum admissible concentrations of toxic metals in soils from the USA, UK, and EU

HM	Previous studies (Pakistan) (Jamali et al. 2009)	Previous studies (India) (Chandra et al. 2009)	USA	UK	EU
Cd	2.4 ± 0.18	10 ± 0.76	20	3	1–3
Cu	39.2 ± 2.1	40.83 ± 2.3	750	135	50–140
Ni	36.0 ± 1.2	42.24 ± 3.3	210	75	30–75
Pb	75.5 ± 4.3	30.81 ± 2.2	150	300	50–300
Zn	235 ± 7.2	143 ± 4.2	1400	300	150–300
Mn	–	1352 ± 10	–	–	–
Fe	–	13,925 ± 12	–	–	–

Table 2.3 Guideline of safe limits for heavy metals in water, soil, and plant

Sample	Standard	Pb	Zn	Cd	Hg
Water (mg L^{-1})	FAO/WHO 2007	5.0	2.0	0.01	0.001
Soil (mg kg^{-1})	USEPA 2010	300	200	3.0	0.30
Plant (mg kg^{-1})	FAO/WHO 2007	5.0	60.0	0.2	0.0016

WHO (2004)

Addition of wastewater, solid waste, and sludge material and emissions from vehicles and industrial discharge contaminate agricultural soil due to the toxic heavy metals present in them (Mahmoud and Ghoneim 2016). The buildup of toxic metals in soil, their presence in wastewater, and uptake by the crops in the Hattar industrial area of Pakistan were well reported (Ramady et al. 2017). Quality of the food is one of the major concerns, and general public takes great interest to know whether edible plants grown in such areas are safe to consume (Wu et al. 2018) (Table 2.2).

A decrease in necessary nutrients in human beings and many health issues can emerge by eating foods contaminated with heavy metals, which include growth retardation, immunological defenses, malnutrition disabilities, impaired psychosocial behaviors, and high risk of gastrointestinal cancer (Pandey 2018). Numerous research reports showed that higher accumulation of heavy metals poses higher risk to human health (Khalid et al. 2018). Health quotient, transfer factor, daily intake of metals, and health risk index are determined to assess the health risks emerged by wastewater containing heavy metals (Shahid et al. 2017). Carcinogenic effects of Pb and Cd have also been observed in humans (Rehman et al. 2018). Cu and Zn are basic micronutrients required for human health, but if these metals exceed permissible limits, they can be very toxic (Steffan et al. 2018). Liver damage and acute intestinal and stomach problems can be caused by Cu surplus, while high lipoprotein density and less immune function can be caused due to higher levels of Zn in the body (Giri et al. 2018) (Table 2.3).

2.2 Effects of Heavy Metals on Soil

Heavy metal pollution of soil is one of the comprehensive and significant issues in several countries (Song et al. 2017). Heavy metal pollution not only adversely affects several parameters related to plant quality and yield but also causes changes in the activity of microbial communities, sizes, and compositions (Sharma et al. 2017). As a result, heavy metals are considered an important source of soil pollution. Heavy metal pollution of soil is due to several metals, especially Zn, Cr, Cu, Pb, Ni, and Cd (Meena et al. 2018). The toxic effects of heavy metals on the biological properties of soil properties and their biochemistry have been well reported. Soil characteristics including clay content, pH, and organic matter play an important role in the influence of metals on soil characteristics (Sparks 2018). Higher concentrations of heavy metals in soil affect the growth of plants (Blum 2018). The heavy metals observed in labile pool include organic bound fraction, Fe/Mn oxides, and metals bound to carbonates. A study showed higher level of Pb, Zn, and Cu in the dumping sites due to industrial discharge, and their labile pool may have negative impacts on the quality of soil (Usman et al. 2017). It has been reported that fungi and actinomycetes were found sensitive to heavy metal pollution and exhibited a negative response to Zn, Cu, and Cr (Kouchou et al. 2017).

2.3 Effects of Heavy Metals on Plants

The ability of plants to accumulate heavy metals in their tissues greatly varies. These heavy metals get accumulated in higher concentrations in edible portion of plants and can poison living organisms if consumed (Augustsson et al. 2018). Higher concentrations of Cd and Pb were observed to be toxic while lower levels were beneficial. Tomato species was found to accumulate heavy metals comparatively lesser than eggplants (Manzoor et al. 2018). Heavy metals that are in soluble form in soil solution are absorbed by root exudates and finally become bioavailable (Etesami and Jeong 2018). Plants need heavy metals for their growth and development in limited amounts, but exceeded concentrations become toxic for plants. The capacity of plants to accumulate essential metals helps a plant to obtain other non-essential elements (Hejna et al. 2018). Metals are non-degradable nature and therefore accumulate in higher concentrations in plant tissues and directly and indirectly affect them. Cell structure damage by oxidative stress and cytoplasmic enzyme inhibition are the direct toxic effects of higher concentration of heavy metals on plants (Sinha et al. 2018). Replacing plant essential nutrients at its cation exchange site is an indirect toxic effect of metals (Van Veld et al. 2018). The growth of plant may be affected indirectly because toxic heavy metals alter the activities and growth of soil microbes. Heavy metals disturb the activities of soil microbial population, which finally reduces the enzymatic activities of plants. The toxic effects of heavy metals may cause death and decline plant growth (Meier et al. 2017). It is observed that some species of plants are highly tolerant to heavy metals. These plants tolerate heavy metals in three ways: (i) restrict metals movement and maintain concentration in the shoot as compared to concentration of the soil solution (exclusion), (ii) the concentration of metals in buds represents their concentration in the soil (inclusion), (iii) accumulation of metals in plant roots and buds (bioaccumulation) (Abbas et al. 2017).

2.4 Heavy Metals Availability as Affected by Soil Properties

The availability of heavy metals is affected by the soil properties. Soil pH is one of the major factors affecting metal availability in the soil (Hou et al. 2017).It was reported that due to higher soil pH, Zn and Cd availability decreased in *Thlaspi caerulescens* roots (Tőzsér et al. 2017). Immobilization and reduction of metals availability has been shown because of ferric oxide and organic matter (Zeng et al. 2017). Water holding capacity and moisture contents of soil have been reported to strongly correlate with toxic heavy metals (Noman et al. 2018). It has also been observed that heavy metals alter the biological properties of soil (Tan et al. 2018). To examine the intensity of soil pollution, it is necessary to monitor changes in biochemical and microbiological properties of soil. The results via these techniques are obtained within less duration compared with other soil properties (Bünemann et al. 2018). Although it has been observed that heavy metals have negative effects on the biological properties of soil, some reports did not establish any positive relationship between them (Choppala et al. 2018). Activities of microbes and their diversity are affected by heavy metals. Inorganic cations and anions, pH, metals chemical forms, organic matter, clay minerals, and soil temperature are the most important factors that affect metal toxicity (Reeve and Fallowfield 2018). Different factors that affect the availability of metals in soil include soil surface area, ligand complexation, types of charge in soil colloids, and soil density (Totsche et al. 2018). The greater surface area and interface provided by colloids of soil help in controlling heavy metals in soil system. Soil particles with high surface area have metal-specific properties that help in reducing soluble portion of metals in polluted soils (Ren et al. 2018). Decreased solubility of Pb, Mo, Cu, Cd, and As did not change the solubility of Zn and Ni due to the amendment of hydroxide with higher surface area (Jafari et al. 2018). Mineral composition, activities of microbes, and soil aeration were reported to affect the availability of metals in soils (Somtners et al. 2018) (Table 2.4).

2.5 Immobilization of Heavy Metals in Soils

The soil acts as a major reservoir for chemical pollutants in ecosystem. The major sources of heavy metals in soils are due to the disposal or manufacturing of industrial and domestic heavy metals. Metals remain in the soil for longer period because

Table 2.4 Fractionation of Cd (mg kg^{-1}) in the soil of wastewater irrigated fields

Wastewater (WW) irrigation	Sampling field	Exchangeable	Carbonate Cd	Oxide Cd	Organic Cd	Residual Cd
WW-irrigated	Field 1	4.3	36.4	86.2	16.4	164.8
	Field 2	0.5	23.4	76.4	14.6	182.5
	Field 3	1.8	25.6	92.6	11.4	191.6
	Field 4	0.16	38.2	71.4	19.8	174.3
	Mean	1.6	30.9	81.6	15.5	178.3
Rainfed	Field 1	0.001	1.6	16.5	0.005	22.6
	Field 2	0.002	0.01	5.5	0.007	11.2
	Field 3	0.001	0.64	11.2	0.008	36.5
	Field 4	0.042	0.12	6.2	0.001	48.1
	Mean	0.01	0.6	9.8	0.005	29.6
	LSD (0.05)	0.001	0.1	1.6	1.2	3.7

Bushra et al. (2018)

they cannot be degraded chemically and biologically compared to other pollutants (Holdaway and Wuyi 2018). To remediate polluted soil, researchers are focusing on the development of new technologies (Gong et al. 2018). Restoration of contaminated soils is vital because utilization of conservative techniques such as landfilling and removal of soil are economically unsuitable and not environment friendly. Sustainable substitutes for the remediation of soil have been investigated. In situ immobilization of metals has been reported a favorable option for soil remediation (O'Connor et al. 2018).

Immobilization of metals by amendments has been developed for inactivation of metals. This process may reduce the danger of groundwater contamination, uptake of heavy metals by plants, and exposure of heavy metals to living organisms (Dahlawi et al. 2018). Organic amendments have been used as obstacles for heavy metals (Pampuro et al. 2018). The toxicity, bioavailability, and mobility of metals are dependent on their chemical nature and pattern of attachment to plants. These factors mainly determine the toxic effects and geochemical paths (Ren et al. 2018). The polluted soils have been remediated by compost and animal manure amendments in agricultural sector (Samuel et al. 2018). These amendments affect the metal speciation and other chemical properties of soil (Kwiatkowska-Malina 2018). Organic matter amendments affect the bioavailability of heavy metals, which rely on the soil type, organic matter, and metals (Egene et al. 2018). Decreased bioavailability of metals converts plant available fractions to fractions of carbonates, oxides of metals, and organic matter (Fu et al. 2018). Organic manures when mixed with polluted soil decreased the mobility of contaminants in the soil (Gondek et al. 2018). Kim et al. (2018) reported that composted cow manure induced Pb and Zn complexes in soil.

2.6 Management of Livestock Manures

Livestock industries have grown tremendously in recent decades to meet the demands of increasing human population. Developed techniques have been introduced for the management of reproduction, breeding, and feeding of livestock production. A large quantity of animal manure has been produced which is linked to the conservation of environment. Livestock manure consists of nutrients which may create eutrophication of lakes and rivers (Won et al. 2017). Death of aquatic organisms, reduced oxygen availability, enhanced algal blooms, and decreased of light penetration are caused by eutrophication (Glibert et al. 2018). Composting is important for the management of animal manure and plays a significant role in nutrient transformations (Jain and Kalamdhad 2018). Agricultural practices for sustainability are supported to maintain conservation of natural resources. The sustainable goal is achieved by recycling nutrients in the ecosystem via best agricultural practices (Ward et al. 2018). Different processes such as denitrification, nitrification, and ammonia volatilization occur during composting. Global warming also occurs due to these processes, which emits N gas and nitrous oxide to the atmosphere. Gaseous emission during composting and storage of manure may result in the reduction of 20–40% of the N (Rotz 2018). Utilization of manure needs composition analysis in relation to the environmental effects. Some poultry manure contain harmful heavy metals as a result of chemicals used in the treatment of diseases in commercial chickens (Codling 2018).

Hou et al. (2018) reported that poultry waste is a major contributor of trace elements that can accumulate in soil after frequent applications. Manure is used as soil amendment for crops that improves the soil properties (Rocco et al. 2018). Menezes-Blackburn et al. (2018) reported the importance of manure for sustainable agricultural practices. It is important to know the effect of animal manure production and quality since it is site specific (Wander 2015). Generation of organic acids, inorganic salts and ammonia and inhibition of plant growth and germination have been reported after the application of immature compost and animal manure (Waqas et al. 2018). Mismanagement of manures increases the chance of air and water pollution (Lecerf et al. 2018). Animal manures are spread into the environment without treatment, and in this way, some nutrients are lost during leaching and volatilization (Malomo et al. 2018). Techniques for the utilization and management of manures are to be explored.

2.7 Mitigating Heavy Metal Pollution in Soils After Manure Amendments

Composting is known to be an economical and environment friendly technique for the treatment of organic wastes, which can transform animal manures into natural fertilizers (Arshad et al. 2018). Adjusting the C:N ratio of lignocellulose compost by adding nitrogen sources in animal residues proved that the composting process was reliable, and the final products were found to be more stable and effective and suitable for safe soil utilization (Fidelis and Rao 2017). It is considered to be most appropriate for recycling of waste, helps in mass and volume reduction, decreases pathogenic microbes, and obtains final end products with good fertilizing property (Asses et al. 2018). During the composting process, factors such as moisture, temperature, chemical nature of the substance, and pH should be maintained (Bayard et al. 2018; Sarker et al. 2020). Li and Takkellapati (2018) reported that the degradation process depends upon the composition of plant materials, i.e., lignin, cellulose, and hemicellulose. It was noticed that the amount of poultry litter may change according to the initial condition of litter material and percentage of lignin (Janssen and Nonnenmann 2017). Composts suitability for land application depends on stability and maturity, which can be assessed by physical and chemical characteristics (Cerda et al. 2017).

During composting, change in temperature, organic carbon, nitrogen pH, C/N, and other physicochemical characteristics are observed. Organic C, water content, and C and N ratio have been found to be reduced during composting (Haghshenas and Ghanbari Malidarreh 2021). It was reported that the compost of poultry manure increased C/N ratio (Onwosi et al. 2017). Based on the temperature range, composting is categorized into three stages (i) mesophilic condition, (ii) thermophilic stage, and (iii) stage of maturation (Hou et al. 2017). To measure the maturity of compost, the chemical and physical characteristics of the compost should be analyzed. Temperature has been known as a simple, easy, and efficient parameter to measure the maturity of compost (Van Fan et al. 2018). The concentrations of NH_4, NO_3, C/N, and pH have been also used to evaluate compost maturity (Azim et al. 2018) Temperature of compost was found correlated with chemical characteristics (Riaz et al. 2018). The quality of compost depends on the compost maturity, composting method, and source of organic matter (Guidoni et al. 2018). Stability of compost is actually the maturity of compost which provides nutrients to the plants. Compost gives a stable form of organic matter and increases nutrient concentration in soils (Barthod et al. 2018).

Heavy metals are toxic in nature and are found in high concentrations in polluted soils (Shifaw 2018). Different human activities contribute to soil contamination, which include pesticide and fertilizer use on a large scale, wastewater irrigation, and dumping of mining and industrial wastes (Shi et al. 2018; Farokhian et al. 2021; Amjadian et al. 2021; Modanlo et al. 2021). Serious economic losses, reduction of crop yield, and inhibition of plant growth are the major impacts of polluted soils (Machado et al. 2017). Exposure to heavy metal contaminated soils greatly affects human health. Microbes cannot decompose heavy metals unlike other organic contaminants (Portier et al. 2017).

Changing the bioavailability and mobility of heavy metals through soil amendments is cheap and a highly applicable method for soil remediation (Lahori et al. 2017). Compost, coal fly ash, biochar, and sewage sludge have been utilized as soil amendments (Lucchini et al. 2014). Composting process leads to the stability and humification of organic (Awasthi et al. 2015). Microbes degrade through different stages and with the help of enzymes convert organic wastes into humic substances, mineral ions, water, and carbon dioxide (Bialobrzewski et al. 2015). The presence of microbes, humic substances, and mineral ions in compost affect heavy metal immobilization, thereby decreasing environmental and ecological risks (Udovic and McBride 2012). Redox reaction, precipitation, complexation, and adsorption are the various types of chemical reactions that are involved in heavy metal immobilization (Park et al. 2011). By using compost, we can decrease the risk of crop failure and reduce human health risk and economic losses.

Compost is referred as a viable option for the management of waste. In this way, parasites and pathogens from waste, organic pollutants, and phytoxic heavy metals are removed (Kulikowska and Gusiatin 2015). Biomass of crop is enhanced, activities of microbes are increased, and physical and soil fertility status of soils become more better due to the application of compost (Calleja-Cervantes et al. 2015). To remediate contaminated soils, application of composted manures is a sustainable, environment friendly, and economical approach. Therefore, compost is considered to be more reliable, safe, and effective soil amendment and is used as soil fertilizer and conditioner by some farmers on agricultural lands in Iran (Sharifi and Renella 2015). In general, compost application is a very common practice for heavy metal fixation and soil improvement. The bioavailability of heavy metals from contaminated soils is reduced by compost amendment (Gul et al. 2016). Therefore, compost technology is recommended as a sustainable approach for soil management (Patra 2018). The reuse of organic waste is the most appropriate way for environmental management (Torkashvand et al. 2015).

2.8 Mobility of Heavy Metals in Manure-Amended Soils

Mineral composition, content of metal oxides, organic matter, pH, and other soil properties are some factors used for monitoring the mobility of heavy metals in the soil (Yang et al. 2016; Yolchiyeva et al. 2020). As mobilization mainly depends on Eh condition but the leaching of Zn, Cu and Cr was due to low pH (Unda-Calvo et al. 2017). In polluted soils, mobility of heavy metals can be possible due to compost addition. It was reported that the application of municipal solid waste compost to a field affected the mobility of heavy metals (Párraga-Aguado et al. 2017). The mobility of heavy metals reduced due to greater amount of Fe and organic carbon present in soil (Clemente et al. 2010). With the application of compost to the polluted soils, serious attention should be given to the change of heavy metals mobility in soils. Different heavy metals may show different responses to compost addition. Under high pH, cationic metals are fixed (Fleming et al. 2013). Soils of agricultural field are usually not contaminated by a single toxic metal but polluted by several types of toxic metals (Cai et al. 2012).

2.9 Bioavailability of Heavy Metals upon Manure Amendment

Bioavailability is an important indicator to monitor the adverse impacts of heavy metals on ecosystem. Bioavailability is generally defined as a process in which chemical substances move from an abiotic environment to a biotic environment (Bolan et al. 2014). Therefore, total organic carbon, soil organic matter, cation or anion exchange capacities, soil pH, soil texture, and soil organisms are considered important factors for the bioavailability of heavy metals (Zhou et al. 2014; Qayyum et al. 2020). The bioavailability of Pb was affected due to the application of composted and fresh olive husk (Dela Fuente et al. 2011). Composted olive husk decreased Pb bioavailability because it forms complexes with humic substances, but fresh olive husk was more toxic for plants. In alkaline soils, addition of green compost decreased extractable Pb and reduced Pb and Zn accumulation (Zhou et al. 2012). Compost application can prevent accumulation of heavy metals in plants and decreases the available portions i.e., adsorbed, exchangeable and extractable and enhances the organic matter bound and total form (non-available portion) of heavy metals to plants (Rosen and Chen 2014).

There are two well-known methods for applying compost to soils: (i) incorporation and (ii) mulching (Bastida et al. 2010). Compost addition to the top few centimeter of soil help to increase the availability of soil microbes and also interacts with the roots of plants and affects soil bulk density, N, and C as compared to surface application (Gale et al. 2017). Mulching is commonly used in the area of agriculture, horticulture, and under dry climatic conditions because it prevents loss of water by evaporation (Kader et al. 2017). Lower accessibility of nutrients to microbes are the drawbacks of mulching, while in case of incorporation, much availability of nutrients to soil microbes is achieved (Bonanomi et al. 2017). Compost with coarse texture is considered better material for mulching because it permits air and water through the soil. Fine-textured compost can be applied as mulch in thick layer (Fuller 2018) (Table 2.5).

2.10 Effects of Manure Amendment on Chemical Properties of the Soil

The effect of compost application on the pH of soil has not been comprehensively reported. The pH of mature composts is neutral or slightly alkalinein nature. Amendment of soil with compost has been reported at high and low soil pH and possesses the buffering ability for soil (Hurel et al. 2017; Hassanpour 2021) depending on soil pH, and the nature of compost. Many chemical reactions, microbial activity, plant growth, movement and uptake of plant nutrients, and metal solubility are affected by soil pH (Ma et al. 2018). Application of broiler litter compost resulted in an increase in soil pH, ammonification process, adding of basic cationic

Table 2.5 Extractable heavy metals (mg kg^{-1}) in contaminated soil amended with cow manure and poultry litter composted with plant residues (Bushra et al. 2019)

Manure/application rate	Plant residues	Manure mixture	Pb	Ni	Cu	Cd	Zn	pH	EC (μS m^{-1})
Control		0.0	27.2	7.2	8.5	2.8	2.41	8.10	636.8
Cow manure, 10 t ha^{-1}		1:0	24.2	6.9	6.5	2.3	1.93	7.15	619.2
	Privet	1:1	21.5	6.6	5.0	1.9	1.72	7.35	623.6
		1:2	21.3	5.7	3.9	1.1	1.15	7.55	626.5
	Cypress	1:1	22.9	5.8	3.6	1.5	1.85	7.26	625.5
		1:2	19.7	4.8	3.2	0.9	1.72	7.45	629.0
Poultry litter, 10 t ha^{-1}		1:0	15.5	5.9	4.0	1.5	1.38	6.73	593.0
	Privet	1:1	14.4	5.3	3.6	1.4	1.19	6.77	595.5
		1:2	12.7	3.5	2.9	1.0	0.94	6.85	600.6
	Cypress	1:1	14.6	4.7	3.8	1.4	0.86	6.82	597.5
		1:2	11.2	4.0	2.4	1.2	0.70	6.86	606.2
Cow manure, 20 t ha^{-1}		1:0	17.5	5.5	3.6	0.8	0.60	7.35	839.2
	Privet	1:1	15.4	5.3	3.2	0.7	0.56	7.65	842.5
		1:2	14.5	4.8	2.8	0.6	0.57	7.85	847.1
	Cypress	1:1	14.4	4.9	2.2	0.5	0.65	7.80	840.5
		1:2	11.5	4.2	1.8	0.5	0.50	7.90	844.0
Poultry litter, 20 t ha^{-1}		1:0	9.8	4.8	1.6	0.5	0.22	7.15	810.0
	Privet	1:1	8.1	4.6	1.1	0.4	0.15	7.41	817.0
		1:2	6.9	3.8	1.1	0.4	0.11	7.45	823.1
	Cypress	1:1	9.2	4.3	0.9	0.3	0.16	7.33	815.5
		1:2	7.8	3.8	0.7	0.3	0.15	7.55	820.0
	LSD (0.05)		0.9	0.3	0.1	0.1	0.05	0.2	12.2

elements and release of ammonia during decomposition stage (Agyarko-Mintah et al. 2017). The release of organic acid in the decomposition and production of pH via nitrification when compost of agro-industrial waste with rice straw was applied on the soil decreased soil pH (Rashad et al. 2011). Alkaline pH under saline conditions affects plant growth (Asadi and Jalilian 2021).

Compost incorporated into the soil can increase the capacity of soil cation exchange. Humic acids are the main constituents of compost that fix cations because they contain carboxylic acid groups (Bass et al. 2016). The compost consists of essential nutrients, which are in organic form. These nutrients are released gradually and leach down less as compared to chemical fertilizers (Rigby et al. 2016). Higher concentration of total C and N was achieved in soil after incorporation of compost of garden waste to soils (Jakubus 2016). Agegnehu et al. (2016) reported that addition of compost to the soil enhanced organic matter and total N and C concentrations. Tien et al. (2017) reported that the use of sewage waste, and dairy manure produced 37–67% C in the soil. In compost, C is mainly present in macro-aggregates (Amendola et al. 2017). Application of dairy manure and municipal solid waste (MSW) compost to soils can increase P, N, and other nutrients in soils (Zornoza et al. 2016). Increased activities of microbes lead to higher N mineralization and denitrification (Nguyen et al. 2017).

2.11 Effects of Manure Amendment on the Physical Properties of Soil

Application of MSWs compost to soils increases soil organic matter, enhances water availability and retention, and reduces water loss through evaporation (Vithanage et al. 2017). In an agricultural field, it was observed that cattle manure when applied to the fields increased 58–86% of available water and enhanced macro- and micro-porosity of soils. Production of perennial crops in arid or semi-arid regions increased by increasing the water retention of soils (Smith et al. 2016). Soil with good structure helps in water and gas movement in soil and reduces soil erosion. Compost with gummy and spongy texture forms aggregates (Sodhi et al. 2009). It was reported that after MSWs application to clay soils, an improvement in the aeration and drainage of soil was observed, and same benefits were observed for cattle manure (Eden et al. 2017). Duong et al. (2011) noticed that compost increased aggregate stability in medium or lighter textured soils than sandy textured soils.

2.12 Effects of Manure Amendment on Biological Properties

The incorporation of compost to soil affects the microbes in soils by supplying nutrients and changing the physical and chemical properties of soil. Compost increases the population of microbes in soil (Sutton-Grier et al. 2010) Compost from various feedstock increases the density of denitrifying, ammonia-oxidizing, and gram-positive bacteria; denitrification and nitrification process; enzymes activities; and soil respiration (Sutton-Grier et al. 2010). These variations affect the availability of nutrients. Community structure and microbial activity changes after compost application suppress the soil-borne plant pathogens (Odlare and Pell 2009) and increase the breakdown of organic pollutants (Yuan et al. 2009).

2.13 Effects of Manure Amendment on Plant Growth and Nutrient Uptake

Compost application to soils enhances the availability of plant nutrients (Agegnehu et al. 2017; Mahboub et al. 2021). When compared to the quantity of P and N added, the contents of these nutrients uptaken by the plants from compost may be smaller in amount than inorganic fertiliser due to the processes of mineralization and immobilization via microbes (Crews et al. 2016). However, at initial stage, slow mineralization process can sustain the release of N over a 2-year period and about 36–44% of compost N was mineralized (Ghidotti et al. 2016). Compost from different sources such as kitchen waste, manure, and fruit residues can increase the retention time of applied N fertilizer in the soil-plant system by encouraging the uptake of N by plant and immobilization process and by reducing leaching of N and gaseous losses (Kuppusamy et al. 2016). A study was conducted in which yard waste compost was incorporated into soils. The results showed that it increased the availability of water, and excellent root system was reported due to the application of compost (Durand et al. 2016). Root development, plant growth, and uptake of nutrients can be stimulated by compost (Oworu et al. 2010). Humic substances are the major and significant part of soil organic matter present in composts and they can increase shoot biomass (Lazcano et al. 2009).Compost can increase nutrient availability and uptake by plant tissues. Direct effects of compost are noticed when nutrients are added with the compost, and indirect effects such as higher microbial activity, enhanced soil structure, or storage of nutrients and water retention are also observed. Improvements in soil structure and water storage capacity promote root growth. Controlling heavy metal pollution in soils and plants using composted manures with plant waste has not been well studied. Co-composting of plant and animal wastes was hypothesized as a better soil amendment for the reduction of heavy metal contamination in soils (Fig. 2.1).

2.14 Mechanism of Heavy Metal Remediation in Manure-Amended Soils

Compost is the end product obtained due to microbial activities. Adding compost to the soil changes the bioavailability and mobility of heavy metals and reduce the harmful effects of polluted soils. These changes in soils occur through several processes (Machado et al. 2017).

2.14.1 Adsorption and Complexation

Immobilization of heavy metals in soils can be achieved by adsorption. The compost has a high ability to adsorb heavy metals (Soares et al. 2016). Various studies have reported on the adsorption ability of compost using isothermal and kinetic adsorption methods (Venegas et al. 2015). Compost is superior to zeolite due to its chemical absorption capacity and reduces up to 85–89% of heavy metal concentrations in water (Simantiraki and Gidarakos 2015). Compost can affect the redox potential, organic matter, pH and sorption of heavy metals in soils (Kargar et al. 2015). Compost contains humus substances that comprise long chain of functional

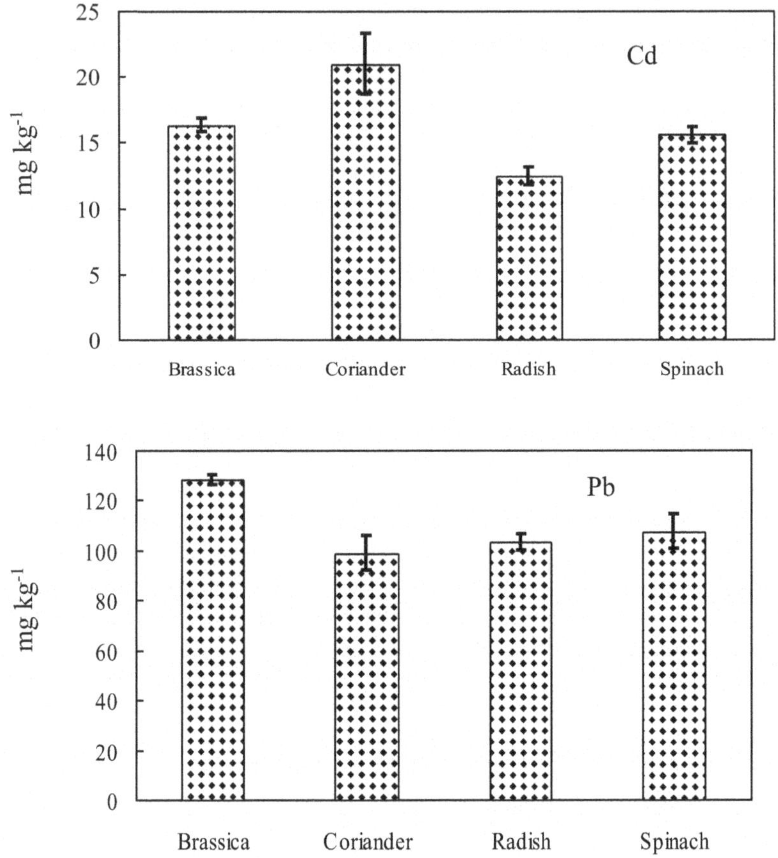

Fig. 2.1 Concentration of Cd and Pb in vegetables irrigated with wastewater (Irshad et al. 2015)

groups of phenols, carboxyl, and carbonyl, which can attach to ions of metals through complexation (Tsang et al. 2014). Composts exhibit affinity toward different types of toxic metals. Antoniadis et al. (2018) reported the affinity of humic substances to heavy metals to be in the following order: Pb > Cu > Cd > Zn. Through mineralization and biosorption, some microbes in compost can immobilize heavy metals (Kushwaha et al. 2018). It was reported that more than 90% of Pb concentration available form was removed by various isolates. Oxides of aluminum, iron and manganese in compost have retaining property of heavy metals in soil (Vaverková 2018). Adsorption of heavy metals by these inorganic substances is almost an irreversible process (Sherlala et al. 2018) (Fig. 2.2).

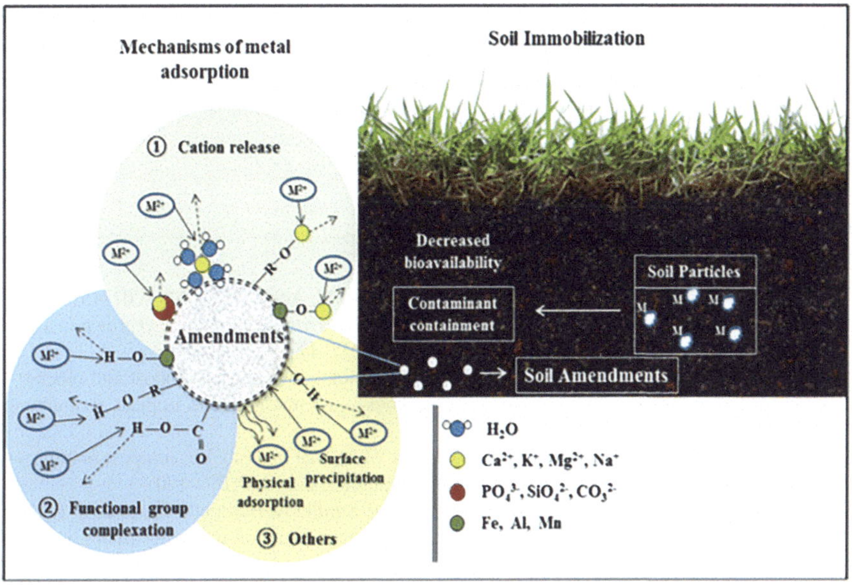

Fig. 2.2 Heavy metal dynamics in soil–plant systems. (Source: Nejad et al. (2017))

2.14.2 Precipitation

Precipitation is a well-known approach for immobilization of heavy metals, and results into low solubility in soil. The most abundant and important component of compost is humic substances. When compost is added to the polluted soil, the humic substances present in compost react rapidly with the surface of heavy metals to generate humic complexes of metals and humate, which quickly precipitate and are insolubilized (Garcia-Mina 2006). Acid-soluble fraction of Cu is reduced by complexation and precipitation reactions due to the compost added to the soil (Lagomarsino et al. 2011). Larger amounts of phosphorus help to immobilize heavy metals in some composts (Katoh et al. 2014). The precipitation of heavy metals by phosphate and humics depend on soil pH (Ye et al. 2017).

2.14.3 Redox Reactions

The reaction that occurs between heavy metals and organic matter in compost changes the toxic state of heavy metals to nontoxic state. Reduction of Cr (VI) to Cr (III) by redox reaction and dissolved organic matter (DOM) contained in compost (Naveen et al. 2017). DOM consists of redox-active functional groups, which gain electrons from electron donors and exhibit electron-transfer capabilities that help in the redox reaction of heavy metals (Yuan et al. 2011). After adding compost to the soil, microbes create an anaerobic environment and decreases the sulfate concentration, thereby decreasing the availability of metals. Heavy metals toxicity and bioavailability are decreased through either indirect or direct reactions of compost (Paul 2014).

References

Abbas T, Rizwan M, Ali S, Zia-ur-Rehman M, Qayyum MF, Abbas F, Ok YS (2017) Effect of biochar on cadmium bioavailability and uptake in wheat (*Triticum aestivum* L.) grown in a soil with aged contamination. Ecotoxicol Environ Saf 140:37–47

Agegnehu G, Bass AM, Nelson PN, Bird MI (2016) Benefits of biochar, compost and biochar–compost for soil quality, maize yield and greenhouse gas emissions in a tropical agricultural soil. Sci Total Environ 543:295–306

Agegnehu G, Srivastava AK, Bird MI (2017) The role of biochar and biochar-compost in improving soil quality and crop performance; a review. Appl Soil Ecol 119:156–170

Agyarko-Mintah E, Cowie A, Van Zwieten L, Singh BP, Smillie R, Harden S, Fornasier F (2017) Biochar lowers ammonia emission and improves nitrogen retention in poultry litter composting. Waste Manag 61:129–137

Alves RN, Maulvault AL, Barbosa VL, Fernandez-Tejedor M, Tediosi A, Kotterman M, Sloth JJ (2018) Oral bioaccessibility of toxic and essential elements in raw and cooked commercial seafood species available in European markets. Food Chem 267:15–27

Amendola C, Montagnoli A, Terzaghi M, Trupiano D, Oliva F, Baronti S, Scippa GS (2017) Short-term effects of biochar on grapevine fine root dynamics and arbuscular mycorrhizae production. Agric Ecosyst Environ 239:236–245

Amjadian E, Zeinodini A, Doğan H (2021) Effect of fertilizer management systems on growth and balance of nutrients in wheat cultivation. Central Asian J Plant Sci Innov 1(2):56–69. https://doi.org/10.22034/CAJPSI.2021.02.01

Antoniadis V, Shaheen SM, Tsadilas CD, Selim MH, Rinklebe J (2018) Zinc sorption by different soils as affected by selective removal of carbonates and hydrous oxides. Appl Geochem 88:49–58

Arshad M, Bano I, Khan N, Shahzad MI, Younus M, Abbas M, Iqbal M (2018) Electricity generation from biogas of poultry waste: An assessment of potential and feasibility in Pakistan. Renew Sust Energ Rev 81:1241–1246

Asadi N, Jalilian S (2021) The effect of methyl jasmonate on the germination of lemon seeds under the influence of salinity stress. Central Asian J Environ Sci Technol Innov 2(3):119–128. https://doi.org/10.22034/CAJESTI.2021.03.03

Asirifi I (2017) Biochar and wastewater use in urban agriculture: soil health and pathogenic properties. Thesis of Department of Biotechnology, Faculty of Agriculture, University for Development Studies, Ghana. http://udsspace.uds.edu.gh/handle/123456789/1235

Asses N, Farhat A, Cherif S, Hamdi M, Bouallagui H (2018) Comparative study of sewage sludge co-composting with olive mill wastes or green residues: process monitoring and agriculture value of the resulting composts. Process Saf Environ Prot 114:25–35

Augustsson A, Uddh-Söderberg T, Filipsson M, Helmfrid I, Berglund M, Karlsson H, Alriksson S (2018) Challenges in assessing the health risks of consuming vegetables in metal contaminated environments. Environ Int 113:269–280

Awasthi MK, Pandey AK, Bundela PS, Khan J (2015) Co-composting of organic fraction of municipal solid waste mixed with different bulking waste: characterization of physicochemical parameters and microbial enzymatic dynamic. Bioresour Technol 182:200–207

Azim K, Soudi B, Boukhari S, Perissol C, Roussos S, Alami IT (2018) Composting parameters and compost quality: a literature review. Org Agric 8(2):141–158

Barthod J, Rumpel C, Calabi-Floody M, Mora ML, Bolan NS, Dignac MF (2018) Adding worms during composting of organic waste with red mud and fly ash reduces CO_2 emissions and increases plant available nutrient contents. J Environ Manag 222:207–215

Bass AM, Bird MI, Kay G, Muirhead B (2016) Soil properties, greenhouse gas emissions and crop yield under compost, biochar and co-composted biochar in two tropical agronomic systems. Sci Total Environ 550:459–470

Bastida F, Hernández T, Garcia C (2010) Soil degradation and rehabilitation: microorganisms and functionality. In: Environmental toxicology. Springer, Berlin, pp 253–270

Bayard R, Benbelkacem H, Gourdon R, Buffière P (2018) Characterization of selected municipal solid waste components to estimate their biodegradability. J Environ Manag 216:4–12

Bialobrzewski I, Miks-Krajnik M, Dach J, Markowski M, Czekala W, Gluchowska K (2015) Model of the sewage sludge-straw composting process integrating different heat generation capacities of mesophilic and thermophilic microorganisms. Waste Manag 43:72–83

Blum A (2018) Plant breeding for stress environments. Sci Hortic 64:72–86

Bolan N, Kunhikrishnan A, Thangarajan R, Kumpiene J, Park J, Makino T (2014) Remediation of heavy metalloids contaminated soils to mobilize. J Hazard Mater 266:141–166

Bonanomi G, Chirico GB, Palladino M, Gaglione SA, Crispo DG, Lazzaro U, Rippa M (2017) Combined application of photo-selective mulching films and beneficial microbes affects crop yield and irrigation water productivity in intensive farming systems. Agric Water Manag 184:104–113

Bünemann EK, Bongiorno G, Bai Z, Creamer RE, De Deyn G, de Goede R, Pulleman M (2018) Soil quality–a critical review. Soil Biol Biochem 120:105–125

Bushra H, Ping A, Pervez A, Faridullah, Irshad M (2018) Characterization of heavy metal in soils as affected by long-term irrigation with industrial wastewater. J Water Reuse Desal 9(1):47–56

Bushra H, Hassan A, Abbasi AM, Ping A, Yang S, Irshad M (2019) Effects of manure amendments on the extractability and bioavailability of heavy metals from contaminated soil after wastewater irrigation. J Water Reuse Desal 10(1):17–29

Cai L, Xu Z, Ren M, Guo Q, Hu X, Hu G (2012) Source identification of eight hazardous heavy metals in agricultural soils of Huizhou, Guangdong Province, China. Ecotoxicol Environ Saf 78:2–8

Calleja-Cervantes ME, Fernández-González AJ, Irigoyen I, Fernández-López M, Aparicio-Tejo PM, Menéndez S (2015) Thirteen years of continued application of composted organic wastes in a vineyard modify soil quality characteristics. Soil Biol Biochem 90:241–254

Calmon C (2018) Ion exchange pollution control. Soil Pollut 31:64–85

Cerda A, Artola A, Font X, Barrena R, Gea T, Sánchez A (2017) Composting of food wastes: status and challenges. Bioresour Technol 40:52–71

Chandra R, Bharagava RN, Yadav S, Mohan D (2009) Accumulation and distribution of toxic metals in wheat (Triticum aestivum L.) and Indian mustard (Brassica campestris L.) irrigated with distillery and tannery effluents. J Hazard Mater 162:1514–1521

Choppala G, Kunhikrishnan A, Seshadri B, Park JH, Bush R, Bolan N (2018) Comparative sorption of chromium species as influenced by pH, surface charge and organic matter content in contaminated soils. J Geochem Explor 184:255–260

Clemente R, Hartley W, Riby P, Dickinson NM, Lepp NW (2010) Trace element mobility in a contaminated soil two years after field-amendment with a greenwaste compost mulch. Environ Pollut 158(5):1644–1651

Codling EE (2018) Accumulation of lead and arsenic by peanut grown on lead and arsenic contaminated soils amended with broiler litter ash or superphosphate. J Plant Nutr 41(12):1615–1623

Crews TE, Blesh J, Culman SW, Hayes RC, Jensen ES, Mack MC, Schipanski ME (2016) Going where no grains have gone before: from early to mid-succession. Agric Ecosyst Environ 223:223–238

Dahlawi S, Naeem A, Iqbal M, Farooq MA, Bibi S, Rengel Z (2018) Opportunities and challenges in the use of mineral nutrition for minimizing arsenic toxicity and accumulation in rice: a critical review. Chemosphere 194:171–188

Dela Fuente C, Clemente R, Martinez-Alcala I, Tortosa G, Pilar Bernal M (2011) Impact of fresh and composted solid olive husk and their water-soluble fractions on soil heavy metal fractionation; microbial biomass and plant uptake. J Hazard Mater 186(2–3):1283–1289

Duong TT, Penfold C, Marschner P (2011) Amending soils of different texture with six compost types: impact on soil nutrient availability, plant growth and nutrient uptake. Plant Soil 354(1–2):197–209

Durand M, Porcheron B, Hennion N, Maurousset L, Lemoine R, Pourtau N (2016) Water deficit enhances C export to the roots in *Arabidopsis thaliana* plants with contribution of sucrose transporters in both shoot and roots. Plant Physiol 170(3):1460–1479

Edelstein M, Ben-Hur M (2018) Heavy metals and metalloids: Sources, risks and strategies to reduce their accumulation in horticultural crops. Sci Hortic 234:431–444

Eden M, Gerke HH, Houot S (2017) Organic waste recycling in agriculture and related effects on soil water retention and plant available water: a review. Agron Sustain Dev 37(2):11

Egene CE, Van Poucke R, Ok YS, Meers E, Tack FMG (2018) Impact of organic amendments (biochar, compost and peat) on Cd and Zn mobility and solubility in contaminated soil of the Campine region after three years. Sci Total Environ 626:195–202

Etesami H, Jeong BR (2018) Silicon (Si): review and future prospects on the action mechanisms in alleviating biotic and abiotic stresses in plants. Ecotoxicol Environ Saf 147:881–896

Farokhian S, Tohidi Nejad E, Mohamadi Nejad G (2021) Studying the effect of bio-fertilizers on the yield of *Sesamum indicum* genotypes under drought stress. Central Asian J Plant Sci Innov 1(1):32–38. https://doi.org/10.22034/CAJPSI.2021.01.04

Fidelis C, Rao BR (2017) Enriched cocoa pod composts and their fertilizing effects on hybrid cocoa seedlings. Int J Recycl Org Waste Agric 6(2):99–106

Fleming M, Tai Y, Zhuang P, McBride MB (2013) Extractability and bioavailability of Pb and As in historically contaminated orchard soil: effects of compost amendments. Environ Pollut 177:90–97

Fu H, Zhang B, Yang J, Liu H, Yang S, Zhao P (2018) Cadmium and lead speciation as affected by soil amendments in calcareous soil. Environ Eng Sci 123:154–161

Fuller WH (2018) Soils in waste treatment and utilization. Environ Pollut 54:68–101

Gale NV, Halim MA, Horsburgh M, Thomas SC (2017) Comparative responses of early-successional plants to charcoal soil amendments. Ecosphere 8(10):24–30

Garcia-Mina JM (2006) Stability, solubility and maximum metal binding capacity in metal–humic complexes involving humic substances extracted from peat and organic compost. Org Geochem 37(12):1960–1972

Ghidotti M, Fabbri D, Hornung A (2016) Profiles of volatile organic compounds in biochar: insights into process conditions and quality assessment. ACS Sustain Chem Eng 5(1):510–517

Giri B, Dey S, Das T, Sarkar M, Banerjee J, Dash SK (2018) Chronic hyperglycemia mediated physiological alteration and metabolic distortion leads to organ dysfunction, infection, cancer progression and other pathophysiological consequences: An update on glucose toxicity. Biomed Pharmacother 107:306–328

Glibert PM, Berdalet E, Burford MA, Pitcher GC, Zhou M (2018) Harmful algal blooms and the importance of understanding their ecology and oceanography. In: Global ecology and oceanography of harmful algal blooms. Springer, Cham, pp 9–25

Gondek K, Mierzwa-Hersztek M, Kopeć M (2018) Mobility of heavy metals in sandy soil after application of composts produced from maize straw, sewage sludge and biochar. J Environ Manag 210:87–95

Gong X, Huang D, Liu Y, Peng Z, Zeng G, Xu P, Wan J (2018) Remediation of contaminated soils by biotechnology with nanomaterials: bio-behavior, applications, and perspectives. Crit Rev Biotechnol 38(3):455–468

Guidoni LLC, Marques RV, Moncks RB, Botelho FT, da Paz MF, Corrêa LB, Corrêa ÉK (2018) Home composting using different ratios of bulking agent to food waste. J Environ Manag 207:141–150

Gul S, Naz A, Khan A, Nisa S, Irshad M (2016) Phytoavailability and leachability of heavy metals from contaminated soil treated with composted livestock manure. Soil Sediment Contam Int J 25(2):181–194

Haghshenas H, Ghanbari Malidarreh A (2021) Response of yield and yield components of released rice cultivars from 1990-2010 to nitrogen rates. Central Asian J Plant Sci Innov 1(1):23–31. https://doi.org/10.22034/CAJPSI.2021.01.03

Hassanpour M (2021) Techno-economic assessment model of screening step of agricultural wastes recycling to animal feed project. Central Asian J Environ Sci Technol Innov 2(1):1–11. https://doi.org/10.22034/CAJESTI.2021.01.01

Hejna M, Gottardo D, Baldi A, Dell'Orto V, Cheli F, Zaninelli M, Rossi L (2018) Nutritional ecology of heavy metals. Animal 12:1–15

Holdaway J, Wuyi W (2018) From soil pollution to "cadmium rice" to public health impacts: An interdisciplinary analysis of influencing factors and possible responses. J Resour Ecol 9(1):1–12

Hou N, Wen L, Cao H, Liu K, An X, Li D, Li C (2017) Role of psychrotrophic bacteria in organic domestic waste composting in cold regions of China. Bioresour Technol 236:20–28

Hou D, Ding Z, Li G, Wu L, Hu P, Guo G, Wang X (2018) A sustainability assessment framework for agricultural land remediation in China. Land Degrad Dev 29(4):1005–1018

Hurel C, Taneez M, Ghirardini AV, Libralato G (2017) Effects of mineral amendments on trace elements leaching from pre-treated marine sediment after simulated rainfall events. Environ Pollut 220:364–374

Irshad M, Naureen R, Faridullah F (2015) Assessing selected heavy metals in vegetables and soils irrigated with wastewater at Haripur, Pakistan. Minerva Biotecnol 27:99–105

Jafari AJ, Kermani M, Kalantary RR, Arfaeinia H (2018) The effect of traffic on levels, distribution and chemical partitioning of harmful metals in the street dust and surface soil from urban areas of Tehran, Iran. Environ Earth Sci 77(2):38

Jain MS, Kalamdhad AS (2018) A review on management of Hydrilla verticillata and its utilization as potential nitrogen-rich biomass for compost or biogas production. Bioresou Technol Rep 1:69–78

Jakubus M (2016) Estimation of phosphorus bioavailability from composted organic wastes. Chem Speciat Bioavailab 28(1–4):189–198

Jamali MK, Kazi TG, Arain MB, Afridi HI, Jalbani N, Kandhro GA, Shah AQ, Baig JA (2009) Heavy metal accumulation in different varieties of wheat (Triticum aestivum L.) grown in soil amended with domes-tic sewage sludge. J Hazard Mater 164:1386–1391

Janssen B, Nonnenmann MW (2017) Public health science in agriculture: Farmers' perspectives on respiratory protection research. J Rural Stud 55:122–130

Kader MA, Senge M, Mojid MA, Ito K (2017) Recent advances in mulching materials and methods for modifying soil environment. Soil Tillage Res 168:155–166

Kargar M, Clark OG, Hendershot WH, Jutras P, Prasher SO (2015) Immobilization of tracemetals in contaminated urban soil amended with compost and biochar. Water Air Soil Pollut 226(6):1–12

Katoh M, Kitahara W, Sato T (2014) Sorption of lead in animal manure compost: contributions of inorganic and organic fractions. Water Air Soil Pollut 225(1):1–12

Khalid S, Shahid M, Bibi I, Sarwar T, Shah A, Niazi N (2018) A review of environmental contamination and health risk assessment of wastewater use for crop irrigation with a focus on low and high-income countries. Int J Environ Res Public Health 15(5):895

Kim HS, Seo BH, Kuppusamy S, Lee YB, Lee JH, Yang JE, Kim KR (2018) A DOC coagulant, gypsum treatment can simultaneously reduce As, Cd and Pb uptake by medicinal plants grown in contaminated soil. Ecotoxicol Environ Saf 148:615–619

Kouchou A, Rais N, Elsass F, Duplay J, Fahli N, Ghachtouli NEL (2017) Effects of long-term heavy metals contamination on soil microbial characteristics in calcareous agricultural lands (Saiss plain, North Morocco). J Mater Environ Sci 8(2):691–695

Kulikowska D, Gusiatin ZM (2015) Sewage sludge composting in a two-stage system: carbon and nitrogen transformations and potential ecological risk assessment. Waste Manag 38:312–320

Kuppusamy S, Thavamani P, Megharaj M, Venkateswarlu K, Naidu R (2016) Agronomic and remedial benefits and risks of applying biochar to soil: current knowledge and future research directions. Environ Int 87:1–12

Kushwaha A, Hans N, Kumar S, Rani R (2018) A critical review on speciation, mobilization and toxicity of lead in soil-microbe-plant system and bioremediation strategies. Ecotoxicol Environ Saf 147:1035–1045

Kwiatkowska-Malina J (2018) Functions of organic matter in polluted soils: the effect of organic amendments on phytoavailability of heavy metals. Appl Soil Ecol 123:542–545

Lagomarsino A, Mench M, Marabottini R, Pignataro A, Grego S, Renella G (2011) Copper distribution and hydrolase activities in a contaminated soil amended with dolomitic limestone and compost. Ecotoxicol Environ Saf 74(7):2013–2019

Lahori AH, Zhang Z, Guo Z, Mahar A, Li R, Awasthi MK, Zhao J (2017) Potential use of lime combined with additives on (im) mobilization and phytoavailability of heavy metals from Pb/Zn smelter contaminated soils. Ecotoxicol Environ Saf 145:313–323

Lazcano C, Arnold J, Zaller JG, Martín JD, Salgado AT (2009) Compost and vermicompost as nursery pot components: effects on tomato plant growth and morphology. Span J Agric Res 4:944–951

Lecerf R, Van Der Velde M, Dentener F (2018) Impact of nitrogen management plans on winter soft wheat yield and nitrogen balance at European scale, simulated with the stics crop model. J Agric 12:45–61

Li T, Takkellapati S (2018) The current and emerging sources of technical lignins and their applications. Biofuels Bioprod Biorefin 12(5):756–787

Libutti A, Gatta G, Gagliardi A, Vergine P, Pollice A, Beneduce L, Tarantino E (2018) Agro-industrial wastewater reuse for irrigation of a vegetable crop succession under Mediterranean conditions. Agric Water Manage 196:1–14

Liu SH, Zeng GM, Niu QY, Liu Y, Zhou L, Jiang LH, Cheng M (2017) Bioremediation mechanisms of combined pollution of PAHs and heavy metals by bacteria and fungi: a mini review. Bioresour Technol 224:25–33

Lucchini P, Quilliam RS, DeLuca TH, Vamerali T, Jones DL (2014) Does biochar application alters heavy metal dynamics in agricultural soil? Agric Ecosyst Environ 184:149–157

Ma C, White JC, Zhao J, Zhao Q, Xing B (2018) Uptake of engineered nanoparticles by food crops: characterization, mechanisms, and implications. Annu Rev Food Sci Technol 9:129–153

Machado J, Campos A, Vasconcelos V, Freitas M (2017) Effects of microcystin-LR and cylindrospermopsin on plant-soil systems: a review of their relevance for agricultural plant quality and public health. Environ Res 153:191–204

Mahboub KA, Haddad A, Alipoor R, Hojati S (2021) Cow manure and sawdust vermicompost effect on nutrition and growth of ornamental foliage plants. Central Asian J Environ Sci Technol Innov 2(2):68–78. https://doi.org/10.22034/CAJESTI.2021.02.03

Mahmoud EK, Ghoneim AM (2016) Effect of polluted water on soil and plant contamination by heavy metals in El-Mahla El-Kobra, Egypt. Solid Earth 7(2):703–711

Malomo GA, Madugu AS, Bolu SA (2018) Sustainable animal manure management strategies and practices. Agric Waste Residues 40(3):20–26

Manzoor J, Sharma M, Wani KA (2018) Heavy metals in vegetables and their impact on the nutrient quality of vegetables: a review. J Plant Nutr 41(13):1744–1763

Meena RAA, Sathishkumar P, Ameen F, Yusoff ARM, Gu FL (2018) Heavy metal pollution in immobile and mobile components of lentic ecosystems: a review. Environ Sci Pollut Res 25(5):4134–4148

Meier S, Curaqueo G, Khan N, Bolan N, Rilling J, Vidal C, Borie F (2017) Effects of biochar on copper immobilization and soil microbial communities in a metal-contaminated soil. J Soils Sediments 17(5):1237–1250

Menezes-Blackburn D, Giles C, Darch T, George TS, Blackwell M, Stutter M, Brown L (2018) Opportunities for mobilizing recalcitrant phosphorus from agricultural soils: a review. Plant Soil 427(1–2):5–16

Modanlo H, Baghi M, Ghanbari Malidarreh A (2021) Sunflower (*Helianthus annuus* L.) grain yield affected by fertilizer and plant density. Central Asian J Plant Sci Innov 1(2):102–108. https://doi.org/10.22034/CAJPSI.2021.02.05

Naveen BP, Mahapatra DM, Sitharam TG, Sivapullaiah PV, Ramachandra TV (2017) Physico-chemical and biological characterization of urban municipal landfill leachate. Environ Pollut 220:1–12

Nejad ZD, Jung MC, Kim KH (2017) Remediation of soils contaminated with heavy metals with an emphasis on immobilization technology. Environ Geochem Health 40:1–27

Nguyen TTN, Xu CY, Tahmasbian I, Che R, Xu Z, Zhou X, Bai SH (2017) Effects of biochar on soil available inorganic nitrogen: a review and meta-analysis. Geoderma 288:79–96

Noman A, Ali Q, Maqsood J, Iqbal N, Javed MT, Rasool N, Naseem J (2018) Deciphering physio-biochemical, yield, and nutritional quality attributes of water-stressed radish (Raphanus sativus L.) plants grown from Zn-Lys primed seeds. Chemosphere 195:175–189

Obosu-Mensah K (2018) Food production in urban areas: a study of urban agriculture in Accra. Ghana, Routledge

O'Connor D, Peng T, Zhang J, Tsang DC, Alessi DS, Shen Z, Hou D (2018) Biochar application for the remediation of heavy metal polluted land: a review of in situ field trials. Sci Total Environ 619:815–826

Odlare M, Pell M (2009) Effect of wood fly ash and compost on nitrification and denitrification in agricultural soil. Appl Energy 86(1):74–80

Onwosi CO, Igbokwe VC, Odimba JN, Eke IE, Nwankwoala MO, Iroh IN, Ezeogu LI (2017) Composting technology in waste stabilization: on the methods, challenges and future prospects. J Environ Manag 190:140–157

Oworu OO, Dada OA, Majekodunmi OE (2010) Influence of compost on growth, nutrient uptake and dry matter partitioning of grain amaranths (*Amaranthus hypochondriacus* L.). Libyan Agric Res Center J Intl 6:375–383

Pampuro N, Caffaro F, Cavallo E (2018) Reuse of animal manure: a case study on stakeholders' perceptions about pelletized compost in Northwestern Italy. Sustainability 10(6):2028

Pandey P (2018) Assessment of quality of leafy vegetables grown in periphery of Allahabad district. Soil Environ 10(2):21–46

Park JH, Lamb D, Paneerselvam P, Choppala G, Bolan N, Chung JW (2011) Role of organic amendments on enhanced bioremediation of heavy metalloid contaminated soils. J Hazard Mater 185(2–3):549–574

Párraga-Aguado I, Álvarez-Rogel J, González-Alcaraz MN, Conesa HM (2017) Metal mobility assessment for the application of an urban organic waste amendment in two degraded semiarid soils. J Geochem Explor 173:92–98

Patra AK (2018) Conservation of microbial biodiversity: a key to sustainable soil health. In: SAARC training manual on integrated nutrient management for improving soil health and crop productivity. ICAR-Indian Institute of Soil Science and SAARC Agriculture Center (SAC), Bhopal/Dhaka, p 14

Paul EA (2014) Soil microbiology, ecology and biochemistry. Chemosphere 46:84–91

Portier CJ, Thigpen TK, Carter SR, Dilworth CH, Grambsch AE, Gohlke J, Maslak T (2017) A human health perspective on climate change. A report outlining the research needs on the human health effects of climate change. Environmental Health Perspectives/National Institute of Environmental Health Sciences, Durham

Qayyum S, Khan I, Meng K, Zhao Y, Peng C (2020) A review on remediation technologies for heavy metals contaminated soil. Central Asian J Environ Sci Technol Innov 1(1):21–29. https://doi.org/10.22034/CAJESTI.2020.01.03

Radojevic M, Bashkin VN (2010) Practical environmental analysis, 2nd edn. RSC Publishing, Cambridge

Ramady H, Alshaal T, Abowaly M, Abdalla N, Taha HS, Al-Saeedi AH, Sztrik A (2017) Nanoremediation for sustainable crop production. In: Nanoscience in Food and Agriculture, vol 5. Springer, Cham, pp 335–363

Rashad FM, Kesba HH, Saleh WD, Moselhy MA (2011) Impact of rice straw composts on microbial population, plant growth, nutrient uptake and root-knot nematode under greenhouse conditions. Afr J Agric Res 6(5):1188–1203

Reeve PJ, Fallowfield HJ (2018) Natural and surfactant modified zeolites: a review of their applications for water remediation with a focus on surfactant desorption and toxicity towards microorganisms. J Environ Manag 205:253–261

Rehman K, Fatima F, Waheed I, Akash MSH (2018) Prevalence of exposure of heavy metals and their impact on health consequences. J Cell Biochem 119(1):157–184

Ren X, Zeng G, Tang L, Wang J, Wan J, Liu Y, Deng R (2018) Sorption, transport and biodegradation–an insight into bioavailability of persistent organic pollutants in soil. Sci Total Environ 610:1154–1163

Riaz U, Murtaza G, Saifullah FM, Farooq M (2018) Comparable effect of commercial composts on chemical properties of sandy clay loam soil and accumulation of trace elements in soil-plant system. Int J Agric Biol 20:85–92

Rigby H, Clarke BO, Pritchard DL, Meehan B, Beshah F, Smith SR, Porter NA (2016) A critical review of nitrogen mineralization in biosolids-amended soil, the associated fertilizer value for crop production and potential for emissions to the environment. Sci Total Environ 541:1310–1338

Rocco C, Seshadri B, Adamo P, Bolan NS, Mbene K, Naidu R (2018) Impact of waste-derived organic and inorganic amendments on the mobility and bioavailability of arsenic and cadmium in alkaline and acid soils. Environ Sci Poll Res 25(26):25896–25905

Rosen V, Chen Y (2014) The influence of compost addition on heavy metal distribution between operationally defined geochemical fractions and on metal accumulation in plant. J Soils Sediments 14(4):713–720

Rotz CA (2018) Modeling greenhouse gas emissions from dairy farms. J Dairy Sci 101(7):6675–6690

Samuel O, Christian S, Frederick O (2018) Effect of pig manure on the microbial remediation of crude oil-polluted soil. Am J Life Sci Res 6(2):76–90

Sarker M, Choudhury S, Islam N, Zeb T, Zeb B, Mahmood Q (2020) The effects of climatic change mediated water stress on growth and yield of tomato. Central Asian J Environ Sci Technol Innov 1(2):85–92. https://doi.org/10.22034/CAJESTI.2020.02.03

Sarwar N, Imran M, Shaheen MR, Ishaque W, Kamran MA, Matloob A, Hussain S (2017) Phytoremediation strategies for soils contaminated with heavy metals: modifications and future perspectives. Chemosphere 171:710–721

Shahid M, Dumat C, Khalid S, Schreck E, Xiong T, Niazi NK (2017) Foliar heavy metal uptake, toxicity and detoxification in plants: a comparison of foliar and root metal uptake. J Hazard Mater 325:36–58

Sharifi Z, Renella G (2015) Assessment of a particle size fractionation as a technology for reducing heavy metal, salinity and impurities from compost produced by municipal solid waste. Waste Manag 38:95–101

Sharma B, Sarkar A, Singh P, Singh RP (2017) Agricultural utilization of biosolids: a review on potential effects on soil and plant grown. Waste Manag 64:117–132

Shen F, Liao R, Ali A, Mahar A, Guo D, Li R, Zhang Z (2017) Spatial distribution and risk assessment of heavy metals in soil near a Pb/Zn smelter in Feng County, China. Ecotoxicol Environ Saf 139:254–262

Sherlala AIA, Raman AAA, Bello MM, Asghar A (2018) A review of the applications of organo-functionalized magnetic graphene oxide nanocomposites for heavy metal adsorption. Chemosphere 193:1004–1017

Shi T, Ma J, Wu X, Ju T, Lin X, Zhang Y, Wu F (2018) Inventories of heavy metal inputs and outputs to and from agricultural soils: a review. Ecotoxicol Environ Saf 164:118–124

Shifaw E (2018) Review of heavy metals pollution in China in agricultural and urban soils. J Health Pollut 8(18):180–198

Simantiraki F, Gidarakos E (2015) Comparative assessment of compost and zeolite utilisation for the simultaneous removal of BTEX, Cd and Zn from the aqueous phase: batch and continuous flow study. J Environ Manag 159:218–226

Sims R et al (2018) A future food system for healthy human beings and a healthy planet. Scientific and Technical Advisory Panel to the Global Environment Facility. Washington, DC

Sinha V, Pakshirajan K, Chaturvedi R (2018) Chromium tolerance, bioaccumulation and localization in plants: an overview. J Environ Manag 206:715–730

Smith P, House JI, Bustamante M, Sobocká J, Harper R, Pan G, Paustian K (2016) Global change pressures on soils from land use and management. Glob Chang Biol 22(3):1008–1028

Soares MA, Marto S, Quina MJ, Gando-Ferreira L, Quinta-Ferreira R (2016) Evaluation of eggshell-rich compost as biosorbent for removal of Pb (II) from aqueous solutions. Water Air Soil Pollut 227(5):1–16

Sodhi GPS, Beri V, Benbi DK (2009) Soil aggregation and distribution of carbon and nitrogen in different fractions under long-term application of compost in rice–wheat system. Soil Tillage Res 103(2):412–418

Somtners L, Van Volk V, Giordano PM, Sopper WE, Bastían R (2018) Effects of soil properties on accumulation of trace elements by crops. In: Land application of sludge. CRC Press, Boca Raton, pp 5–24

Song B, Zeng G, Gong J, Liang J, Xu P, Liu Z, Ye S (2017) Evaluation methods for assessing effectiveness of in situ remediation of soil and sediment contaminated with organic pollutants and heavy metals. Environ Int 105:43–55

Sparks DL (2018) Kinetics and mechanisms of chemical reactions at the soil mineral/water interface. In: Soil physical chemistry, 2nd edn. CRC Press, Boca Raton, pp 135–192

Srivastava V, Sarkar A, Singh S, Singh P, de Araujo AS, Singh RP (2017) Agroecological responses of heavy metal pollution with special emphasis on soil health and plant performances. Front Environ Sci 5:64–70

Steffan JJ, Brevik EC, Burgess LC, Cerdà A (2018) The effect of soil on human health: an overview. Eur J Soil Sci 69(1):159–171

Sutton-Grier AE, Kenney MA, Richardson CJ (2010) Examining the relationship between ecosystem structure and function using structural equation modelling: a case study examining denitrification potential in restored wetland soils. Ecol Model 221(5):761–768

Sun J, Pan L, Tsang DC, Zhan Y, Zhu L, Li X (2018) Organic contamination and remediation in the agricultural soils of China: A critical review. Sci Total Environ 615:724–740

Tan LC, Nancharaiah YV, van Hullebusch ED, Lens PN (2018) Selenium: environmental significance, pollution, and biological treatment technologies. In: Anaerobic treatment of mine wastewater for the removal of selenate and its co-contaminants. CRC Press, Boca Raton, pp 9–71

Tien YC, Li B, Zhang T, Scott A, Murray R, Sabourin L, Topp E (2017) Impact of dairy manure pre-application treatment on manure composition, soil dynamics of antibiotic resistance genes, and abundance of antibiotic-resistance genes on vegetables at harvest. Sci Total Environ 581:32–39

Torkashvand AM, Alidoust M, Khomami AM (2015) The reuse of peanut organic wastes as a growth medium for ornamental plants. Int J Recycl Org Waste Agric 4(2):85–94

Tóth G, Hermann T, Da Silva MR, Montanarella L (2016) Heavy metals in agricultural soils of the European Union with implications for food safety. Environ Int 88:299–309

Totsche KU, Amelung W, Gerzabek MH, Guggenberger G, Klumpp E, Knief C, Ray N (2018) Microaggregates in soils. J Plant Nutr Soil Sci 181(1):104–136

Tőzsér D, Magura T, Simon E (2017) Heavy metal uptake by plant parts of willow species: a meta-analysis. J Hazard Mater 336:101–109

Tsang DCW, Yip ACK, Olds WE, Weber PA (2014) Arsenic and copper stabilisation in a contaminated soil by coal fly ash and green waste compost. Environ Sci Pollut Res 21(17):54–68

Udovic M, McBride MB (2012) Influence of compost addition on lead and arsenic bioavailability in reclaimed orchard soil assessed using Porcellioscaber bioaccumulation test. J Hazard Mater 205:144–149

Unda-Calvo J, Martínez-Santos M, Ruiz-Romera E (2017) Chemical and physiological metal bioaccessibility assessment in surface bottom sediments from the Deba River urban catchment: harmonization of PBET, TCLP and BCR sequential extraction methods. Ecotoxicol Environ Saf 138:260–270

Usman AR, Almutairi AA, Elmaghraby S, Al-Farraj AS (2017) Levels, solid-phase fractions and sources of heavy metals at site received industrial effluents: a case study. Chem Speciat Bioavailab 29(1):78–88

Van Fan Y, Lee CT, Klemeš JJ, Chua LS, Sarmidi MR, Leow CW (2018) Evaluation of effective microorganisms on home scale organic waste composting. J Environ Manag 216:41–48
Van Veld PA, Sanders BM, Fowler BA, Lars F, Di Giulio RT, Marius B, Stegeman JJ (2018) Molecular responses to environmental contamination: enzyme and protein systems as indicators of chemical exposure and effect. In: Biomarkers. CRC Press, Boca Raton, pp 235–336
Vaverková MD (2018) Evaluation of the addition of immobilizing agents on selected physicochemical properties of soil contaminated with heavy metals. Polish J Soil Sci 51(1):59
Venegas A, Rigol A, Vidal M (2015) Viability of organic wastes and biochars as amendments for the remediation of heavy metal-contaminated soils. Chemosphere 119:190–198
Vithanage M, Herath I, Joseph S, Bundschuh J, Bolan N, Ok YS, Rinklebe J (2017) Interaction of arsenic with biochar in soil and water. A critical review. Carbon 113:219–230
Wander M (2015) Managing manure fertilizers in organic systems. Artikkeli Haettu 22(20):17
Waqas M, Nizami AS, Aburiazaiza AS, Barakat MA, Rashid MI, Ismail IMI (2018) Optimizing the process of food waste compost and valorizing its applications: a case study of Saudi Arabia. J Clean Prod 176:426–438
Ward A, Sharpley A, Miller K, Dick W, Hoorman J, Fulton J, LaBarge GA (2018) An assessment of in-field nutrient best management practices for agricultural crop systems with subsurface drainage. J Soil Water Conserv 73(1):5A–10A
WHO U (2004) Why are 4 million newborn babies dying each year. Lancet 364:399–401
Won S, Shim SM, You BG, Choi YS, Ra C (2017) Nutrient production from dairy cattle manure and loading on arable land. Asian Australas J Anim Sci 30(1):125
Wu J, Hou Y, Wen Y (2018) Tourist behavior and conservation awareness on eating wild edible plants in mountainous protected areas: a case study in Northwest China. J Sustain For 5:489–503
Yang X, Liu J, McGrouther K, Huang H, Lu K, Guo X, Wang H (2016) Effect of biochar on the extractability of heavy metals (Cd, Cu, Pb, and Zn) and enzyme activity in soil. Environ Sci Pollut Res 23(2):974–984
Ye S, Zeng G, Wu H, Zhang C, Dai J, Liang J, Zhang C (2017) Biological technologies for the remediation of co-contaminated soil. Crit Rev Biotechnol 37(8):1062–1076
Yolchiyeva F, Hacjiyeva S, Huseyinli A, Hasanova A (2020) Ecological problems of water resources in Azerbaijan and their impact on human health. Central Asian J Environ Sci Technol Innov 1(2):71–76. https://doi.org/10.22034/CAJESTI.2020.02.01
Yuan SY, Su LM, Chang BV (2009) Biodegradation of phenanthrene and pyrene in compost-amended soil. J Environ Sci Health A 44(7):648–653
Yuan T, Yuan Y, Zhou S, Li F, Liu Z, Zhuang L (2011) A rapid and simple electrochemical method for evaluating the electron transfer capacities of dissolved organic matter. J Soils Sediments 11(3):467–473
Zafarzadeh A, Rahimzadeh H, Mahvi AH (2018) Health risk assessment of heavy metals in vegetables in an endemic esophageal cancer region in Iran. Health Scope 7(3):45–58
Zeng G, Wan J, Huang D, Hu L, Huang C, Cheng M, Jiang D (2017) Precipitation, adsorption and rhizosphere effect: the mechanisms for phosphate-induced Pb immobilization in soils. A review. J Hazard Mater 339:354–367
Zhou YF, Haynes RJ, Naidu R (2012) Use of inorganic and organic wastes for insitu immobilisation of Pb and Zn in a contaminated alkaline soil. Environ Sci Pollut Res 19(4):1260–1270
Zhou H, Zhou X, Zeng M, Liao B-H, Liu L, Yang W-T (2014) Effects of combined amendments on heavy metal accumulation in rice (*Oryza sativa* L.) planted on contaminated paddy soil. Ecotoxicol Environ Saf 101:226–232
Zia MH, Watts MJ, Niaz A, Middleton DR, Kim AW (2017) Health risk assessment of potentially harmful elements and dietary minerals from vegetables irrigated with untreated wastewater, Pakistan. Environ Geochem Health 39(4):707–728
Zornoza R, Moreno-Barriga F, Acosta JA, Muñoz MA, Faz A (2016) Stability, nutrient availability and hydrophobicity of biochars derived from manure, crop residues, and municipal solid waste for their use as soil amendments. Chemosphere 144:122–130

Chapter 3
Potential of Inorganic Fertilizers for Sustainable Development in Agriculture

Abida Parveen, Muhammad Arslan Ashraf, Iqbal Hussain, Shagufta Perveen, Rizwan Rasheed , Abdol Ghaffar Ebadi, and Sumaira Thind

3.1 Introduction to Fertilizers and Soil Fertility

After the second agricultural revolution, various practices of fertilizers to the agricultural soil have now become common. It is well documented that a nutrient-rich and fertile soil strongly impacts crop growth and development, and nutrient additions greatly enhance agricultural output. Primarily, the nutrient concentration of soil is hardly at a plausible stage to fulfil the basic requirements of plant growth; farmers adopt many strategies to compensate with soil addition of inorganic and organic remedies that are plentiful in essential nutrients (Ahmad et al. 2009). Unfortunately, improper addition of organic fertilizer affects the soil environment by nitrate leaching (Dong et al. 2005), After leaching in the root environment, NO_3-N uptake becomes restricted for plants, ultimately causing groundwater contamination (Ball Coelho et al. 2006) and eventually causing organic farming to fail (Pang and Lety 2000). Also, nutrient mobility in the vadose zone depends on the nutrient profile in that zone (Woodard et al. 2002), greatly dependent on soil water (Chinkuyu and Kanwar 2001). Nutrient mobility is also influenced by soil physical properties, conductivity, permeability, water status, and chemical properties (Kung et al. 2000), which change with a massive addition of organic fertilizers (Fares et al. 2008). Further, eco-friendly properties associated with the organic fertilizers, and the low economic cost with rare side effects, cause organic fertilizers to be

A. Parveen (✉) · I. Hussain · S. Thind · M. A. Ashraf · S. Perveen · R. Rasheed
Department of Botany, Government College University, Faisalabad, Pakistan

A. Ebadi
Department of Agriculture, Islamic Azad University Jouybar Branch, Iran, Iran

Q. Mahmood (ed.), *Sustainable Plant Nutrition under Contaminated Environments*, Sustainable Plant Nutrition in a Changing World,
https://doi.org/10.1007/978-3-030-91499-8_3

extensively used as a substitute for synthetic fertilizers (El-Hawary et al. 2002). Unfortunately, timing and rate of release of the nutrient are very slow when using organic fertilizer as per the requirements of plants. Also, organic fertilizers depend on a microbially assisted conversion process that converts these into inorganic forms that are readily absorbed by plants. Organic forms convert into absorbable inorganic forms through a very slow process via soil microbes. For example, nitrogen (N) is a key limiting factor for the growth of plants that is converted from the organic form to ammonium through microbial activity; the ammonium quickly oxidizes to nitrate (Gaskell and Smith 2007). Farmers add organic matter without prior knowledge of vital factors such as the dependence of decomposition on soil temperature, moisture, and microbial activity including the nature of organic material without thinking about how agricultural practices are well managed (Scagel 2005).

Inorganic fertilizers are an immediately available source of nutrients because they release quickly into the soil after application as compared to organic fertilizers (Gaskell and Smith 2007). The global feeding challenge will reach 9.6 billion humans by 2050. The huge dietary requirements are changing economic values by introducing diversified farming systems. To improve this goal, there is a dire need to increase the quantity and quality of food crops by addition of seeds, pesticides, and fertilizers. Fertilizers have a key role in the productivity of modern agriculture (Erisman et al. 2008). Moreover, a decline rate has been witnessed in the agricultural productivity of inappropriate regions. Unfortunately, consumption of nutrient-deficient dietary products motivates us for the search of a sustainable key solution to introduce crops with high yield potential and with strong nutritional qualities (Bindraban et al. 2015).

Keeping this in view, adoption of relatively rapid nutrient source, its easy availability and inorganic or synthetic fertilizers will increase quantity and quality of food while enhanced agriculture yield and productivity with inorganic fertilizers (Bindraban et al. 2015). In this context, inorganic fertilizers or use of synthetic fertilizers are therefore an efficient source of nutrients.

3.2 Interactive Effects of Inorganic Fertilizers on the Crop Plant

The positive effect of various inorganic fertilizers on the growth and productivity of crop plants has been reported earlier in this chapter. In this context, application of phosphate and potassium sulfate was documented to improve the yield of rice and wheat (Shaukat et al. 2020). Similarly, the addition of phosphorus increases the growth and productivity of cotton (Iqbal et al. 2020). However, soil application of sulfate enhance growth and along with the improved nutritional potential of ryegrass (Ikoyi et al. 2020). The comparative influence of different inorganic fertilizers on various crops is discussed in Table 3.1.

Table 3.1 Interactive effect of inorganic nutrition regarding growth and development of crop plants

Sample no.	Crop	Inorganic fertilizer applied	Growth medium	Biological effects	References
1	Maize	NPK	Soil	Improvements in the soil nutrient availability. Significant increase in nutritional status. Enhanced maize grain yield.	Faloye et al. (2020)
2	Rice and wheat	Urea, diammonium phosphate, and potassium sulfate	Soil (silt + loam)	Organ-mineral treatment promoted early tillering and anthesis. Maximum growth and yield improved.	Shaukat et al. (2020)
3	Sweet potato	NPK + organo Mineral fertilizer	Soil	Increase in the ether, crude fiber and vitamin A. Increase in the energy contents of the tuber.	Kareem et al. (2020)
4	Maize	NPK with interactive others	Sandy soil	Maize yield and biomass of grain increased. Availability of P to soil increased and crop development.	Faloye et al. (2017)
5	Sugar cane	NPK + amalgamation of others	Saline soil	Increase 100% in the diameter of a plant with inorganic fertilizer. Raised uptake of nutrients. Increase in the growth of sugarcane. Highest absorption of N, P and K.	Djajadi et al. (2020)
6	Soybean	NPK	Alkaline soil	Availability of P in soil increased. Total biomass and yield increased with the combined effect of NPK fertilizer and biochar.	Shamim et al. (2015)

(continued)

Table 3.1 (continued)

Sample no.	Crop	Inorganic fertilizer applied	Growth medium	Biological effects	References
7	Cucumber	NPK and micro fertilizer	Soil	Increase in the growth and yield. Increase in the fruit weight and length. Absorption of nutrient along with essential elements.	Marliah et al. (2020)
8	Cotton	Phosphorus	Soil	Improvement in the yield rate. Increased in biomass production. Enhanced nutrient accumulation.	Iqbal et al. (2020)
9	Ryegrass	Sulfate	Soil (in a greenhouse)	Increase in sulfate concentration and shoot dry biomass. Improvement in growth rate. Increase in accumulation of N, P, Mg, and K.	Ikoyi et al. (2020)
10	Radish	NPK + soil mixing with others	Black soil	Improvement in growth rate and number of leaves per plant. Increase in root biomass and yield.	Kushwah et al. (2020)
11	Okra	NPK + biofer combined	Soil	Increase in plant height, number of branches and leaves per plant. Increase in stem girth. Total yield enhanced.	Kumar et al. (2019)
12	Wheat	Liquid amide N +magnesium + sulfur	Soil (clay loam)	Combines N, mg, and S improved the photosynthetic activity of crop. Higher chlorophyll contents in leaves due to combined nutrient supply. Improvement in the dry matter content. Grain yield significantly enhanced.	Pranckietene et al. (2020)

(continued)

Table 3.1 (continued)

Sample no.	Crop	Inorganic fertilizer applied	Growth medium	Biological effects	References
13	Barley + Wheat	Potassium + magnesium + sulfur	Soil	Significantly increased grain yield. Increase protein contents in both bowls of cereal. Prevents soil element deficiency.	Hlisnikovsky et al. (2019)
14	Guava	Calcium + Magnesium + iron	Sandy soil	Improvement in fruit yield. Significant increase in sugar contents of a fruit.	Kumar et al. (2019)

3.3 Importance of Inorganic over Traditional Fertilizer

Inorganic fertilizers are nonliving, man-made, derived from a by-product of the petroleum industry, and called synthetic or commercial fertilizers. These chemical fertilizers contain elements almost the same as traditional fertilizers, but they release nutrients very quickly in soil that are readily absorbed by plants (Sabry 2015). Also, these chemical or synthetic fertilizers are easily available, with high nutrient values, and quickly absorbed by plant roots. Overfertilization leads to nutrient deficiency and both groundwater and surface water contamination, with changes in soil acidic and basic properties with a significant loss in the microbial population in the soil environment (Chen 2006). Unfortunately, the addition of the huge amounts of organic or traditional farming has various limitations, viz. low nutrient content, very slow decomposition rate, and changes in different nutrient compositions that are linked to its organic texture properties. From the perspective of microbial activity, organic fertilizers increase at a higher rate by changing soil texture as compared to inorganic fertilizers (Dinesh et al. 2010; Gonzalez et al. 2010). Similarly, it is reported that organic fertilizers having the potential to increase plant growth and productivity last only a single season because they add nutrients into the soil in very small quantities (Arancon et al. 2004, 2005). Also, it is documented that some biological plant growth-promoting mechanisms take place, such as soil microbial activity, and some biochemical mechanisms can overreact in changed soil practices. However, microbial biomass increases rapidly in soil by application of organic matter. Consequently, increased soil microbial biomass reduces nutrient availability in the plant soil root system because at this stage microbial immobilization takes place in the whole soil environment (Geisseler et al. 2010; Inselsbacher et al. 2010). Of particular importance, it is regarded that organic or traditional farming may compromise crop growth and productivity because of changes in soil nutrient status, decrease in biological activities, and changes in soil biochemical properties as compared to inorganic fertilizers (Pimental et al. 2005). Consequently, inorganic fertilizers are a source of rapid nutrition as compared with organic fertilizer, so preferences are made in terms of plant nutrient requirement as a basis. A brief account of various inorganic or synthetic fertilizers is discussed next.

3.4 Macro and Micronutrient Fertilization

Mineral nutrient farming is a very innovative agriculture management practice that is crucial for crop growth and productivity. Some of those nutrients, known as macronutrients, are required in relatively large amounts (Kulcheski et al. 2015). Macronutrients include nitrogen (N), phosphorus (P), potassium (K), calcium (Ca), magnesium (Mg), and sulfur (S), having a place in essential elements. Others, required in small amounts, are called micronutrients: these include copper (Cu), iron (Fe), manganese (Mn), molybdenum (Mo), nickel (Ni), zinc (Zn), boron (B), silicon (Si), and sodium (Na), which have properties to improve growth by adding countless benefits to plants (Datnoff et al. 2007). Nutrients in plants have important functions, such as osmotic regulation, membrane permeability, a vital component of key structures, and crucial roles in cell catabolic and anabolic functions simultaneously, particularly in plant growth and developmental stages.

As an estimate, by 2030, the global decline in productivity, increase in stunting, loss of cognitive skills, and increase in many chronic disorders could cost up to $35 trillion (Bereuter and Glickman 2015). The addition of micronutrients to the soil contributes positively to decreasing their deficiencies in humans (Cakmak 2008). Similarly, selenium- (Se) fortified soil enhanced crop growth and yields and improved Se uptake in human beings (Ekholm et al. 2005). Similarly, nutrient deficiency in soil is a worldwide issue, as crop species respond differently according to the requirement for each nutrient (Monreal et al. 2015). Also, nutrients strengthen our physical and mental health by stimulating the immune system. Similarly, important metabolic functions in the human body performed by enzymes, proteins, and biological compounds can fail in their functions with deficiency of micronutrients (Akeredolu et al. 2011).

It is documented that soil application of micronutrients does not show any antagonistic response, whereas the more plausible effects of these nutrients lead to enhanced agriculture growth and yield in crop plants such as soybean (*Glycine max*), tobacco (*Nicotiana* sp.), wheat (*Triticum aestivum* L.), and cauliflower (*Brassica oleracea*). Nevertheless, the soil media of these crops received nitrogen, phosphorus, and potash in particular quantity already. Similarly, it is reported that soil application with Zn increased rice growth and productivity; in contrast, the uptake of Zn is not antagonistically linked with the presence of a varying concentration of nitrogen, phosphorus, and potassium already in the soil mixture (Katyal and Ponamperuma 1974).

Crops depend upon plant metabolic functions that are regulated by nitrogen, phosphorus, potassium, magnesium, calcium, sulfur, iron, copper, and chloride because without these micronutrients the cell environment is unable to perform optimal functions. Iron, copper, manganese, and chloride are involved as co-factors in various metabolic reactions and ultimately carry out the process of photosynthesis. All these micronutrients regulate the activity of specific enzymes, such as DNA/RNA polymerase, N-metabolizing enzymes, superoxide dismutase, catalase, dehydrogenases, oxidases, and ATPases, which have a vital role in cellular functions (Broadley et al. 2012). A key role of micronutrients as co-factor in multiple enzymatic and nonenzymatic reactions is pivotal: the most important among nutrients is Zn, which has a role in the biosynthesis of plant growth promoters such as auxin, involved in growth enhancement (Fageria 2002). Application of Zn also has a stress mitigation function because Zn modulates the key functions of NADPH oxidases by changing the permeability of membrane lipids (Cakmak 2000). Similarly, it is also reported that Zn mediates key functions that regulate reactive oxygen species (ROS) under a stressful environment, protecting the cell environment by generating a signaling mechanism that provides defense in the cell under drought conditions (Golldack et al. 2014). Above all, micronutrients upon application to the soil improve the growth and yield of crops and behave as a resistance mechanism under various biotic and abiotic stresses from a range of 10–70% (Dimkpa and Bindraban 2016). A brief account of different macro- and micronutrients given in Table 3.2.

3.5 Synthetic Fertilizer of Nitrogen (N)

Nitrogen is present abundantly in the atmosphere and is relatively rare in the soil medium. During the early twentieth-century, ammonia production was based on the Haber–Bosch process, with a particular mixture of hydrogen and nitrogen at optimum conditions (Engelstad 1985). As an estimate, more than 21 million tons of synthetic nitrogen fertilizer is used in the United States in food production in various agricultural practices (Sabry 2015). A dramatic increase is observed by the addition of soil with synthetic nitrogen from a range of 11.6 Tg in 1961 to 104 Tg in 2006 In addition, nitrate fertilizer increases amino acid and protein levels to ultimately improve crop growth and productivity (Bernier et al. 1993). It is also reported that dramatic alteration in plant hormones (Crawford 1995) reduces growth attributes (Scheible et al. 1997) and changes in root composition (Stitt and Feil 1999). Some nitrogen fertilizers are discussed in Table 3.3.

3.6 Synthetic NPK Fertilizer

After the Green Revolution, inorganic or chemical fertilizers, which are a source of nitrogen (N), phosphorus (P), and potassium (K), enhanced the yield potential of crop plants. The application of these nutrients to the soil causes changes in environmental factors that alter soil physical properties (Øvreås and Torsvik 1998).

Table 3.2 Effects of macro- and microfertilization on growth and development of crop plants

Sample no.	Crop	Inorganic fertilizer applied	Growth medium	Biological effects	References
1	Maize	Potassium	Silica sandy soil	K deficiency increased sugar accumulation in leaves. K deficiency significantly decreased starch content in leaves.	Martineau et al. (2017)
2	Sunflower	Sulfur	Sandy loam	Induces an increase in uptake of K^+ and Ca^{2+}. Improves leaf turgor pressure. Influences lipid peroxidation.	Aziz et al. (2019)
3	Sunflower	Potassium	Soil	Achene and straw yield significantly increased. Increased oil yield.	Shehzad et al. (2020)
4	Sugarcane	Potassium, calcium and magnesium	Oxisol soil	Stimulates the sugar transporters in phloem. Increase the length and diameter of stalk and number of stalks. Mg uptake is suppressed by potassium.	Rhodes et al. (2018)
5	Soybean	Potassium	Hydroponic	Improved functioning of stomata. Potassium deficiency causes decrease in rubisco activity.	Xiao-guang et al. (2015)
6	Potato	Potassium	Sandy loam	Increases the number of aerial stems in each plant. Improves plant height. Increases number of tubers and tuber weight.	Zelelew et al. (2016)
7	Oat	Potassium	Sandy clay loam	Increases plant height and biomass. Improved chlorophyll content. Increased alanine and aspartate amino transferase activity was observed in shoot and root.	Ahanger et al. (2015)
8	Sesame	Nitrogen and potassium	Loamy sand	Increased number of capsules per plant. Improved the harvest index.	Kale et al. (2019)

(continued)

Table 3.2 (continued)

Sample no.	Crop	Inorganic fertilizer applied	Growth medium	Biological effects	References
9	Potato	Calcium	Sandy loam	Improves the calcium content of potato tuber. NPK concentration was increased. Increases average tuber size.	Hamdi et al. (2015)
10	Maize	Zinc	Soil	Increases photosynthetic activity. Increased chlorophyll pigments, enhanced stomatal conductance and transpiration rate. Increased yield.	Liu et al. (2016)

Moreover, the rate of decline in agricultural productivity and consumption of nutrient-deficient dietary products motivates the search for a sustainable key solution to introduce crops with high yield potential with strong nutritional qualities (Bindraban et al. 2015). NPK-fertilizer got the attention of farmers as a modern fertilizer that serves as three limiting fertilizers and controls several physiological and metabolic functions inside the cell (Kulcheski et al. 2015). It is documented that the growth and yield of wheat and cereals have rapidly increased since 1970 because of the massive use of synthetic fertilizers along with agronomic amendments (Iqtidar et al. 2006; Khan et al. 2009). Although synthetic fertilizers do not have a significant role related to soil properties, these still have importance in several ways over organic fertilizers (Hammad et al. 2010; Dong et al. 2006; Hussain and Khan 2000). Currently, soil components such as microbial biomass carbon (MBC), soil community structure, and enzymatic functions have been used as a tool to study soil qualities in diversified agricultural systems (Mäder et al. 2002; Badalucco et al. 2010). In this context, inorganic fertilizers (NPK) have been observed to have positive as well as negative effects on the structures of MBC and microbial mass (Böhme et al. 2005). Incorporation of nutrients enhances the nutritional values without any effect on the carbon quantity of the cell that is used for the growth of microorganisms. So, the addition of organic manures brings the highest level of MBC in the soil compared to that without any addition of inorganic nutrients (Goyal et al. 1999).

Inorganic fertilization has a surprising impact on the growth and productivity of oats, maintaining market standards for a high quality of oat for human consumption as well as horse feed, and provides macronutrients such as N, P, and KCl in the soil (May et al. 2004). Similarly, the addition to the soil with NPK-fertilizer at different rates enhances the nutritional values of crops and increases productivity (Olson and Kurtz 1982; Dibb and Thompson 1985). Similarly, it is reported by Arab et al. (2015) that the addition to soil of an increased amount of NPK up to 100% surprisingly enhanced growth, viz. the number of branches, flowers, and capitulum as well

Table 3.3 Effect of nitrogen-nutrition regarding growth and development of crop plants

Sample number	Crop	Inorganic fertilizer applied	Growth medium	Biological effects	References
1	Maize	Nitrogen	Soil	Fresh and dry forage yield increase. Overall plant growth improved. Enhanced tasseling stage.	Amin (2010)
2	Wheat	Nitrogen	Smonitza soil	Increased the sedimentation value. Improves the grain quality. Increased wet gluten content.	zecevic et al. (2010)
3	Rice	Nitrogen	Sandy loam soil	Increased grain number and leaf area index. Increased yield specially in cold areas.	Ju et al. (2021)
4	Sesame	Nitrogen	Fine sand	Photosynthetic rate is improved. Causes extension of internodes. Reduced biomass under N-deprived conditions.	Sehgal et al. (2021)
5	Potato	Nitrogen	Nitisol soil	Improve yield and yield components. Increase the number of main stems per plant.	Setu and Mitiku (2020)
6	Barley	Nitrogen	Soil	Protein content was increased. Germination energy was slightly decreased by increasing N rates.	Terefe et al. (2018)
7	Sweet pepper	Nitrogen	Soil	Increased plant height at vegetative and flowering stages. Increased the average fruit weight and volume. Improved vitamin C.	Aminifard et al. (2012)
8	Cotton	Nitrogen	Loam soil	Causes root elongation. Increase boll biomass. Provided high seed cotton yield. Improved rate of photosynthesis.	Chen et al. (2020)

(continued)

Table 3.3 (continued)

Sample number	Crop	Inorganic fertilizer applied	Growth medium	Biological effects	References
9	Sunflower	Nitrogen	Silt loam, sandy-clay-loam and clay-loam	Significantly increased the total dry matter production (TDM). Enhanced the leaf area index. Improved source to sink translocation of photo-assimilates. Increased number of achenes per head.	Ahmad et al. (2017)
10	Tomato	Nitrogen	Soil	Increased number of flower cluster^{-1}. Improved number of fruits and fruit weight. Increased fruit yield (t ha^{-1}).	Haque et al. (2011)

as phosphorus and potassium content as compared to that without NPK-fertilizer addition. Similarly, it is reported that the addition of nitrogen enhances vegetative growth because N is an essential part of nucleic acids and enhanced growth (Lampkin 1990). The addition of phosphorus in soil media involved in chlorophyll structures participates in various physiological processes, such as cell division, meristematic tissues, photosynthesis, different metabolic functions of a cell, and constituents of carbohydrates, fats, and proteins (Singh and Singh 2003).

Acknowledgments We are thankful to the Department of Languages, the University of Faisalabad for their generous help in improving language quality.

References

Ahanger MA, Agarwal RM, Tomar NS, Shrivastava M (2015) Potassium induces positive changes in nitrogen metabolism and antioxidant system of oat (Avena sativa L cultivar Kent). J Plant Interact 10(1):211–223

Ahmad AA, Fares A, Abbas F, Deenik JL (2009) Nutrient concentrations within and below root zones from applied chicken manure in selected Hawaiian soils. Environ Sci Heal B 44(8):828–843

Akeredolu I, Oguntona BE, Okafor C, Osisanya OJ (2011) Iron, zinc, and copper malnutrition among primary school children in Lagos, Nigeria. Food Sci Nutr 2:1063–1070. https://doi.org/10.4236/fns.2011.210142

Aminifard MH, Aroiee H, Ameri A, Fatemi H (2012) Effect of plant density and nitrogen fertilizer on growth, yield and fruit quality of sweet pepper (Capsicum annum L.). Afr J Agric Res 7(6):859–866

Arab A, Zamani GR, Sayyari MH, Asili J (2015) Effects of chemical and biological fertilizers on morpho-physiological traits of marigold (*Calendula officinalis* L.). J Med Plants Res:60–68
Arancon NQ, Edwards CA, Bierman P, Welch C, Metzger JD (2004) Influences of vermicomposts on field strawberries:1. Effects on growth and yields. Bioresour Technol 93:145–153
Arancon NQ, Edwards CA, Bierman P, Metzger JD, Lucht C (2005) Effects of vermicomposts produced from cattle manure, food waste and paper waste on the growth and yield of peppers in the field. Pedobiologia 49:297–306
Aziz A, Ashraf M, Sikandar S, Asif M, Akhtar N, Shahzad SM, Babar BH (2019) Optimizing sulfur for improving salt tolerance of sunflower (Helianthus annuus L.). Soil Environ 38:222–233
Badalucco L, Rao M, Colombo C, Palumbo G, Laudicina VA, Gianfreda L (2010) Reversing agriculture from intensive to sustainable improves soil quality in a semiarid south Italian soil. Biol Fertil Soils 46(5):481–489
Ball Coelho BR, Roy RC, Bruin AJ (2006) Nitrogen recovery and partitioning with different rates and methods of side dressed manure. Soil Sci Soc Am J 70(2):464–473
Bereuter D, Glickman D (2015) Healthy food for a healthy world: leveraging agriculture and food
Bernier G, Havelange A, Houssa C, Petitjean A, Lejeune P (1993) Physiological signals that induce flowering. Plant Cell 5(10):1147
Bindraban PS, Dimkpa CO, Nagarajan L, Roy AH, Rabbinge R (2015) Revisiting fertilisers and fertilisation strategies for improved nutrient uptake by plants. Biol Fertil Soils 51:897–911. https://doi.org/10.1007/s00374-015-1039-7
Bohme L, Langer U, Bohme F (2005) Microbial biomass, enzyme activities and microbial community structure in two European long-term field experiments. Agric Ecosyst Environ 109:141–152
Broadley M, Brown P, Cakmak I, Rengel Z, Zhao F (2012) Functions of nutrient: micronutrients. In: Marschner P (ed) Marschner's mineral nutrition of higher plants, 3rd edn. Elsevier, Oxford, pp 243–248
Cakmak I (2000) Tansley review no. 111-possible roles of zinc in protecting plant cells from damage by reactive oxygen species. New Phytol 146:185–205. https://doi.org/10.1046/j.1469-8137.2000.00630
Cakmak I (2008) Enrichment of cereal grains with zinc: agronomic or genetic biofortification? Plant and Soil 302:1–17. https://doi.org/10.1007/s11104-007-9466-3
Chen JH (2006) The combined use of chemical and organic fertilizers and/or biofertilizer for crop growth and soil fertility. In International workshop on sustained management of the soil-rhizosphere system for efficient crop production and fertilizer use (Vol. 16, No. 20, pp. 1–11). Land Development Department Bangkok Thailand
Chen J, Liu L, Wang Z, Zhang Y, Sun H, Song S, Li C (2020) Nitrogen fertilization increases root growth and coordinates the root–shoot relationship in cotton. Front Plant Sci 11:880
Chinkuyu AJ, Kanwar RS (2001) Predicting soil nitrate-nitrogen losses from incorporated poultry manure using the gleams model. Trans ASAE 44(6):1643
Crawford NM (1995) Nitrate: nutrient and signal for plant growth. Plant Cell 7(7):859
Datnoff LE, Elmer WH, Huber DM (2007) Mineral nutrition and plant disease. American Phytopathological Society (APS Press)
Dibb DW, Thompson Jr WR (1985) Interaction of potassium with other nutrients. Potassium in agriculture:515–533
Dimkpa CO, Bindraban PS (2016) Fortification of micronutrients for efficient agronomic production: a review. Agron Sustain 36(1):7
Dinesh R, Srinivasan V, Hamza S, Manjusha A (2010) Short-term incorporation of organic manures and biofertilizers influences biochemical and microbial characteristics of soils under an annual crop [*Turmeric (Curcuma longa* L.)]. Bioresour Technol 101:4697–4702
Djajadi D, Syaputra R, Hidayati SN, Khairiyah Y (2020) Effect of vermicompost and nitrogen on N, K, Na uptakes and growth of sugarcane in saline soil. Agrivita J Agric Sci 42(1):110–119
Dong YH, Ouyang Z, Liu SL (2005) Nitrogen transformation in maize soil after application of different organic manures. Int J Environ Sci 17(2):340–343

Dong J, Hengsdijk H, Ting-Bo D, Boer W, Qi J, Xing CW (2006) Long-term effects of manure and inorganic fertilizers on yield and soil fertility for a winter wheat maize system in Jiangsu, China. Pedosphere 16:25–32
Ekholm P, Eurola M, Venalainen ER (2005) Selenium content of foods and diets in Finland. In: Eurola M (ed) Proceedings twenty years of selenium fertilization. Agri-food Res Rep 69, pp 39–45
El-Hawary MI, Hawarytalman IEI, El- Ghamary AM, El-Naggar E (2002) Effect of application of biofertilizer on the yield and NPK uptake of some wheat genotypes as affected by the biological properties of soil. PJBS 5(11):1181–1185
Engelstad OP (1985) Fertilizer technology and use, 3rd edn. Soil Science Society of America, Inc., Madison, p 18
Erisman JW, Sutton MA, Galloway JN, Klimont Z, Winiwarter W (2008) How a century of ammonia synthesis changed the world. Nat Geosci 1:636–639. https://doi.org/10.1038/ngeo325
Fageria NK (2002) Influence of micronutrients on dry matter yield and interaction with other nutrients in annual crops. Pesq agropec bras Brasília 37:1765–1772
Faloye OT, Alatise MO, Ajayi AE, Ewulo BS (2017) Synergistic effects of biochar and inorganic fertiliser on maize (*zea mays*) yield in an alfisol under drip irrigation. Soil Tillage Res 174:214–220
Faloye OT, Ajayi AE, Alatise MO, Ewulo BS, Horn R (2020) Nutrient uptake, maximum yield production, and economic return of maize under deficit irrigation with biochar and inorganic fertiliser amendments. Biochar:1–14
Fares A, Abbas F, Ahmad A, Deenik JL, Safeeq M (2008) Response of selected soil physical and hydrologic properties to manure amendment rates, levels, and types. J Soil Sci 173(8):522–533
Gaskell M, Smith R (2007) Nitrogen sources for organic vegetable crops. HortTechnology 17(4):431–441
Geisseler D, Horwath WR, Joergensen RG, Ludwig B (2010) Pathways of nitrogen utilization by soil microorganisms – a review. Soil Biol Biochem 42:2058–2067
Golldack D, Li C, Mohan H, Probst N (2014) Tolerance to drought and salt stress in plants: unravelling the signalling networks. Front Plant Sci 5:151. https://doi.org/10.3389/fpls.2014.00151
González M, Gomez E, Comese R, Quesada M, Conti M (2010) Influence of organic amendments on soil quality potential indicators in an urban horticultural system. Bioresour Technol 101:88978901
Goyal S, Chander K, Mundra MC, Kapoor KK (1999) Influence of inorganic fertilizers and organic amendments on soil organic matter and soil microbial properties under tropical conditions. Biol Fertil Soils 29(2):196–200. https://doi.org/10.1007/s003740050544
Hamdi W, Helali L, Beji R, Zhani K, Ouertatani S, Gharbi A (2015) Effect of levels calcium nitrate addition on potatoes fertilizer. Int Res J Eng Technol 2(3):2006–2013
Hammad HM, Khaliq A, Ahmad A, Aslam M, Khaliq T, Wajid SA, Hussain A, Usman M, Nasim W, Farhad W, Sultana R (2010) Influence of organic manures on weed dynamics and wheat productivity under low rainfed area. J Crop Environ 1(1):13–17
Haque ME, Paul AK, Sarker JR (2011) Effect of nitrogen and boron on the growth and yield of tomato (Lycopersicon esculentum M.). Int J Stress Manag 2(3):277–282
Hlisnikovský L, Čermák P, Kunzová E, Barłóg P (2019) The effect of an application of potassium, magnesium and sulphur on wheat and barley grain yield and protein content
Hussain N, Khan MB (2000) Effect of farmyard manure on growth and yield of wheat (*Triticum aestivum*) on saline soil. J Res Sci 11:31–34
Ikoyi I, Fowler A, Storey S, Doyle E, Schmalenberger A (2020) Sulfate fertilization supports the growth of ryegrass in soil columns but changes microbial community structures and reduces abundances of nematodes and *arbuscular mycorrhiza*. Sci Total Environ 704:135315
Inselsbacher E, Hinko-Najera UN, Stange FC, Gorfer M, Schüller E, Ripka K, Zechmeister-Boltenstern S, Hood-Novotny R, Strauss J, Wanek W (2010) Short-term competition between crop plants and soil microbes for inorganic N fertilizer. Soil Biol Biochem 42:360–372

Iqbal B, Kong F, Ullah I, Ali S, Li H, Wang J, Zho Z (2020) Phosphorus application improves the cotton yield by enhancing reproductive organ biomass and nutrient accumulation in two cotton cultivars with different phosphorus sensitivity. J Agron 10(2):153

Iqtidar H, Ayyaz KM, Ahmad KE (2006) Bread wheat varieties as influenced by different nitrogen levels. J Zhejiang Univ Sci 7:70–78

Ju C, Zhu Y, Liu T, Sun C (2021) The effect of nitrogen reduction at different stages on grain yield and nitrogen use efficiency for nitrogen efficient rice varieties. Agronomy 11(3):462

Kale PD, Thaokar A, Gawali KA, Sarda A, Nagmote A, Jatav SK (2019) Response of sesame (Sesamum indicum L.) to nitrogen and potassium fertilization. J Pharmacogn Phytochem 8:411–414

Kareem I, Akinrinde EA, Oladosu Y, Eifediyi EK, Abdulmaliq SY, Alasinrin SY, Adekola OF (2020) Influence of organic, inorganic and organo-mineral fertilizers on yield and quality of sweet potato (*Ipomoea batatas*). JASEM 24(1):111–118

Katyal JC, Ponnamperuma FN (1974) Zinc deficiency: a widespread nutritional disorder of rice in Agusan del Norte [Philippines]. The Philippine Agriculturist

Khan A, Jan MT, Marwat KB, Arif M (2009) Organic and inorganic nitrogen treatments effects of plant and yield attributes of maize in a different tillage system. Pak J Bot 41:99–108

Kulcheski FR, Côrrea R, Gomes IA, de Lima JC, Margis R (2015) NPK macronutrients and microRNA homeostasis. Front Plant Sc 6:451

Kung KJ, Kladivko EJ, Gish TJ, Steenhuis TS, Bubenzer G, Helling CS (2000) Quantifying preferential flow by the breakthrough of sequentially applied tracers silt loam soil. Soil Sci Soc Am J 64(4):1296–1304

Kushwah I, Sharma R, Kushwah S, Singh O (2020) Influence of organic manures and inorganic fertilizers on growth, yield and profitability of radish (*Raphanus sativus* L.). Ann Plant Soil Res 22(1):14–18

Lampkin N (1990) In: Ipswich UK (ed) Organic farming. Press Books, pp 701–710

Liu H, Gan W, Rengel Z, Zhao P (2016) Effects of zinc fertilizer rate and application method on photosynthetic characteristics and grain yield of summer maize. J Plant Nutr Soil Sci 16(2):550–562

Mäder P, Fliessbach A, Dubois D, Gunst L, Fried P, Niggli U (2002) Soil fertility and biodiversity in organic farming. Science 296(5573):1694–1697

Marliah A, Anhar A, Hayati E (2020) Combine organic and inorganic fertilizer increases yield of cucumber (*Cucumis sativus* L.). In IOP Conference Series: Earth and Environmental Science (Vol. 425, No. 1, p. 012075). IOP Publishing

Martineau E, Dome JC, Bosc A, Dannoura M, Gibon Y, Bénard C, Jordan-Meille L (2017) The role of potassium on maize leaf carbon exportation under drought condition. Acta Physiol Plant 39(10):1–13

Monreal CM, DeRosa M, Mallubhotla SC, Bindraban PS, Dimkpa CO (2015) Nanotechnologies for increasing the crop use efficiency of fertilizer-micronutrients. Biol Fertil Soils. https://doi.org/10.1007/s00374-015-1073-5

Olsen RA, Kurtz LT (1982) Crop nitrogen requirements, utilization and fertilization. p. 567-604. In F. J. Stevenson (ed.) nitrogen in agricultural soils. Agron. Mongr. 22. ASA, CSSA, and SSSA, Madison, WI

Øvreas L, Torsvik VV (1998) Microbial diversity and community structure in two different agricultural soil communities. Microb Ecol 36(3):303–315. Parham JA, Deng SP, Raun WR, J

Pang XP, Letey J (2000) Organic farming challenge of timing nitrogen availability to crop nitrogen requirements. Soil Sci Soc Am J 64(1):247–253

Pimentel D, Hepperly P, Hanson J, Douds D, Seidel R (2005) Environmental, energetic, and economic comparisons of organic and conventional farming systems. Bioscience 55:573–582

Pranckietiene I, Dromantiene R, Smalstiene V, Jodaugiene D, Ilona V, Aurelija P, Marks M (2020) Effect of liquid amide nitrogen fertilizer with magnesium and sulphur on spring wheat chlorophyll content, accumulation of nitrogen and yield. J Elem 25(1):139–152

Rhodes R, Miles N, Hughes JC (2018) Interactions between potassium, calcium and magnesium in sugarcane grown on two contrasting soils in South Africa. Field Crop Res 223:1–11
Sabry AK (2015) Synthetic fertilizers; role and hazards. Fertil Technol 1:110–133
Scagel CF (2005) Inoculation with ericoid mycorrhizal fungi alters fertilizer use of highbush blueberry cultivars. Hort Science 40(3):786–794
Scheible WR, Gonzalez-Fontes A, Lauerer M, Muller-Rober B, Caboche M, Stitt M (1997) Nitrate acts as a signal to induce organic acid metabolism and repress starch metabolism in tobacco. Plant Cell 9(5):783–798
Sehgal A, Smith CA, Walne CH, Chastain D, Shankle M, Reddy KR (2021) Developing functional relationships between sesame growth, development, and nitrogen nutrition during early season. Agrosyst Geosci Environ 4(3):e20198
Setu H, Mitiku T (2020) Response of potato to nitrogen and phosphorus fertilizers at Assosa, western Ethiopia. Agron J 112(2):1227–1237
Shamim MIA, Dijkstra FA, Abuyusuf M, Hossain AI (2015) Synergistic effects of biochar and NPK fertilizer on soybean yield in alkaline soil. Pedosphere 25(5):713–719
Shaukat M, Ahmad A, Khaliq T, Afzal I (2020) Organic and inorganic fertilization change the phenophases of rice (*Oryza sativa* L.) and wheat (*Triticum aestivum* L.) by modulating the photosynthetic capacity of leaves. Pak J Agric Sci 57(1):289–294
Shehzad MA, Nawaz F, Ahmad F, Ahmad N, Masood S (2020) Protective effect of potassium and chitosan supply on growth, physiological processes and antioxidative machinery in sunflower (Helianthus annuus L.) under drought stress. Ecotoxicol Environ Saf 187:109841
Singh AK, Singh Y (2003) Leaf nutrient status, growth and flower yield in rose as influenced by organic and inorganic sources. J Ornam Hortic 6(3):229–233
Stitt M, Feil R (1999) Lateral root frequency decreases when nitrate accumulates in tobacco transformants with low nitrate reductase activity: consequences for the regulation of biomass partitioning between shoots and root1. Plant 215(2):143–153
Terefe D, Desalegn T, Ashagre H (2018) Effect of nitrogen fertilizer levels on grain yield and quality of malt barley (Hordeum vulgare L.) varieties at Wolmera District, Central Highland of Ethiopia. Int J Agric Sci 4, 4:29–43
Woodard KR, French EC, Sweat LA, Graetz DA, Sollenberger LE, Macoon B, Van Horn HH (2002) Nitrogen removal and nitrate leaching for forage systems receiving dairy effluent. J Environ Qual 31(6):1980–1992
Zecevic V, Knezevic D, Boskovic J, Micanovic D, Dozet G (2010) Effect of nitrogen fertilization on winter wheat quality. Cereal Res Commun 38(2):243–249
Zelelew DZ, Lal S, Kidane TT, Ghebreslassie BM (2016) Effect of potassium levels on growth and productivity of potato varieties. Am J Plant Sci 7(12):1629–1638

Chapter 4
Wastewater Irrigation and Plant Growth: An Insight into Molecular Studies

Nadia Riaz, Muhammad Saqib Khan, Maria Sabeen, Bibi Saima Zeb, Shahida Shaheen, and Tahir Hayat

4.1 Introduction

Due to rapid socioeconomic development and increasing global populations, water scarcity has become a significant global issue. Generating huge volumes of wastewater (treated or untreated) discharged by industry into water bodies does not typically meet the recommended requirements. Only 2% of wastewater is reported to be treated in developing countries and discharged into freshwater bodies (Minhas and Samra 2003), while the remaining 98% is lost or used in wastewater irrigation. Irrigating agricultural land with wastewater can help farmers to save fertilizer costs (Zhang and Shen 2019) as wastewater is a stable and reliable water source and is not influenced by climate and other seasonal variations, unlike freshwater resources (Balkhair et al. 2013). Humans and environment are vulnerable to untreated wastewater, irrespective of its continuous availability, when polluted with biological contaminants and heavy metals (Keraita et al. 2012; Gola et al. 2016). Untreated wastewater is extensively used worldwide, particularly in the water scarcity regions of developing countries, to irrigate agricultural crops. Nearly 86% of irrigated agricultural lands are in China, Pakistan, India, Iran, and Mexico, where waste water treatment is lower than 75% (Thebo et al. 2017). This study exclusively offers global estimates on how treated and untreated wastewater impacts irrigated crops. Agricultural activities in some developed countries such as South Africa and Mexico depend on wastewater (Mendoza-Espinosa et al. 2019). The consumption of contaminated foods grown on irrigated soil is responsible for the ill health effects.

N. Riaz (✉) · M. S. Khan · M. Sabeen · B. S. Zeb · S. Shaheen · T. Hayat
Department of Environmental Sciences, COMSATS University Islamabad, Abbottabad, Pakistan
e-mail: nadiariazz@cuiatd.edu.pk

Q. Mahmood (ed.), *Sustainable Plant Nutrition under Contaminated Environments*, Sustainable Plant Nutrition in a Changing World,
https://doi.org/10.1007/978-3-030-91499-8_4

4.2 Heavy Metal Contamination

Heavy metals (HM) are widely spread among several environmental toxins. Such contaminants might originate due to human or natural sources, but human involvement plays a crucial role in their increase due to continuous discharge of untreated wastewater from industrial, sewage, or mining activities; dumping of metallic waste; and use of agrochemicals (Zhang et al. 2010). Heavy metals are transition metals having atomic mass >0.002 kg, weight ≅ 5 N/m^3, and density >5 g/cm^3. Such elements are considered necessary to living organisms but may become toxic above a certain concentration threshold. Arsenic (As), iron (Fe), lead (Pb), cobalt (Co), mercury (Hg), nickel (Ni), cadmium (Cd), copper (Cu), molybdenum (Mo), zinc (Zn), manganese (Mn), and selenium (Se) are few examples. These elements, when exposed to polluted soil or water, can alter the basic physiological and metabolic routes of plant processes. HM-induced toxic effects are primarily due to their persistent and non-biodegradable nature, shape and route of exposure, and poor biotransformation rates and bioaccumulation capabilities, resulting in significant damage including defective cellular processes (Martínez-Alcalá and Bernal 2020; Singh et al. 2020). Classical examples of the most dangerous HM include asbestos, arsenic, and mercury, which can pose a possible danger even at lower levels when exposed (Jan et al. 2015). Untreated urban and industrial wastewater from several manufacturers is released worldwide into water systems forced to be used by farmers for irrigation purposes, which in turn contaminates the soil and then bioaccumulates in agronomic crops, causing significant harm to both the environment and human health (Abegunrin et al. 2016; Mendoza-Espinosa et al. 2019). Major effects because of such practice have serious consequences resulting in poor soil quality, surface and groundwater contamination and heavy metals (HM) enrichment (Teklehaimanot et al. 2015; Gola et al. 2016; Zhang and Shen 2019). Consumption of food crops contaminated with heavy metal is a significant route of human access to the food chain. Studies have shown that a significant amount of heavy metals are accumulated in wastewater-irrigated soils resulting in health risks (Khan et al. 2008; Sridhara Chary et al. 2008; Sabeen et al. 2019) (Fig. 4.1).

4.3 Effects of HMs on Human Health

Agricultural activities are being affected worldwide, mainly because of water scarcity along with the resultant HM contamination due to the application of contaminated wastewaters for irrigation of agricultural fields. Such irrigation practices might have significant impact on the contamination of soil and water, thereby badly affecting the quality of the cultivated crops on contaminated soils. Such crops serve a major exposure route to humans due to bioaccumulation of HM. The scientific community worldwide is concerned about the effects of metal accumulation in crops and entry into the food chain (Jan et al. 2015; Ahmed et al. 2018). Exceeding

Fig. 4.1 Heavy metal accumulation in food crops irrigated with contaminated water

Table 4.1 Association of human health risk with different heavy metals

No	Heavy metal	Associated risk	References
1	Cadmium (Cd)	Kidneys, bone, and lungs damage	(Casalino et al. 2002; de Vries et al. 2007; Bernard 2008; Moulis and Thévenod 2010; Liu et al. 2014a; Zhang et al. 2014)
2	Nickel (Ni)	Allergic dermatitis, eczema, and respiratory disorders	(Christensen et al. 1999; Yeganeh et al. 2013; Ahlström et al. 2018)
3	Lead (Pb)	CNS effects and related disorders (hypertension, disruption of nervous systems, brain damage), kidney damage (renal dysfunction), blood disorders (anemia), miscarriages, and infertility	(Liu et al. 2014a; Shvachiy et al. 2018)
4	Arsenic (As)	Responsible for Birth defects, carcinogenic and can effect different body organs (lungs, skin, liver, bladder, kidneys)	(Banerjee et al. 2013; Sun et al. 2014)

the permissible limits of HM and trace elements (like Cu and Zn) and their accumulation are well documented and have been associated with serious health risks upon consumption of HM-contaminated crops (Khan et al. 2008; Zhao et al. 2014; Liao et al. 2015; Chen et al. 2018a). There are several factors responsible for metal toxicity including concentration, oxidation state, mode of exposure, gender and age group, level of consumption, and genetics (Reichman 2002; Tran and Popova 2013; Ghori et al. 2019; Martínez-Alcalá and Bernal 2020; Singh et al. 2020). The impacts of consumption of different heavy metals are outlined in Table 4.1. Respiratory issues and skin problems like allergic contact dermatitis and eczema can be caused by the exceeding limits of nickel (Christensen et al. 1999; Yeganeh et al. 2013; Ahlström et al. 2018). Accumulation of Cd in humans above a certain level can cause kidney, bone, and pulmonary damage, Alzheimer's disease, and many other adverse disorders (Casalino et al. 2002; de Vries et al. 2007; Bernard 2008; Fowler 2009; Moulis and Thévenod 2010; Liu et al. 2014a; Zhang et al. 2014). Pb exposure may cause mental disorder, kidney dysfunction, and blood system impairment (Liu et al. 2014a; O'Connor et al. 2018; Shvachiy et al. 2018). Daily intake and metal accumulating in food are critical factors in determining the risk of metals to human health from food consumption.

4.4 Effects of HMs on Plants

Substantial quantities of HM might accumulate in agricultural crops grown in contaminated soils or irrigated using untreated wastewater. Major route of HM exposure to human is by consumption of contaminated crops. HM like Pb, Cr, Cu, As, and Cd (at higher levels exceeding the permissible levels) might accumulate in certain plant tissues/parts, especially leafy parts, when grown on contaminated industrial soils or when irrigated using untreated industrial wastewater (Liu et al. 2005; Yebpella et al. 2011; Tasrina et al. 2015). An Indian scholar (Kawatra and Bakhetia 2008) reported that HM (e.g., Pb, Cd, and Ni) usually transfer to edible parts of plant through roots (Kumar Sharma et al. 2007). In a study conducted in Bangladesh by Goni et al. (2014), they found maximum concentrations of Fe, Zn, Cr, Cu, and Pb in soils around the industrial area.

Irrespective of the fact that HM may also serve as vital plant nutrients but HM presence exceeding certain limit, might be toxic to plant itself. Some examples of such HM include cadmium (Cd), chromium (Cr), lead (Pb), and nickel (Ni). Cadmium can readily be absorbed by many plant species from contaminated soils, but its long-term use (0.01 mg/L) can cause toxicity to beans, beets, and turnips. Chromium (Cr) toxicity has not been yet investigated in plants, but its reference dose in wastewater is up to 0.1 mg/L in aerated soils. Lead (Pb) can hinder plant cell growth at high concentrations (>5.0 mg/L) (Rowe and Abdel-Magid 1995). Nickel (Ni) is lethal to some plants at a concentration of 0.5–1.0 mg/L. High Cd buildup in plants, for example, results in reduced photosynthetic rate, reduced water and nutrient intake, yellowing of leaves, slowed development, burning of tips, and plant necrosis (Mohanpuria et al. 2007). Elevated levels of Cr hinder plant growth, accelerating the yellowing of leaves, disturb nutrients balance, damage the tops and roots, and hinder chlorophyll anabolism (Mohanpuria et al. 2007; Moulis and Thévenod 2010; Tran and Popova 2013). Cr toxicity affects plant growth by modifying the germination process; reducing the growth of root, stem, and leaves; declining the production of total dry matter and yield of plant, etc. Metabolic variations by Cr exposure influence enzymes and metabolites and their capacity to release reactive oxygen species (ROS) (Shanker et al. 2005). Ni toxicity in plants causes deterioration in water content of dicotyledonous and monocotyledonous plants. The water content of dicotyledonous and monocotyledonous plants worsens as a result of Ni toxicity in plants. Because of its toxicity, plants' normal growth, photosynthetic activity, cell division, DNA damage, chlorosis, and necrosis all declined (Yang et al. 1996; Zornoza et al. 1999; Yusuf et al. 2011). Underdeveloped growth of roots and tops, dark green foliage, or dark brown to purple leaves on some plants reveals high levels of iron (Fe) accumulation (Wheeler et al. 1985; Fageria and Rabelo 1987; Ghori et al. 2019). Arsenic (As) is not essential for plants and is usually toxic. Metalloid prevents proliferation and root extension. Arsenic can seriously inhibit plant growth by slowing or stopping expansion and accumulation of biomass and by compromising plant reproductive ability by losses in fertility, yield, and fruit production (Mascher et al. 2002; Finnegan and Chen 2012; Abbas et al. 2018; Singh et al. 2020). Lead causes oxidative stress in growing plants (Liu et al. 2003; Verma and Dubey 2003; Sharma and Dubey 2005).

Plant exposure to heavy metals over an extended period is linked to oxidative stress due to the generation/accumulation of high level of methylglyoxal (MG) and reactive oxygen species (ROS). The major ROS are singlet oxygen ($1O_2$), superoxide (O^{-}_2), hydroxyl radicals (•OH), and hydrogen peroxide (H_2O_2), which cause homeostasis disruption in cellular compartments between ROS generation and sequestration (Rughani et al. 2015; Chandrakar et al. 2020). This is associated with numerous degrading effects such as lipid peroxidation, sulfhydryl group binding, membrane structure breakage or disturbance, ion leakage, redox imbalance, inactivation of the antioxidant defense mechanism, and amino acid, protein, and DNA oxidation (Liu et al. 2014b). Long-term plant exposure to various HMs results in overproduction and accumulation of MG and ROS, producing oxidative stress within the plant cell which is highly toxic to the cellular macromolecules. In this sense, plant cells have a mechanism of defense. There is significant evidence that cells detoxify the deleterious effects of HMs by various mechanisms such as compartmentalization, chelation, hyperaccumulation, synthesis of various compatible solutes and signaling molecules, and activation of a plant defense system.

4.5 Health Risk Assessment and Genetic Changes in Plants Consequent to Wastewater Irrigation

The existence of significant amounts of valuable nutrients in wastewater helps reduce the additional need for chemical fertilizers (Schwartz and Boyd 1994; Jiménez and Asano 2008), but it has been documented that long-term use of wastewater for irrigation causes potential health hazards due to metal bioaccumulation in agricultural crops (Farahat and Linderholm 2015; Christou et al. 2017). The presence of dangerous metals beyond acceptable limit is unavoidable and may be detrimental to animal and human health (Jaishankar et al. 2014; Iqbal et al. 2016; Chen et al. 2018b). Industrial wastewater containing contaminants such as metals, when used for irrigation, will seriously affect the quality of food. Metal bioaccumulation enhances genetic content, chromosomal aberrations (CA), and mutation rates (Haq et al. 2017; Silveira et al. 2017). Eco-toxicological effects due to HM accumulation

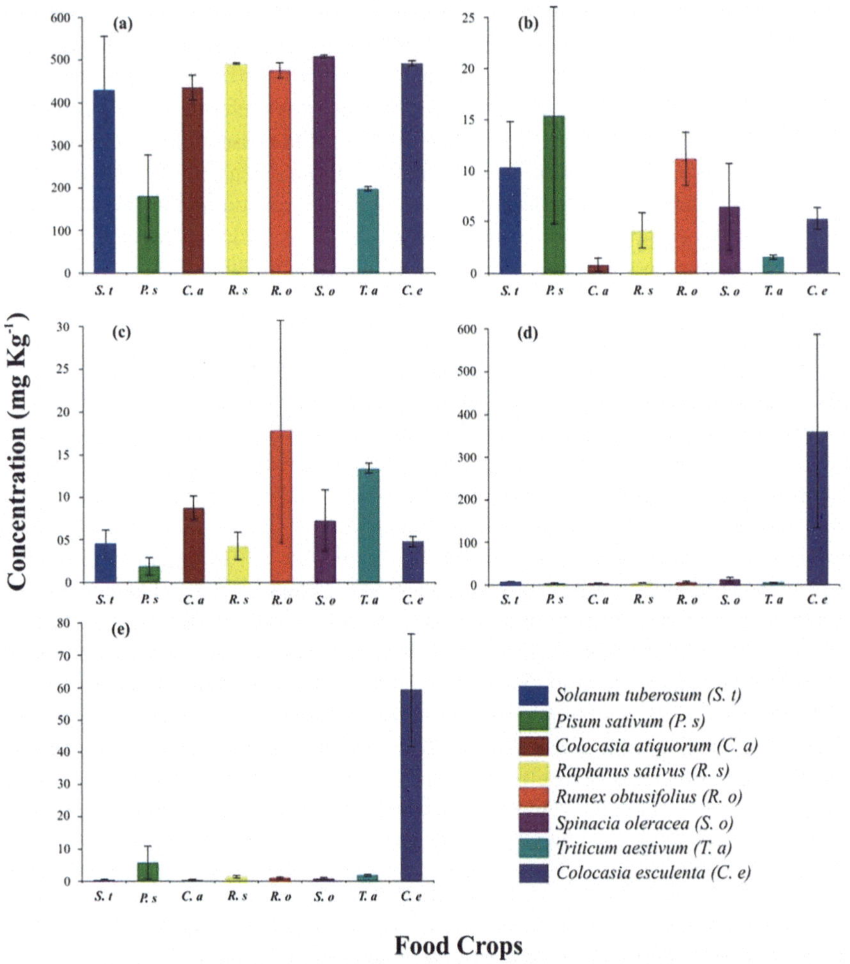

Fig. 4.2 Accumulation of Fe (**a**), Pb (**b**), Ni (**c**), Cr (**d**), and Cd (**e**) in food crops with long-term wastewater irrigation. (Sabeen et al. 2019)

can be measured using different techniques (Baderna et al. 2011). To compare the possible human health risks associated with industrial wastewater irrigation, Sabeen et al. (2019) investigated the absorption and accumulation of HM in eight food crops (as shown in Fig. 4.2) irrigated with industrial wastewater. Researchers have reported differing concentrations of HM between different crops, finding that crops contaminated with HM may present a serious health risk to local consumers.

Effects of HM accumulation and toxicity in plants and related effects have been well established, but not many studies on plant response in terms of genetic changes in edible crops have been published. In recent years, increasing stresses on food safety have increased the risk of intake of food products contaminated with pesticides, metals, or toxins among researchers. The related metal concentration and the analysis of the genetic capacity of organisms are of utmost importance in shedding light on our understanding of metal absorption, accumulation, and cytotoxicity (Kumar Sharma et al. 2007; Saleh 2015; Sabeen et al. 2019).

4.5.1 Genotoxicity Assays

Genotoxic assays are useful for predicting genetic alterations by evaluating gene mutation and chromosome damage caused by the contaminants. Plant bioassays are very simple to perform, inexpensive, and quick and good genetic toxicity analysts (Grant 1994; Steinkellner et al. 1998). Chromosome aberration analyses, mutation assays, unique locus mutation assays, and cytogenetic studies have been used to observe the damages caused by metal in plants. Many types of plants can tolerate higher concentrations of metals or other harmful substances being native to metalliferous soils. Metals such as Pb, Cu, Mn, and Cd may affect the DNA of the plant cells (Silveira et al. 2017).

Allium cepa test, developed and acknowledged by Levan (1949), is efficient to track the atmosphere and a very sensitive test to investigate CA triggered by many chemical toxins. The procedure is used to examine the adverse effects of toxins on cell division normality and is a simple technique for fostering intricate information about how toxic substances alone or in a combination induce chromosomal changes (Fiskesjö 1988; Mishra 1993).

With a simple method such as randomly amplified polymorphic DNA (RAPD), genotoxicity can be easily measured by using a small amount of DNA from PCR amplification. The outcome can vary due to band changes, bands skipped, or new bands emerging. Those bands then determine genetic resemblances or differences. This can also be used to monitor genomic DNA to detect any form of damage or change in DNA structure (e.g., rearrangements, point mutation, small insertion or deletion of DNA and ploidy changes) and to propose it as a basis for new biomarker assays to detect DNA damage and genetic changes in bacterial, plant, and animal cells (Atienzar and Jha 2006).

In advanced plants, both natural and unnatural stresses are complemented by a complex chemical and physiological phenomenon of oxidative stress and activated by the development and accumulation of more reactive oxygen species (ROS). ROS damages and induces malfunction of all essential cell organelles (Atienzar and Jha 2006; Kumari et al. 2016). In terms of protecting the plant from free radical damage, antioxidants also safeguard the plant's protection and development. An increase in the quantity of free radicals demands the use of additional antioxidants to counter them. In order to protect the body system from ROS, an extremely classy and complicated antioxidant protection mechanism is formed in the plants that involves various components working interactively and synergistically to neutralize free radicals (Atienzar and Jha 2006).

4.5.2 Genotoxicity Evaluation of HM in Plants

Metal accumulation in plants is subsequently linked to genetic changes that affect the quality and safety of food. There is minimal evidence available on vegetable genotoxicity investigations of plants (plant root damage and growth inhibition) that are exposed to HM due to irrigation from wastewater. Higher plants are well known in monitoring studies as excellent genetic models for detecting mutagens in the field. The advantage of measuring the impact of genotoxic chemicals directly on DNA is mainly related to the sensitivity and fast reaction times. Recently, advances in molecular biology in eco-genotoxicology have resulted in the production of many selective and sensitive assays for DNA analysis. *Allium cepa* assay is well known for the assessment of the impact of various genotoxic materials (Levan 1949). *Allium cepa* is one of those plants that are used to determine genetic alterations such as chromosomal variations and irregular mitotic cycles. This was used to test a variety of chemical compounds that helped with the environmental monitoring of its expanding application. *Allium cepa* is inexpensive, has low chromosomal count, is easily treated, and facilitates the evaluation of various aspects and genetic toxicity. The micronucleus is used to assess mutagenicity with the use of certain chemicals, while the mitotic index and gene aberrations are used to detect the effects of cytotoxicity. In addition, *A. cepa* test also provides valuable information for estimating the mode of action of a specific pollutant and its effects on its genome (Rathnasamy et al. 2013; Sabeen et al. 2020).

Because of its clear, diversified, and low chromosomal count, onion is an appropriate resource for investigating the genotoxic effects of HM induction, as well as other benefits such as well-defined mitotic phases, karyotype stability, rapid response to hazardous chemicals, and so on. In this regard, it has been stated elsewhere that genotoxicity of pollutants like metal can be best explored by growing root tips, which can ultimately give a good portion of the plant for investigation (Sabeen et al. 2020). Many researchers have documented clear evidence of HM toxicity and mutagenic and carcinogenic existence (Nagao 1978; Bhat et al. 2017). Chromosomal aberrations (CA) symbolize a change in the arrangement or number of the chromosomes. Changes in the structure of the chromosome may be due to breaks, inhibition, and/or replication of DNA modification. Different chromosomal aberrations, like breaks and bridges of chromosome, demonstrate the clastogenic action. Aneuploidy and polyploidy are numerical CA caused by aneugenic agents or naturally occurring irregular chromosomal segregation. In the absence of telomeres, chromosomes are "sticky," which may be associated with other broken chromosomal endings (Nefic et al. 2013). The altered MI is mainly related to cytotoxic effects are primarily responsible for the altered MI relative to regulation of karyokinesis (Leme and Marin-Morales 2009). C-mitosis is distinguished by the progression of metaphase to anaphase, thereby leading to polyploidy. Increased dose of HM directly affects MI. Sabeen et al. (2020) reported a gradual decrease in MI that was associated with an increase in HM dose. Cytological observations are shown in Fig. 4.3 as evidence of diverse types of abnormalities studied. Increased or reduced

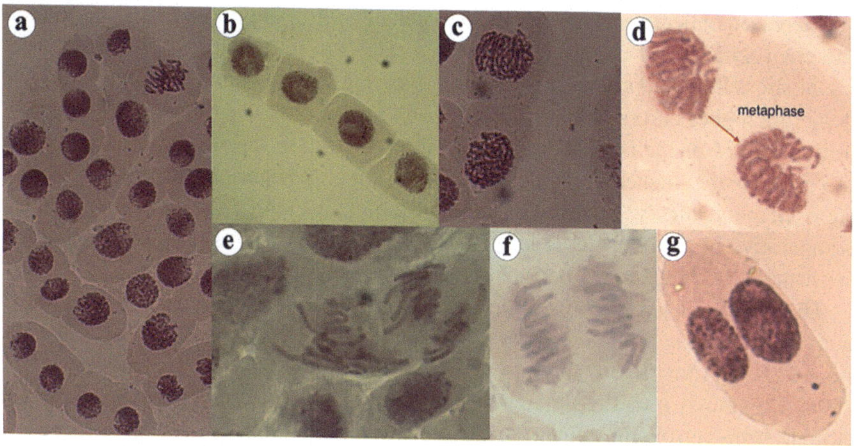

Fig. 4.3 Various normal mitotic stages. (**a**, **b**) prophase, (**c**, **d**) metaphase, (**e**) interphase, (**f**) anaphase, and (**g**) telophase. (Sabeen et al. 2020)

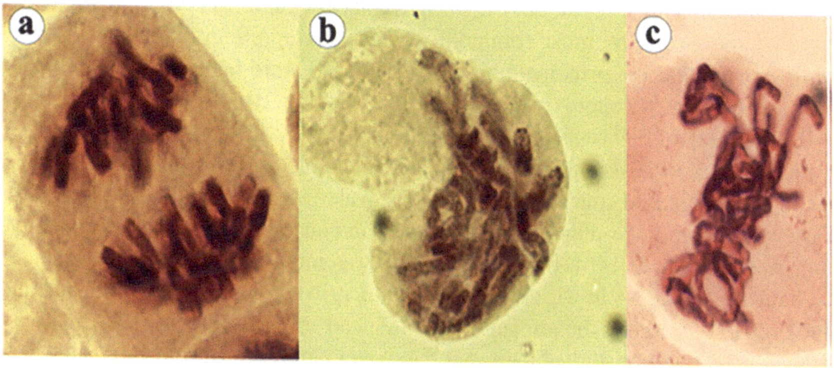

Fig. 4.4 Diverse types of chromosome aberrations have been observed. (**a**) Sticky anaphase with chromosome fragments, (**b**) sticky anaphase, and (**c**) laggards. (Sabeen et al. 2020)

mitotic index (MI) determines the cytotoxic level of a waste/sludge. The MI can be calculated by the following Eq. (4.1):

$$\text{Mitotic Index}(\text{MI}) = \frac{\text{Number of deviding cells}}{\text{Total number of cells}} \times 100 \quad (4.1)$$

Bridges, anaphase, breakage, and ring chromosomes are examples of clastogenic aberrations, whereas chromosomal losses, delays, multipolarity, adhesion, and c-mitosis are examples of physiological aberrations (Bhat et al. 2017; Sabeen et al. 2020). Several studies have reported that chromosomal defects in *A. cepa* can be due to long-term exposure to specific chemical effluents (Dixit and Nerle 1985) (Fig. 4.4).

Increased or decreased mitotic index is considered an important factor to monitor the environment and cytotoxic substances (Hoshina and Marin 2009). Increased metal stress hindered cell division, thus increasing the incidence of cell aberrations depicting higher chromosome changes (Sabeen et al. 2020). Changing the concentration of HM or the type of wastewater used to irrigate various agricultural crops provided clear evidence of the negative effects on crop health. By examining leachate from metal and dye residues that created a high level of Cr, Ni, and other metals, comparative biomonitoring of leachates from two factories using allium bioassay confirmed the cytogenetic abnormalities (Chandra et al. 2005).

The effects of a 100% fly-ash concentration with coal genotoxicity combined with a comet assay on *A. cepa* root tip cells including HMs such as Pb, Zn, Cu, Cd, As, and Ni on root growth and mitotic indices were investigated (Chakraborty et al. 2008). Previously it was reported that root tip cells of A. *cepa* can be affected by the effluent from petroleum refinery through aberrations in micronucleus and chromosome (Hoshina and Marin 2009). Upon washing, refinery effluent can also conflict with the consistency of the river water. The *A. Cepa* test revealed that paint and textile effluents with an average concentration of 6.93 % caused chromosomal abnormalities (Samuel et al. 2010). Bridges, pieces, fragments, vagrant and sticky chromosomes were the most commonly found aberrations in this event. Another study showed that the principal aberrations induced were laggards and concluded that paint effluent is genotoxic to A. *cepa* (Njoku et al. 2015).

RAPD assay of ballast water with *A. cepa* root cells showed genotoxicity caused by the presence of HMs and other physicochemical factors (Olorunfemi et al. 2012, 2014). With increased concentration of ballast water, the MI decreased significantly. Wastewater from coke plants on *Vicia faba* can cause genotoxicity and oxidative damage (Liu et al. 2014b). In plants, toxicity was induced by free-radical-damage mechanism. Mutagens in wastewater from tannery were confirmed by Masood and Malik (2013). To assess the impact of HM-induced soil genotoxicity, the sensitivity of the *A. cepa* test was studied using MI parameters, mitosis, germination, and chromosomal aberration (CA) in an allium assay (de Souza et al. 2013). The results showed that A. *cepa* bioassay was prone to genotoxicity of the soil and can be used as an indicator for initial bioscreening. *A. cepa* genotoxicity assay for oil refinery wastewater was performed, and significant damage to the chromosomes, stickiness, stray, and clumped were detected (Fazili and Ahmad 2014).

A. cepa bioassay is a proven method for measuring domestic sewage sludge toxicity. Environmental toxicity resulted in effective chromosomal frequencies as well as micronucleic and nuclear aberrations, as proven by nucleolar modification analysis (Mazzeo and Marin-Morales 2015). Herbicides induced a substantial increase in nucleoli and micronuclei count, perceived as resulting from the removal of excessive nucleolar content resulting from polyploidization. Domestic sludge toxicity using A. *cepa* test system showed that raw sludge showed toxicity after 1 year of natural attenuation, while sludge/soil mixture did not show any toxicity after 1 year of natural attenuation (Mazzeo et al. 2015). Leather industry effluents without treatment for irrigation purposes should not be used because seed germination in *V. radiata L* had been affected by these effluents (Yadav et al. 2019). The growing roots of

Table 4.2 Heavy metal-associated risk to plants

No	Heavy metal	Associated risk	References
1	Cadmium (Cd)	Slowdown of photosynthetic rate, reduced uptake of water and nutrients, yellowing of leaves, retarded growth, burning of tips, and plant necrosis	(Mohanpuria et al. 2007; Moulis and Thévenod 2010; Tran and Popova 2013)
	Chromium (Cr)	Hinders plant growth, metabolic variations, accelerates the yellowing of leaves, disturbs nutrients balance, damages the tops and roots, and hinders chlorophyll anabolism	(Shanker et al. 2005; Kumari et al. 2016; Singh et al. 2020)
	Nickel (Ni)	Deteriorates water content, decreases normal growth, photosynthetic activity, cell division, DNA damage, chlorosis, and necrosis	(Yang et al. 1996; Zornoza et al. 1999; Yusuf et al. 2011)
	Lead (Pb)	Hinders plant cell growth	(Liu et al. 2003; Verma and Dubey 2003; Sharma and Dubey 2005)
	Iron (Fe)	Underdeveloped growth of roots and tops, dark green foliage or dark brown to purple leaves on some plants	(Wheeler et al. 1985; Fageria and Rabelo 1987; Ghori et al. 2019)
	Arsenic (As)	Inhibits plant growth	(Mascher et al. 2002; Finnegan and Chen 2012; Abbas et al. 2018; Singh et al. 2020)

A. cepa were observed with genotoxic results such as chromosomal aberrations (stickiness, loss of chromosome, C-mitosis, and vaginal chromosome) and nuclear anomalies such as micro-nucleated and binucleated cells. *Vicia faba L.* and *Allium cepa L.* exposed to untreated wastewater were used to study chromosomal aberrations (Tables 4.2 and 4.3).

Chromosomal aberrations caused by industrial waste/sludge were also studied using bioassays of *Vicia faba*; genotoxicity was evaluated by chromosome aberration (CA) and mitotic indices (MI); and cytotoxicity and mutagenicity were determined using nuclear aberration (NA) and micronuclei (MN) (Kristen 1997).Tannery solid waste (TSW) leachates including key components such as chromium and nickel were tested for genotoxic effects on *V. faba* somatic cells, with the conclusion that Cr and Ni were responsible for the genetic abnormalities (Chandra et al. 2004). These findings are supported elsewhere that DNA damage in the tested plants was induced by exceeded limits of Cr (Raj et al. 2014). Cr as a major toxic element was responsible for the genotoxicity and oxidative damage in Faba bean (*Vicia faba L.*) when irrigated using wastewater from coke plants (Liu et al. 2014b). In an analysis, Raj et al. (2014) observed that the seed germination of mung bean (*Vigna radiata L.*) was inhibited (90%) when it was grown in soils using tannery effluents. Excessive levels of chromium (Cr) (4.48 and 3.81 mg/L) in untreated tannery effluent were deemed genotoxic due to a cessation in growth. It was found that the use of treated wastewater for irrigation presents environmental health threat to both humans and animals. Sabeen et al. (2020) examined the impact of bioaccumulation of metals on

Table 4.3 Genotoxicity evaluation studies using untreated industrial effluents

No	Plant	Contaminant/wastewater	Reported risk	References
1	Faba bean (*Vicia faba* L.)	Chromium and nickel/ tannery waste leachate	Genetic abnormalities	(Chandra et al. 2004)
2	*Allium cepa* L. root cells	Coal fly ash (with HM contents Zn, Pb, Cu, Ni, Cd and As)	Genotoxic effects	(Chakraborty et al. 2008)
3	*Allium cepa* L. root cells	Ballast water	DNA polymorphism/ damage	(Olorunfemi et al. 2012; Olorunfemi et al. 2014)
4	Faba bean (*Vicia faba* L.)	Chromium/coke plant	Genotoxicity and oxidative damage	(Liu et al. 2014b)
5	Mung bean, *Vigna radiata* L.	Chromium/tannery	Growth inhibition	(Raj et al. 2014)
6	*V. radiata* L. and root growth of *A. cepa*	Leather effluents	Genotoxic effects	(Yadav et al. 2019)
7	*Allium cepa* L	Known HMs conc. in synthetic wastewater (Cr, Cd, Fe, Pb, and Ni)	Decreased mitotic indices	(Sabeen et al. 2020)

selected vegetables (tomato and chili). Cytotoxic effects have been observed as demonstrated by reduced MI at exposure to HM. Chili extracts were significantly affected by decreasing MI from 19 to 12 when compared to tomato extracts with reduced MI from 20 to 16.

4.6 Conclusions

Wastewater irrigation is becoming a practice in several countries around the world. There are several reasons for this, including climate change, population growth, scarcity of freshwater irrigation, geopolitical conflicts over water resources (especially between India and Pakistan), and lack of knowledge of the poor impacts of wastewater irrigation on farmers. In Pakistan, the situation is not different from that of the other countries affected by the water crisis, where wastewater is used without any treatment for irrigation. Agricultural soils irrigated from industrial wastewater containing hazardous substances such as HM are responsible for triggering mutations. Irrigation of agricultural land using industrial sewage should be strongly discouraged, as contaminants can infiltrate the food chain and pose a threat to human health. Daily ingestion and accumulation of metals in food are critical aspects from the perspective of metals and human health risk assessment by food consumption, daily ingestion and buildup of metals in food are critical aspects. However, the accumulation of metals in food crops and related risks in areas near industrial sites in Pakistan and other countries have not yet been fully explored. Therefore, industrial effluents should be carefully monitored and appropriately controlled prior to discharge to minimize possible threats to both aquatic and human lives. The basic principles behind the mechanism(s) of toxicity and tolerance caused by HM at the physiological, biochemical, and molecular levels still depend on the degree of scientific drive.

References

Abbas G, Murtaza B, Bibi I, Shahid M, Niazi NK, Khan MI, Amjad M, Hussain M (2018) Arsenic uptake, toxicity, detoxification, and speciation in plants: physiological, biochemical, and molecular aspects. Int J Environ Res Public Health 15:59

Abegunrin TP, Awe GO, Idowu DO, Adejumobi MA (2016) Impact of wastewater irrigation on soil physico-chemical properties, growth and water use pattern of two indigenous vegetables in Southwest Nigeria. Catena 139:167–178. https://doi.org/10.1016/j.catena.2015.12.014

Ahlström M, Thyssen J, Menné T, Midander K, Julander A, Lidén C, Johnsen C, Johansen J (2018) Short contact with nickel causes allergic contact dermatitis: an experimental study. Br J Dermatol 79(5):1127–1134

Ahmed M, Matsumoto M, Kurosawa K (2018) Heavy metal contamination of irrigation water, soil, and vegetables in a multi-industry district of Bangladesh international. J Environ Res 12:531–542. https://doi.org/10.1007/s41742-018-0113-z

Atienzar FA, Jha AN (2006) The random amplified polymorphic DNA (RAPD) assay and related techniques applied to genotoxicity and carcinogenesis studies: a critical review. Mutat Res 613:76–102. https://doi.org/10.1016/j.mrrev.2006.06.001

Baderna D, Maggioni S, Boriani E, Gemma S, Molteni M, Lombardo A, Colombo A, Bordonali S, Rotella G, Lodi M, Benfenati E (2011) A combined approach to investigate the toxicity of an industrial landfill's leachate: chemical analyses, risk assessment and in vitro assays. Environ Res 111:603–613. https://doi.org/10.1016/j.envres.2011.01.015

Balkhair KS, El-Nakhlawi FS, Ismail SM, Al-Solimani SG (2013) Treated wastewater use and its effect on water conservation, vegetative yeild, yield components and water use efficiency of some vegetable crops grown under two different irrigation systems in western region, Saudi Arabia. Eur Sci J 9:21

Banerjee M, Banerjee N, Bhattacharjee P, Mondal D, Lythgoe PR, Martínez M, Pan J, Polya DA, Giri AK (2013) High arsenic in rice is associated with elevated genotoxic effects in humans. Sci Rep 3:2195. https://doi.org/10.1038/srep02195

Bernard A (2008) Cadmium & its adverse effects on human health. Indian J Med Res 128:557

Bhat SA, Singh J, Singh K, Vig AP (2017) Genotoxicity monitoring of industrial wastes using plant bioassays and management through vermitechnology: a review agriculture and natural. Resources 51:325–337. https://doi.org/10.1016/j.anres.2017.11.002

Casalino E, Calzaretti G, Sblano C, Landriscina C (2002) Molecular inhibitory mechanisms of antioxidant enzymes in rat liver and kidney by cadmium. Toxicology 179:37–50. https://doi.org/10.1016/S0300-483X(02)00245-7

Chakraborty R, Mukherjee AK, Mukherjee A (2008) Evaluation of genotoxicity of coal fly ash in Allium cepa root cells by combining comet assay with the Allium test. Environ Monit Assess 153:351. https://doi.org/10.1007/s10661-008-0361-z

Chandra S, Chauhan LKS, Pande PN, Gupta SK (2004) Cytogenetic effects of leachates from tannery solid waste on the somatic cells of Vicia faba. Environ Toxicol 19:129–133. https://doi.org/10.1002/tox.20005

Chandra S, Chauhan LKS, Murthy RC, Saxena PN, Pande PN, Gupta SK (2005) Comparative biomonitoring of leachates from hazardous solid waste of two industries using Allium test. Sci Total Environ 347:46–52. https://doi.org/10.1016/j.scitotenv.2005.01.002

Chandrakar V, Yadu B, Xalxo R, Kumar M, Keshavkant S (2020) Mechanisms of plant adaptation and tolerance to metal/metalloid toxicity. In: Hasanuzzaman M (ed) Plant ecophysiology and adaptation under climate change: mechanisms and perspectives II. Springer, Singapore. https://doi.org/10.1007/978-981-15-2172-0

Chen H, Yang X, Wang P, Wang Z, Li M, Zhao F-J (2018a) Dietary cadmium intake from rice and vegetables and potential health risk: a case study in Xiangtan, southern China. Sci Total Environ 639:271–277. https://doi.org/10.1016/j.scitotenv.2018.05.050

Chen L, Zhou S, Shi Y, Wang C, Li B, Li Y, Wu S (2018b) Heavy metals in food crops, soil, and water in the Lihe River Watershed of the Taihu Region and their potential health risks when ingested. Sci Total Environ 615:141–149. https://doi.org/10.1016/j.scitotenv.2017.09.230

Christensen JM, Kristiansen J, Nielsen NH, Menné T, Byrialsen K (1999) Nickel concentrations in serum and urine of patients with nickel eczema. Toxicol Lett 108:185–189. https://doi.org/10.1016/S0378-4274(99)00088-0

Christou A, Karaolia P, Hapeshi E, Michael C, Fatta-Kassinos D (2017) Long-term wastewater irrigation of vegetables in real agricultural systems: concentration of pharmaceuticals in soil, uptake and bioaccumulation in tomato fruits and human health risk assessment. Water Res 109:24–34. https://doi.org/10.1016/j.watres.2016.11.033

de Souza PR, da Costa TC, Vargas VMF (2013) Investigation of sensitivity of the Allium cepa test as an alert system to evaluate the genotoxic potential of soil contaminated by heavy metals water. Air Soil Pollut 224:1460. https://doi.org/10.1007/s11270-013-1460-1

de Vries W, Römkens PF, Schütze G (2007) Critical soil concentrations of cadmium, lead, and mercury in view of health effects on humans and animals. In: Reviews of environmental contamination and toxicology. Springer, Cham, pp 91–130

Dixit G, Nerle S (1985) Cytotoxic effects of industrial effluents on Allium cëpa L. Geobios 12:237–240

Fageria N, Rabelo N (1987) Tolerance of rice cultivars to iron toxicity. J Plant Nutr 10:653–661

Farahat E, Linderholm HW (2015) The effect of long-term wastewater irrigation on accumulation and transfer of heavy metals in Cupressus sempervirens leaves and adjacent soils. Sci Total Environ 512-513:1–7. https://doi.org/10.1016/j.scitotenv.2015.01.032

Fazili NA, Ahmad M (2014) In vitro analysis of the phytotoxic and genotoxic potential of Aligarh wastewater and Mathura refinery wastewater. Toxicol Rep 1:981–986. https://doi.org/10.1016/j.toxrep.2014.06.004

Finnegan P, Chen W (2012) Arsenic toxicity: the effects on plant metabolism. Front Physiol 3. https://doi.org/10.3389/fphys.2012.00182

Fiskesjö G (1988) The allium test — an alternative in environmental studies: the relative toxicity of metal ions. Mutat Res 197:243–260. https://doi.org/10.1016/0027-5107(88)90096-6

Fowler BA (2009) Monitoring of human populations for early markers of cadmium toxicity: a review. Toxicol Appl Pharmacol 238:294–300. https://doi.org/10.1016/j.taap.2009.05.004

Ghori N-H, Ghori T, Hayat M, Imadi S, Gul A, Altay V, Ozturk M (2019) Heavy metal stress and responses in plants. Int J Environ Sci Technol 16:1807–1828

Gola D, Malik A, Shaikh ZA, Sreekrishnan T (2016) Impact of heavy metal containing wastewater on agricultural soil and produce: relevance of biological treatment. Environ Process 3:1063–1080. https://doi.org/10.1007/s40710-016-0176-9

Goni MA, Ahmad JU, Halim MA, Mottalib MA, Chowdhury DA (2014) Uptake and translocation of metals in different parts of crop plants irrigated with contaminated water from DEPZ area of Bangladesh. Bull Environ Contam Toxicol 92:726–732. https://doi.org/10.1007/s00128-014-1264-z

Grant WF (1994) The present status of higher plant bioassays for the detection of environmental mutagens. Mutat Res 310:175–185. https://doi.org/10.1016/0027-5107(94)90112-0

Haq I, Kumar S, Raj A, Lohani M, Satyanarayana GNV (2017) Genotoxicity assessment of pulp and paper mill effluent before and after bacterial degradation using Allium cepa test. Chemosphere 169:642–650. https://doi.org/10.1016/j.chemosphere.2016.11.101

Hoshina MM, Marin MMA (2009) Micronucleus and chromosome aberrations induced in onion (Allium cepa) by a petroleum refinery effluent and by river water that receives this effluent. Ecotoxicol Environ Saf 72:2090–2095. https://doi.org/10.1016/j.ecoenv.2009.07.002

Iqbal HH, Taseer R, Anwar S, Qadir A, Shahid N (2016) Human health risk assessment: heavy metal contamination of vegetables in Bahawalpur, Pakistan. Bull Environ Stud 1:10–17

Jaishankar M, Tseten T, Anbalagan N, Mathew BB, Beeregowda KN (2014) Toxicity, mechanism and health effects of some heavy metals. Interdiscip Toxicol 7:60–72

Jan AT, Azam M, Siddiqui K, Ali A, Choi I, Haq QM (2015) Heavy metals and human health: mechanistic insight into toxicity and counter defense system of antioxidants. Int J Mol Sci 16:29592–29630. https://doi.org/10.3390/ijms161226183

Jiménez B, Asano T (2008) Water reuse: an international survey of current practice, issues and needs. IWA, London

Kawatra BL, Bakhetia P (2008) Consumption of heavy metal and minerals by adult women through food in sewage and tube-well irrigated area around Ludhiana city (Punjab, India). J Hum Ecol 23:351–354. https://doi.org/10.1080/09709274.2008.11906089

Keraita BN, Akatse J, Kinane M, Osei CK, Mateo-Sagasta J, Beernaerts I, Koo-Oshima S, Youdeowei A, Fredrix M, Neate PJ (2012) On-farm practices for the safe use of wastewater in urban and peri-urban horticulture. FAO, Rome

Khan S, Cao Q, Zheng YM, Huang YZ, Zhu YG (2008) Health risks of heavy metals in contaminated soils and food crops irrigated with wastewater in Beijing, China. Environ Pollut 152:686–692. https://doi.org/10.1016/j.envpol.2007.06.056

Kristen U (1997) Use of higher plants as screens for toxicity assessment. Toxicol In Vitro 11:181–191. https://doi.org/10.1016/S0887-2333(97)00005-2

Kumar Sharma R, Agrawal M, Marshall F (2007) Heavy metal contamination of soil and vegetables in suburban areas of Varanasi, India. Ecotoxicol Environ Safety 66:258–266. https://doi.org/10.1016/j.ecoenv.2005.11.007

Kumari V, Yadav A, Haq I, Kumar S, Bharagava RN, Singh SK, Raj A (2016) Genotoxicity evaluation of tannery effluent treated with newly isolated hexavalent chromium reducing Bacillus cereus. J Environ Manag 183:204–211

Leme DM, Marin-Morales MA (2009) Allium cepa test in environmental monitoring: a review on its application. Mutat Res 682:71–81. https://doi.org/10.1016/j.mrrev.2009.06.002

Levan A (1949) The influence on chromosomes and mitosis of chemicals, as studied by the Allium test. Hereditas 35:325–337. https://doi.org/10.1111/j.1601-5223.1949.tb03345.x

Liao QL, Liu C, Wu HY, Jin Y, Hua M, Zhu BW, Chen K, Huang L (2015) Association of soil cadmium contamination with ceramic industry: a case study in a Chinese town. Sci Total Environ 514:26–32

Liu J, Li K, Xu J, Zhang Z, Ma T, Lu X, Yang J, Zhu Q (2003) Lead toxicity, uptake, and translocation in different rice cultivars. Plant Sci 165:793–802. https://doi.org/10.1016/S0168-9452(03)00273-5

Liu W-h, Zhao J-z, Ouyang Z-y, Söderlund L, Liu G-h (2005) Impacts of sewage irrigation on heavy metal distribution and contamination in Beijing, China. Environ Int 31:805–812. https://doi.org/10.1016/j.envint.2005.05.042

Liu G, Yu Y, Hou J, Xue W, Liu X, Liu Y, Wang W, Alsaedi A, Hayat T, Liu Z (2014a) An ecological risk assessment of heavy metal pollution of the agricultural ecosystem near a lead-acid battery factory. Ecol Indic 47:210–218

Liu Y, Wang Y, Lv Y (2014b) Evaluation of genotoxicity and oxidative damage induced by coke plant wastewater. Int J Environ Technol Manag 17:450–458. https://doi.org/10.1504/ijetm.2014.064587

Martínez-Alcalá I, Bernal MP (2020) Environmental impact of metals, metalloids, and their toxicity. In: Metalloids in plants, pp 451–488. https://doi.org/10.1002/9781119487210.ch21

Mascher R, Lippmann B, Holzinger S, Bergmann H (2002) Arsenate toxicity: effects on oxidative stress response molecules and enzymes in red clover plants. Plant Sci 163:961–969. https://doi.org/10.1016/S0168-9452(02)00245-5

Masood F, Malik A (2013) Mutagenicity and genotoxicity assessment of industrial wastewaters. Environ Sci Pollut Res 20:7386–7397. https://doi.org/10.1007/s11356-013-1756-0

Mazzeo DEC, Marin-Morales MA (2015) Genotoxicity evaluation of environmental pollutants using analysis of nucleolar alterations. Environ Sci Pollut Res 22:9796–9806. https://doi.org/10.1007/s11356-015-4134-2

Mazzeo DEC, Fernandes TCC, Levy CE, Fontanetti CS, Marin-Morales MA (2015) Monitoring the natural attenuation of a sewage sludge toxicity using the Allium cepa test. Ecol Indic 56:60–69. https://doi.org/10.1016/j.ecolind.2015.03.026

Mendoza-Espinosa LG, Burgess JE, Daesslé L, Villada-Canela M (2019) Reclaimed water for the irrigation of vineyards: Mexico and South Africa as case studies. Sustain Cities Soc 51:101769. https://doi.org/10.1016/j.scs.2019.101769

Minhas PS, Samra JS (2003) Quality assessment of water resources in Indo-Gangetic basin part in India. Central Soil Salinity Research Institute, Karnal

Mishra K (1993) Cytotoxic effects of distillary waste on Allium cepa L. Bull Environ Contam Toxicol 50:199–204. https://doi.org/10.1007/BF00191722

Mohanpuria P, Rana NK, Yadav SK (2007) Cadmium induced oxidative stress influence on glutathione metabolic genes of Camellia sinensis (L.) O. Kuntze. Environ Toxicol 22:368–374. https://doi.org/10.1002/tox.20273

Moulis J-M, Thévenod F (2010) New perspectives in cadmium toxicity: an introduction. Springer, Cham

Nagao M, Takashi S, Taijiro M (1978) Environmental mutagens and carcinogens. Annual review of genetics 12(1)117–159

Nefic H, Musanovic J, Metovic A, Kurteshi K (2013) Chromosomal and nuclear alterations in root tip cells of allium cepa L. induced by alprazolam. Med Arch 67:388–392. https://doi.org/10.5455/medarh.2013.67.388-392

Njoku K, Akinola M, Tommy I (2015) Genotoxicity of industrial paint effluent on the root meristem of Allium Cepa. IOSR J Environ Sci Toxicol Food Technol 9:11–17. https://doi.org/10.9790/2402-09211117

O'Connor D, Hou D, Ye J, Zhang Y, Ok YS, Song Y, Coulon F, Peng T, Tian L (2018) Lead-based paint remains a major public health concern: a critical review of global production, trade, use, exposure, health risk, and implications. Environ Int 121:85–101. https://doi.org/10.1016/j.envint.2018.08.052

Olorunfemi D, Duru E, Okieimen F (2012) Induction of chromosome aberrations in Allium cepa L. root tips on exposure to ballast water. Caryologia 65:147–151. https://doi.org/10.1080/00087114.2012.711676

Olorunfemi DI, Okieimen EA, Ovwemuvwose J (2014) DNA integrity of onion (Allium cepa L.) root cells exposed to ballast water. Stud Univ Vasile Goldis Seria Stiintele Vietii (Life Sci Ser) 24:305–309

Raj A, Kumar S, Haq I, Kumar M (2014) Detection of tannery effluents induced DNA damage in mung bean by use of random amplified polymorphic DNA markers. ISRN Biotechnol 2014:727623. https://doi.org/10.1155/2014/727623

Rathnasamy S, Mohamed KB, Sulaiman SF, Akinboro A (2013) Evaluation of cytotoxic, mutagenic and antimutagenic potential of leaf extracts of three medicinal plants using Allium cepa chromosome assay. Int Curr Pharm J 2:131–140. https://doi.org/10.3329/icpj.v2i8.15588

Reichman S (2002) The responses of plants to metal toxicity: a review forusing on copper, manganese & zinc. Australian Minerals & Energy Environment Foundation, Melbourne

Rowe DR, Abdel-Magid IM (1995) Handbook of wastewater reclamation and reuse. CRC Press, Boca Raton

Rughani G, Chandra J, Chandrakar V, Keshavkant S (2015) Lipid peroxidation and in-situ localization of ROS in assorted seeds exposed to salinity and artificial ageing. Appl Sci Rep 12:123–127. https://doi.org/10.15192/PSCP.ASR.2015.12.3.123127

Sabeen M, Mahmood Q, Ebadi AG, Bhatti ZA, Faridullah, Irshad M, Kakar A, Bilal M, Arshad HM, Shahid N (2019) Health risk assessment consequent to wastewater irrigation in Pakistan. Soil Sci Soc Pakistan 39:67–76. https://doi.org/10.25252/SE/2020/71758

Sabeen M, Mahmood Q, Ahmad Bhatti Z, Faridullah, Irshad M, Bilal M, Hayat MT, Irshad U, Ali Akbar T, Arslan M, Shahid N (2020) Allium cepa assay based comparative study of selected vegetables and the chromosomal aberrations due to heavy metal accumulation. Saudi J Biol Sci 27:1368–1374. https://doi.org/10.1016/j.sjbs.2019.12.011

Saleh B (2015) Detection of genetic variations in marine algae Ulva lactuca (Chlorophyta) induced by heavy metal pollutants. J Stress Physiol Biochem 11(3):26–37

Samuel OB, Osuala FI, Odeigah PG (2010) Cytogenotoxicity evaluation of two industrial effluents using Allium cepa assay. Afr J Environ Sci Technol 4

Schwartz MF, Boyd CE (1994) Effluent quality during harvest of channel catfish from watershed ponds. Prog Fish Cult 56:25–32

Shanker AK, Cervantes C, Loza-Tavera H, Avudainayagam S (2005) Chromium toxicity in plants. Environ Int 31:739–753. https://doi.org/10.1016/j.envint.2005.02.003

Sharma P, Dubey RS (2005) Lead toxicity in plants. Braz J Plant Physiol 17:35–52. https://doi.org/10.1590/S1677-04202005000100004

Shvachiy L, Geraldes V, Amaro-Leal Â, Rocha I (2018) Intermittent low-level lead exposure provokes anxiety, hypertension, autonomic dysfunction and neuroinflammation. Neurotoxicology. https://doi.org/10.1016/j.neuro.2018.08.001

Silveira GL, Lima MGF, Reis GB, Palmieri MJ, Andrade-Vieria LF (2017) Toxic effects of environmental pollutants: comparative investigation using Allium cepa L. and Lactuca sativa L. Chemosphere 178:359–367. https://doi.org/10.1016/j.chemosphere.2017.03.048

Singh S, Yadav V, Arif N, Singh VP, Dubey NK, Ramawat N, Prasad R, Sahi S, Tripathi DK, Chauhan DK (2020) Chapter 12 - Heavy metal stress and plant life: uptake mechanisms, toxicity, and alleviation. In: Tripathi DK, Pratap Singh V, Chauhan DK, Sharma S, Prasad SM, Dubey NK, Ramawat N (eds) Plant life under changing environment. Academic Press, pp 271–287. https://doi.org/10.1016/B978-0-12-818204-8.00001-1

Sridhara Chary N, Kamala CT, Samuel Suman Raj D (2008) Assessing risk of heavy metals from consuming food grown on sewage irrigated soils and food chain transfer. Ecotoxicol Environ Saf 69:513–524. https://doi.org/10.1016/j.ecoenv.2007.04.013

Steinkellner H, Mun-Sik K, Helma C, Ecker S, Ma T-H, Horak O, Kundi M, Knasmüller S (1998) Genotoxic effects of heavy metals: comparative investigation with plant bioassays. Environ Mol Mutagen 31:183–191. https://doi.org/10.1002/(sici)1098-2280(1998)31:2<183::aid-em11>3.0.co;2-8

Sun H-J, Rathinasabapathi B, Wu B, Luo J, Pu L-P, Ma LQ (2014) Arsenic and selenium toxicity and their interactive effects in humans. Environ Int 69:148–158. https://doi.org/10.1016/j.envint.2014.04.019

Tasrina R, Rowshon A, Mustafizur A, Rafiqul I, Ali M (2015) Heavy metals contamination in vegetables and its growing soil. J Environ Anal Chem 2:2

Teklehaimanot GZ, Kamika I, Coetzee MAA, Momba MNB (2015) Seasonal variation of nutrient loads in treated wastewater effluents and receiving water bodies in Sedibeng and Soshanguve, South Africa. Environ Monit Assess 187:595. https://doi.org/10.1007/s10661-015-4795-9

Thebo AL, Drechsel P, Lambin EF, Nelson KL (2017) A global, spatially-explicit assessment of irrigated croplands influenced by urban wastewater flows. Environ Res Lett 12:074008. https://doi.org/10.1088/1748-9326/aa75d1

Tran TA, Popova LP (2013) Functions and toxicity of cadmium in plants: recent advances and future prospects. Turk J Bot 37:1–13

Verma S, Dubey RS (2003) Lead toxicity induces lipid peroxidation and alters the activities of antioxidant enzymes in growing rice plants. Plant Sci 164:645–655. https://doi.org/10.1016/S0168-9452(03)00022-0

Wheeler B, Al-Farraj M, Cook R (1985) Iron toxicity to plants in base-rich wetlands: comparative effects on the distribution and growth of Epilobium hirsutum L. and Juncus subnodulosus Schrank. New Phytol 100:653–669

Yadav A, Raj A, Purchase D, Ferreira LFR, Saratale GD, Bharagava RN (2019) Phytotoxicity, cytotoxicity and genotoxicity evaluation of organic and inorganic pollutants rich tannery wastewater from a Common Effluent Treatment Plant (CETP) in Unnao district, India using Vigna radiata and Allium cepa. Chemosphere 224:324–332

Yang X, Baligar VC, Martens DC, Clark RB (1996) Plant tolerance to nickel toxicity: I. influx, transport, and accumulation of nickel in four species. J Plant Nutr 19:73–85. https://doi.org/10.1080/01904169609365108

Yebpella G, Magomya A, Udiba U, Gandu I, Amana S, Ugboaja V, Usman N (2011) Assessment of Cd, Cu, Mn and Zn levels in soil, water and vegetable grown in irrigated farm along river Kubani, Zaria, Nigeria. J Appl Environ Biol Sci 1:84–89

Yeganeh M, Afyuni M, Khoshgoftarmanesh A-H, Khodakarami L, Amini M, Soffyanian A-R, Schulin R (2013) Mapping of human health risks arising from soil nickel and mercury contamination. J Hazard Mater 244:225–239

Yusuf M, Fariduddin Q, Hayat S, Ahmad A (2011) Nickel: an overview of uptake, essentiality and toxicity in plants. Bull Environ Contam Toxicol 86:1–17. https://doi.org/10.1007/s00128-010-0171-1

Zhang Y, Shen Y (2019) Wastewater irrigation: past, present, and future. WIREs Water 6:e1234. https://doi.org/10.1002/wat2.1234

Zhang M-K, Liu Z-Y, Wang H (2010) Use of single extraction methods to predict bioavailability of heavy metals in polluted soils to rice. Commun Soil Sci Plant Anal 41:820–831

Zhang W-L, Du Y, Zhai M-M, Shang Q (2014) Cadmium exposure and its health effects: a 19-year follow-up study of a polluted area in China. Sci Total Environ 470-471:224–228. https://doi.org/10.1016/j.scitotenv.2013.09.070

Zhao Q, Wang Y, Cao Y, Chen A, Ren M, Ge Y, Yu Z, Wan S, Hu A, Bo Q (2014) Potential health risks of heavy metals in cultivated topsoil and grain, including correlations with human primary liver, lung and gastric cancer, in Anhui province, Eastern China. Sci Total Environ 470:340–347

Zornoza P, Robles S, Martin N (1999) Alleviation of nickel toxicity by ammonium supply to sunflower plants. Plant Soil 208:221–226. https://doi.org/10.1023/A:1004517414730

Chapter 5
Nutrient Uptake and Plant Growth Under the Influence of Toxic Elements

Javed Nawab, Junaid Ghani, Sardar Khan, Muhammad Amjad Khan, Abid Ali, Ziaur Rahman, Mehboob Alam, Abd El-Latif Hesham, and Ming Lei

5.1 Plants and Toxic Elements' Interaction

Vegetables are considered as primary source of essential nutrients and basic need of human diets, and their contamination with toxic elements is susceptible to human health through consumption (Nawab et al. 2018a, 2019; Pruvot et al. 2006; Radwan and Salama 2006; Lim et al. 2008). Vegetables are subject to high concentration of toxic elements, while use of fertilizers for high crop production and extensive use of

J. Nawab (✉)
Department of Environmental Sciences, Abdul Wali Khan University Mardan, Mardan, Pakistan
e-mail: javednawab@awkum.edu.pk

J. Ghani
School of Environmental Studies, China University of Geosciences, Wuhan, People's Republic of China

S. Khan · M. A. Khan
Department of Environmental Sciences, University of Peshawar, Peshawar, Khyber Pakhtunkhwa, Pakistan

A. Ali
Department of Zoology, Abdul Wali Khan University Mardan, Mardan, Pakistan

Z. Rahman
Department of Microbiology, Abdul Wali Khan University Mardan, Mardan, Pakistan

M. Alam
Department of Horticulture, The University of Agriculture Peshawar, Peshawar, Pakistan

A. E.-L. Hesham
College of Resources and Environment, Hunan Agricultural University, Changsha, People's Republic of China

M. Lei
Department of Genetics, Faculty of Agriculture, Beni-Suef University, Beni-Suef, Egypt

Q. Mahmood (ed.), *Sustainable Plant Nutrition under Contaminated Environments*, Sustainable Plant Nutrition in a Changing World,
https://doi.org/10.1007/978-3-030-91499-8_5

wastewater for irrigation could lead to increase in metal contamination (Gil et al. 2004). The rate and accumulation level of toxic elements (TEs) and their affects vary with different kinds of hyper-accumulative plant species and, also within the same plants species (John and Van Laerhoven 1976). Toxic elements can be transferred to different parts of the plants at high concentration (Gawęda 2007). The high concentrations of metals could be found in various parts of plants including vegetative and non-vegetative parts of plants cultivated in wastewater irrigated soil (Khan et al. 2008). Garate et al. (1993) reported that lettuce plant tissues have high capability to uptake the toxic elements. On the other hand, toxicity of elements mostly affected the lettuce plant roots and shoots with increasing duration of metal concentration exposure and could negatively affect the plants growth and biomass (López-Millán et al. 2009). Photosynthetic process and other plants minerals can also be substantially influenced by toxicity of elements (López-Millán et al. 2009). The soil pH plays a significant role to supply the important nutrients to plants, such as low pH in soil means more nutrients transfer to plants (Lutz et al. 1972). Furthermore, some factors including type of soil, cultivation practices, storage condition, and climate can also affect nutrient uptake to plants (Srikumar and Ockerman 1990).

It has been previously reported that toxic elements did not show any toxicity and harmful effects in vegetables plant tissues (Intawongse and Dean 2006), while it has been described in literature that toxic element concentrations and harmful symptoms can be found in various parts of the same plant. However, it has been reported that high metal concentrations have been observed in plant roots than shoot (Vanassche and Clijsters 1990). Tomato is a very essential food which is important from nutritional point of view as well as economical (FAOSTAT 2007). The physiology and growth of tomato can be affected due to availability of toxic elements at high concentrations and result to enhance the necrotic symptoms and chlorosis in tomato leaves (López-Millán et al. 2009). Tomato fruit is considered as source of energy enriched with nutrients, vitamins as well as minerals (Giovanelli and Paradise 2002) and can be consumed both in raw and processed forms (Martinez-Valvercle et al. 2002). It has been found in previous studies that toxic elements such as Cd resulted to accumulate in different parts of tomato (Donma and Donma 2005). Similarly, other edible food crop such as potato is an essential vegetable, enriched with crucial elements such as K, Ca, Zn, Fe, and Zn as well as energy vitamins, carbohydrates, and dietary fibers (Finglas and Faulks 1984). Toxicity of metals also effects the roots such as browning of roots, changes in photosynthesis process in plants, and variation in mineral concentrations (López-Millán et al. 2009). In the previous studies, it is concluded that leafy vegetables are more susceptible and affected by high metal concentration and toxic elements such as Cr were highly observed in lettuce stem and exceeded the permissible limit of FAO (0.05 mg/kg) and EU (1.0 mg/kg). Likewise, the Pb concentrations were also exceeded the FAO acceptable limit (0.3 mg/kg), EU limit (0.3 mg/kg), Indian standard (2.5 mg/kg), and SEPA permissible limit (9.0 mg/kg) in many vegetable plants. The detail regarding the effects of TEs on food plant nutrients is shown in Fig. 5.1.

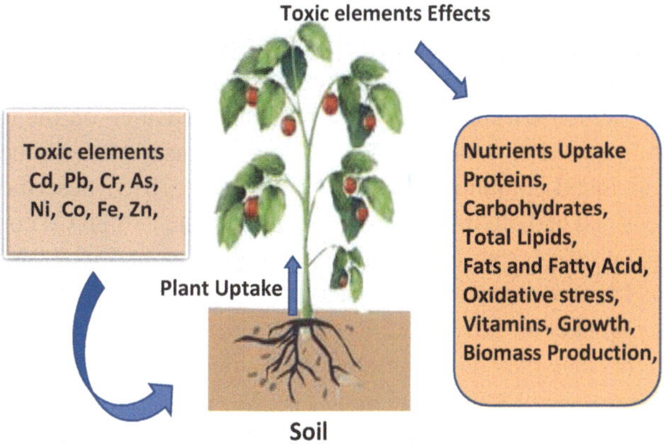

Fig. 5.1 Bioaccumulation of toxic elements in food plant and nutrients

5.2 Phytoremediation and Toxic Elements' Tolerance

Plants are considered as the major source of high toxic element accumulators grown on contaminated soil. Based on uptake of toxic elements, plants are categorized as accumulator plants on the basis of accumulation as well as hyper-accumulator plants and excluder plants. Contamination of toxic elements could possibly accumulate and transfer from soil to plant, water to plant, and air to plant (Nawab et al. 2016a, b, c). However, high accumulation of metals in plants may occur through soil-plant interface. In previous studies, there was a strong relationship observed between toxic elements in soil and vegetable crops (Khan et al. 2015, 2020; Nawab et al. 2015, 2018a, b). In another study, concentration of toxic elements was different in various parts of the plant, and the bioaccumulation of metals and absorption rate were found higher in plant roots, compared with different other parts of the plant (Verma and Dubey 2003). The presence of elevated toxic element contamination has shown adverse effects on both plants and animals, and the potential toxicity rate may vary from species to species plant and, also from metal to metal. The past studies showed that plant leaves' size and structure are strongly influenced by toxic elements, and metal toxicity symptoms are first observed on plant leaves; hence, it is important to determine length to width ratio of plant organs to assess tolerance rate of plant to toxic elements (Zhang et al. 2014). In addition, the bioavailability and variation of toxic elements in soil could be influenced by the amount of exchangeable metals. Likewise, the heavy metal phyto-accumulation may also affected by solubility and soil type in plants (Castro et al. 2009). The heavy metal uptake by plants enhances as the toxic element concentrations increase in the soil. It has been reported in previous literature that there is correlation between heavy metal toxicity and its relevant speciation in soil. However, it is quite difficult to observe particular species due to complexity of metals in nature, its function, and distribution level in the soil (Czupyran and Levy 1989). Moreover, the bioavailability of exchangeable TEs and some carbonate-bound TEs are highly available in soil than other fractions (Wong et al. 2002). According to Galal and Shehata (2015), the bioavailability of toxic elements varies in different sites; authors reported that concentrations of the elements were found high in plants cultivated on the roadside soil. Liu et al. (2005) stated that metal concentration such as Cd was higher than As in plant grown in metal-contaminated soil. Irfan et al. (2013) described that plant growth was retarded and elicited the toxic symptoms due to high Cd concentrations, and toxicological effects can occur on seed germination and its growth at low Cd concentration. Likewise, toxic effects can affect seedling growth due to high Cu concentration (Iqbal et al. 2018).

Micronutrients also play an important role to interact with toxic elements in soil and affect the transport of metals and uptake by plants (Hernández et al. 1998; Thys et al. 1991). Essential nutrients (Nitrogen and Sulfur) provide defensive mechanisms for the tolerance of heavy metal such as protein and glutathione synthesis (Hossain et al. 2012; Zechmann and Müller 2010). Moreover, plants' growth and their physiological functions can also be affected by essential nutrients. The tolerance of toxic elements by plants may depend and attributed to their various chemical forms (Hossain et al. 2012).

Various plant species are recommended to assess the toxicity of toxic elements in soil, depending on physiochemical properties of soil. Plant grown in contaminated soil is used for assessment of metal toxicity in adequate pH environment such as pH level ranging from 6 to 8 (Chapman et al. 2010, 2012; Sheppard et al. 1993) because vegetable is not considered as good for acid-tolerant plant species (Environment Canada 2005; Chapman et al. 2012). The *Brassica juncea* (BjCdR15) and *Arabidopsis* (TGA3) plant species are used to tolerate high concentration of metals due to some processes evolved such as BjCdR15/ TGA3 control the phytochelatin synthesis (Farinati et al. 2010). Whereas, *Allium cepa* has shown Pb tolerance in contaminated soil (Dang et al. 1990). Oat plant is also used for assessment of toxic elements among higher plants (Chang et al. 1997) under inadequate low pH conditions in soil (Loureiro et al. 2006; Bilski and Foy 1987; Small and Jackson 1949). Mellem et al. (2009) reported that *Amaranthus dubius* plant is more capable of removing the toxic elements by roots in contaminated soil, but toxic elements such as Cr, Ni, As, Pb, and Cu could not uptake and absorb in plant shoot. Environmentally friendly techniques such as bioassay and transgenic techniques have been recommended to evaluate the genotoxic potential effects of contaminants on plants such as micronucleus (MCN) induction and random-amplified polymorphic DNAs (RAPDs) techniques (Angelis et al. 2000; Arkhipchuk et al. 2000; DeWolf et al. 2004; Liu et al. 2005). An advanced detoxification technique evolved by selective plants can influence and efficiently reduce the uptake and toxicity of metal, as well as other mechanisms including chelation, secretion of metals, and compartmentalization (Pourrut et al. 2013). The toxic elements interact with genetic materials through DNA and cause genotoxicity to living organisms (Hossain and Huq 2002). Some reactive oxygen species (ROS) that are induced by heavy metal, mainly play key role in peroxidation of fatty acid. Plants adopt various defensive strategies including antioxidant enzymes (Gill and Tuteja 2010). Furthermore, addition of compost and sewage sludge in soils reduce the bioavailability of toxic metals to plants and avoid contamination in food chain (Smith 2009). Phytochelatins play significant role in detoxification of toxic elements in soil (Cobbett 2000). Similarly, metal tolerance is affected by involving different types of genes, which mediate transferring of metal glutathione conjugate (Kim et al. 2006). However, protein in plant can also take active part in metal tolerance and homeostasis of TEs (Suzuki et al. 2002).

5.3 Influence of Toxic Elements on Plant Growth

Contamination of toxic elements in the environment has adverse impacts on the growth of plants due to their phytotoxicity, even low metal concentration also inhibits the growth of plant (Chibuike and Obiora 2014). The plant growth severely reduces at high metal concentrations. Toxic elements can negatively affect the photosynthetic process, retard the plants growth, and induce oxidative stresses. Elevated and maximum concentration of toxic elements can hamper the plant growth by hindering the photosynthetic process in plant and also disrupting the coordination of essential elements and their effective mechanism (Singh et al. 2016). Maximum metal concentration was identified in the leaves of plants cultivated on highly contaminated soil. As a result, photosynthetic processes, respiration in plant leaves, and protein synthesis were negatively affected and damaged. Rate of high metal concentration may substantially affect the plants' metabolic processes and growth of cultivated crops in contaminated soil (Chibuike and Obiora 2014). Hagemeyer 2004 reported that presence of toxic elements negatively affects the plant growth, cell division, and various developmental mechanisms. Cultivated plant height has been greatly reduced from 18% to 77% in contaminated soil. Element such as Cu plays an important role to enhance the plant growth and different physiological functions; however, when its concentration increases and thereby affects the plant growth due to its toxicity and interfere the plant physiological functions (Zhao et al. 2012). Increase of Cu concentration results in inhibiting the root growth and causing ion leakage from the cells by disrupting the plasma membrane (Tan et al. 2014). Whereas, Cd at high concentration has substantial negative effect on the growth of shoot (Khan et al. 2018), and reduction in growth could be observed because of metal induction of chromosomal aberration (Shi et al. 2014).

Influence and stress of TEs on various plants species are presented in Table 5.1. It has been known from previous literature that TEs such as Pb, Cd, and Cr reduced the shoot length and adversely affected fresh and dry weight of *Sorghum bicolor* (Bacaha et al. 2015). According to Nawab et al. (2018b), the presence of elevated TEs including Cr, Ni, As, Zn, Pb, and Cd effectively reduced the *Pisum sativum* and *Capsicum annuum* plants' root and shoot growth in control treatments. Similarly, Nawab et al. (2019) also showed that TEs noticeably reduced spinach growth and biomass in control treatments mine-degraded soil. In another study, the presence of 10 mg Cd kg^{-1} significantly reduced the growth of *Hordeum vulgare*

Table 5.1 Effect of TEs stress on plant growth

TEs	Plants	Impacts on plant growth	References
Pb, Cd, Cr	*Sorghum bicolor*	Reduced shoot length, fresh and dry (weights)	Bacaha et al. (2015)
Cr, Ni, As, Zn, Pb, Cd	*Pisum sativum, Capsicum annuum*	Effectively reduced the pea and chilly plants' roots and shoots growth	Nawab et al. (2018b)
Cr, As, Ni, Cd, Zn, Pb	*Spinacia oleracea*	Decreased and hindered the growth and spinach biomass	Nawab et al. (2019)
Cd	*Hordeum vulgare*	Hindered the growth at concentration of 10 mg kg^{-1}.	Hernández-allica et al. (2008)
Ni, Co	*Cicer arietinum*	Reduced the growth, plant biomass, seed germination and chlorophyll content. Damaged the root and shoot. Suppressed the root nodules and adversely affected the plant yield	Khan and Khan (2010)
Cd, Pb, Zn	*Spinacia oleracea*	Significantly decreased the biomass shoot and root (fresh and dry) weights	Alia et al. (2015)
Cu	*Lemna gibba*	Reduced the plant growth when treated with high concentration	Babu et al. (2003)
Pb	*Lycopersicon esculentum*	Gradual decline was observed in plant growth	Opeolu et al. (2010)
Fe	*Eugenia uniflora L.*	Decreased plant root and shoot growth, caused by oxidative stress	de Oliveira Jucoski et al. (2013)
Cr, Ni, Cd, Cu,	*Alfalfa plant (cv. Malone).*	Significantly hampered ability of the seed to germinate and reduced growth of root and shoot at 0.04 mg/L dose	Peralta-Videa et al. (2001)
Mo	*Cicer arietinum*	Declined the root and shoot length at metal concentration more than 7.5 ppm and altered the plant anatomy at concentration level more than 1.5 ppm	Datta et al. (2011)
Zn	*Camellia sinensis L.*	Reduced the fresh plant root, shoot, and dry weight. Disorganization of cellular organelles. Hindered net photosynthetic process, transpiration rate, stomatal conductance, and chlorophylls a and b contents	Mukhopadhyay et al. (2013)

(Hernández-Allica et al. 2008). Moreover, Khan and Khan (2010) evaluated an experiment to study the adverse effect of Ni and Co at different concentration, such as low level (0, 10, 50 ppm) and high level (100, 200, and 400 ppm) on chick pea *(Cicer arietinum)*. They showed that the growth and biomass of chick pea plant decreased with high toxicity of Ni and Co concentrations, and reduction was observed in seed germination and chlorophyll content in plant (Table 5.1). While it has been reported in previous studies that toxicity of elements such as Cd, Pb, and Zn potentially affected biomass of *S. oleracea* root and shoot (Alia et al. 2015).

Likewise, the high reduction was observed in growth of *L. gibba*, when treated with high Cu concentration (Babu et al. 2003) and high concentration of Pb caused decline in the growth of *Lycopersicon esculentum* (Opeolu et al. 2010).

A previous study showed that Fe at concentration level of 1.0 and 2.0 mM reduced the *Eugenia uniflora L.* root and shoot growth, caused oxidative stress, and increased peroxidation of lipid in leaves (de Oliveira Jucoski et al. 2013). Peralta-Videa et al. (2001) studied that the dose of 0.04 mg/L of TE treatments Cr^{6+}, Ni^{2+}, Cu^{2+}, and Cd^{2+} significantly reduced the capability of seed germination and growth of alfalfa plant (*cv. Malone*) root and shoot. In other previous research, Datta et al. (2011) observed reduction in *Cicer arietinum* root and shoot length, when exposed to high concentration of Mo more than 7.5 ppm, and plant anatomy was also altered at more than 1.5 ppm concentration. Similarly, Mukhopadhyay et al. (2013) identified that Zn (30 μM) concentration decreased the root and shoot (fresh and dry weights) in Tea plant (*Camellia sinensis L.*) and hindered net photosynthesis process, transpiration rate, and chlorophylls a and b contents (Table 5.1). Therefore, the findings of these studies suggest that toxicity of TEs at high concentrations hinder the plant's growth (root and shoot) and negatively alter the plant's structure.

It has been reported in past studies that TEs at elevated and even at low concentration may have negative effects on the plant's growth. The presence of TEs in contaminated soil significantly affects metabolic and photosynthetic process and induces oxidative stress in vegetable crops, thereby hindering the growth and biomass (John et al. 2009). Previous studies show that elevated TE concentration including Cr (30.5), Ni (23.6), Zn (32.7), As (0.25), Cd (0.34), and Pb (1.54 mg/kg) effectively reduced the *Pisum sativum* and *Capsicum annuum* plants' root and shoot growth in contaminated soil. These TEs (Cr, Ni, Zn, As, Cd, and Pb) remarkably reduced spinach (*Spinacia oleracea*) growth by low concentration of 3.18, 3.62, 9.18, 0.22, 0.26, and 0.52 mg/kg, respectively, as shown in Fig. 5.2 (Nawab et al. 2018b, 2019). Potential toxic elements such as As, Cd, Co, Cu, Pb, and Zn at concentration of 8, 0.18, 0.1, 16, 12, and 100 mg/kg reduced the growth of rice (*Oryza sativa* L) grown in contaminated soil and decreased the biomass of grains, leaves, and stems (Khan et al. 2014). Whereas, high reduction was observed in ryegrass shoots due to phytotoxicity of Cu and Pb and metal stress at high concentration of 30 and 500 mg/kg, respectively, and as a result, chlorosis and low rye grass biomass were produced (Karami et al. 2011). Similarly, Younis et al. (2016) identified that elevated Cd at 100 mg/kg concentration reduced the spinach plant (*Spinacia oleracea*) biomass on fresh and dry weight basis, and its toxicity also decreased the photosynthetic pigments, amino acid, and protein contents in spinach leaves. Hernández-Allica et al. (2008) found that Cd at concentration of 10 mg/kg highly reduced the growth of *Hordeum vulgare* (Fig. 5.2). These findings show that growth of different plant species could be hampered and highly reduced under low and high TE stress and phytotoxicity.

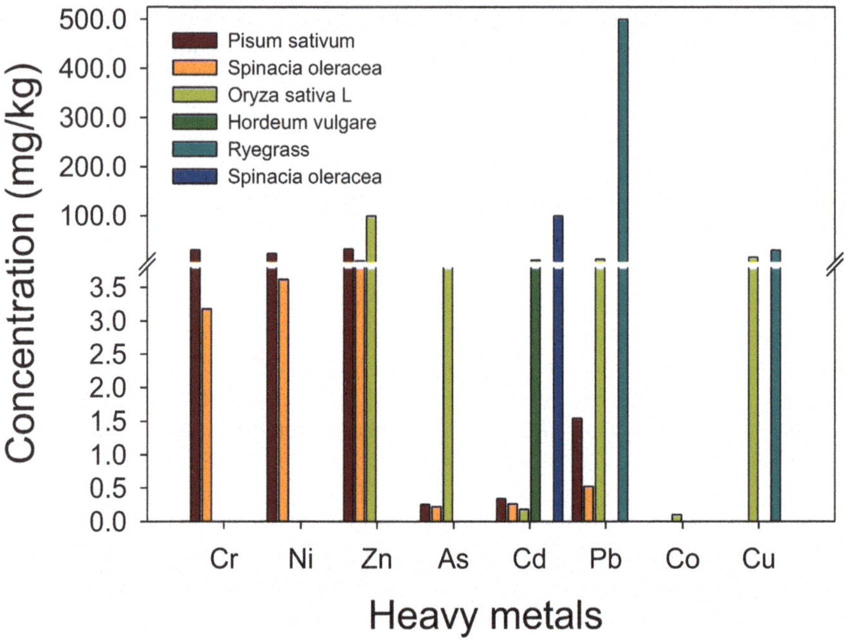

Fig. 5.2 Shows high and low concentrations of TEs in different plants

5.4 Impacts of Copper, Zinc, and Iron on Plants

Essential elements such as Cu, Zn, and Fe are important for plants, and these elements serve as micronutrients and are required for plant growth; however, their excess uptake by the plants have toxic effects (Reeves and Baker 2000; Monni et al. 2000). These TEs play physiological and biochemical functions in plants. They take part in redox reaction and are important part of several enzymes. The plants uptake these metals from soil and then accumulate in different parts depending on the metal. The accumulation rate of Cu and Fe is high in roots and moderate in upper parts, while Zn is uniformly distributed throughout the plant body (Siedlecka 1995).

Cu is an important TE for plants and plays a pivotal role in photosynthesis as a constituent of electron donor in photo system (Mahmood and Islam 2006). Plants like other organisms are sensitive to both the excess availability and deficiency of these metals. These metals are necessary for plant growth and functions, but at higher concentrations, they can cause toxicity. Cu in excess amount in soil is reported to be cytotoxic, generating stress and causing injuries to the plants, resulting in retarded growth and chlorosis. It also induces oxidative stress (Lewis et al. 2001). Cu acts as cofactor of oxygenase, oxidase, and enzymes which are involved to eliminate super oxide radicals. Several enzymes such as carbonic anhydrase, RNA polymerase, and alcohol dehydrogenase contain Zn. Zn takes part in maintaining ribosome integrity, carbohydrate formation, and catalyzing oxidation processes in plants. Fe is one of the essential elements playing a key role in many metabolic processes, photosynthesis, development of chloroplast, and synthesis of chlorophyll. Fe is a component of proteins such as cytochrome, hemoglobin, and myoglobin. It also catalyzes redox reactions (Nagajyoti et al. 2010).

Zn at a higher concentration is phytotoxic and decreases the development and growth, induces oxidative changes, and effects metabolism. Zn is also reported to alter enzyme catalytic efficiency. Its toxicity causes chlorosis in leaves (Ebbs and Kochian 1997). The cholorosis may result from deficiency of elements such as Zn^{+2} and Fe^{+2} and ions. Excess amount of Zn causes Cu and Mn deficiencies in plant shoots. These deficiencies mostly result by transferring of these metals from root to the plant shoot. Zn also causes P deficiency which results in turning the leaves purplish red color (Ebbs and Kochian 1997; Lee et al. 1996). Fe toxicity appears in plants wherein roots uptake high concentrations of Fe and transport to the leaves. Fe^{+2} in excess leads to the production of free radical which damages cellular structure, membrane, proteins, and DNA (de Dorlodot et al. 2005).

5.5 Toxic Elements and Plant Morphology

The toxic effects of elements on plant morphology are markedly known. The presence of trace elements stimulates some toxic symptoms and increases negative effects on plants' structure (Rout and Das 2009). The toxic effects of TEs vary in different cellular structure and function of organisms (Emamverdian et al. 2015). It has been reported in other study that some of structural changes occur in plants under the condition of high metal stress show severe symptoms including reduction in leaf thickness as well as structural changes in mitochondria and absence of palisade structure. Rout and Das (2009) stated that plant morphology is strongly affected due to high metal concentrations in plants. The adverse effects of toxic elements show harmful symptoms in the structure of plant components (Lindsey and Lineberger 1981). Toxic elements can damage the plant cellular metabolism, change the cell membrane structure (Parrotta et al. 2015), and disrupt the chloroplast, result in reducing the photosynthesis process (Mahmood et al. 2010). Moreover, high concentration of Fe mediates free radicals, thereby alter and disrupt cellular structure and stimulate cell membrane damage, DNA and protein (Mackenzie et al. 2008).

5.6 Influence of Toxic Elements on Plant Nutrients

Balance diet and proper nutrition are the basic requirements of good health in our daily life. Food plants especially the fruits and vegetables are the main sources of essential nutrients to fulfil our energy requirements. Fruits and vegetables include a diverse group of plant foods that vary greatly in content of energy and nutrients. Fruits and vegetables also supply vitamins and minerals to the diet and are sources of phytochemicals that function as antioxidants, phytoestrogens, and anti-inflammatory agents and through other protective mechanisms. Thus, a varied good dietary management is vital to fulfill daily requirements of nutrients for our balance diet (Slavin and Lloyd 2012). According to Stangeland et al. (2009), the impact of toxic element concentration on nutritional components of plants depends on soil, food crops, plant organs, and exposure duration. It is impossible to determine the essential dietary contents in food plants, because variation in serving size significantly depends upon people's knowledge, availability of seasonal food plants, different seasons, habits and food traditions, and financial condition of particular communities. Excessive toxic elements may have negative affect by altering the various antioxidative enzyme functions in plants and disrupt the plant cells and disturb the plant oxidative stress (Dutta et al. 2018; Sharma et al. 2012), resulting to the plant death.

The consumption of toxic elements at high rate can cause potential health problems such as subsequent reduction of essential required nutrients in the body (Jaishankar et al. 2014; Tchounwou et al. 2012). High toxic metals have a strong negative impact to disorder genetic characteristics such as mutation and recombination of different plant species (Chibuike and Obiora 2014). Toxic elements cause instability of genes in plants, thereby damaging the DNA (Dutta et al. 2018). Previous literature has shown that plants are continuously exposed to several environmental stresses including metals contamination either directly or indirectly via the induction of oxidative stress and overproduction of reactive oxygen species (ROS), resulting to damages in oxidative state of proteins, nucleic acid, and lipids and causing various consequent physiological disorders including nutrient deficiency, retardation of photosynthesis and growth in plants, reduction of nutrient transport, and genotoxicity (Dutta et al. 2018). Likewise, ROS may stimulate inactivation of enzyme and damage DNA, along with the serious disorders as mentioned above (Sharma et al. 2012). It has been reported in literature that food plants grown on metal-contaminated sites can affect both plant nutritional mechanisms and other dietary qualities (Khan et al. 2019). Women and children are more susceptible to Cd uptake at high rate in their bodies than men because they require more Fe than men (Vahter et al. 2007). Hence, enough nutrition is important for pregnant women and children to have good growth conditions and optimum development in their earlier stages of life (Marangoni et al. 2016).

5.6.1 Influence of Toxic Elements on Nitrogen and Phosphorous Contents

Green leafy vegetables are important part of human diet and contain rich sources of essential nutrients. TE contamination in plants (food crops) and their adverse effects cannot be ignored. However, contaminated food crops may have probably potential effects on human health risks via consumption (Khan et al. 2008). Phytotoxicity of toxic elements may affect the plants and thereby result in yield depression, decrease the uptake of nutrients, reduction in plant growth and chlorosis, and reduce the N_2 fixation in leguminous plants (Guala et al. 2010). Nitrogen (N) is considered as the most important nutrient and major component of nutrients, proteins, vitamins, hormones, and nucleic acids. N has capability to avoid or reduce the toxicity of metals, and it enhances the photosynthetic activities by increasing chlorophyll synthesis in plants and increases the antioxidant enzymes activity. Moreover, N element can synthesize different metabolites such as glutathione (GSH) and proline (Sharma and Dietz 2006; Lin et al. 2011). It has been reported that TEs primarily increase protease activity (Chaffei et al. 2003), and hence they decrease the enzyme activities involved in nitrate NO_3 such as nitrate reductase (NR) and nitrite reductase (NiR). Whereas, Lin et al. (2010) stated in previous research that there is a complex interaction between essential nutrients and metal accumulation in plants and show either synergistic or antagonistic effects. In addition, high elevated TEs have been observed to affect physiological process of N metabolism, including plant functions, growth, and development. High concentration of heavy metal Cd affects nitrogen metabolism by hindering glutamine synthetase (GS) activity, NO_3 reductase, NO_3 transportation and uptake, thus hampering to affect N assimilation processes (Hernández et al. 1998).

Another essential nutrient phosphorus (P) is a major constituent of nucleic acid and cell membrane and plays a key role in phosphorylation reaction. P may alleviate the TE toxicity, thereby altering to reduce the mobilization of metals and form metal phosphate compound (Sarwar et al. 2010). The P interact with various metabolic components such as sugar phosphates, nucleic acids and nucleotides, phospholipids, and coenzymes. Plant roots can be developed by addition of P to soil, increase in tolerance induction, capability of immobilization in the soil, and various physiological processes in plants (Onasanya et al. 2009; Ahmad et al. 2018). Shahid et al. (2012) stated that uptake of P to plant can influence by high Pb concentration. Moreover, high concentrated Pb in soil decreased the soil pH, as a result may enhance availability of P to plants. Similarly, the P concentrations were determined higher in celery and Chinese cabbage plants than the plant-free Pb-contaminated soils. This may be attributed to plants' root activity by releasing organic acids in the soil, resulting to enhance the soil P accumulation to the plant tissues (He et al. 2018). Additionally, P can also enhance glutathione (GSH) content and prevent membrane damage and potentially confer the tolerance to plants (Wang et al. 2009). N and P could enhance the maize plants growth, thereby increase metal accumulation due to the exertion of additional nutrients in the plant biomass. In a previous study, translocation factor values for both N and P were found efficiently higher, and the values for the metals were low. The results showed that N and P nutrient concentrations are generally higher in plants than those of metals in the plant tissues, while proportion of TEs became more concentrated in the roots than in the plant. Therefore, availability of essential nutrients (N and P) in plants and tissues is required to adapt to the environment than toxic metals stress (Mendoza et al. 2015).

The effect of TE stress on plants essential nutrients including N and P is shown in Table 5.2. In the previous study, Sharmin et al. (2012) reported that Cr metal

Table 5.2 Effect of TEs stress on plants' essential nutrients (nitrogen and phosphorus)

TEs	Effect on nitrogen	Effect on phosphorus	References
Cr	Metal toxicity altered the nitrogen metabolism and nitrate reductase in *Miscanthus sinensis* plant	*Zea mays* plant exhibited toxicity due to major changes in proteomics and affected mitochondrial oxidative phosphorylation	Sharmin et al. (2012), Wang et al. (2009)
Cd	Affected nitrogen metabolism by hindering nitrate transportation, accumulation, and nitrate reductase	Significant variation occurred in trehalose-6-phosphate phosphatase	Lea and Miflin (2004)
As	Increased N concentration in *P.vulgaris L.* plants after exposed to arsenite	Enhanced P uptake in tomato plant when exposed to low As level	Carbonell-Barrachina et al. (1997), Carbonell et al. (1998)
Cu	Excessive concentration inhibits root length, induces changes in auxin levels, and damages nitric oxide function	Phosphorylation process activated and metal stress stimulated the four isoforms of mitogen-activated protein kinase (MAPK) in *Medicago sativa*	Zobel et al. (2007), Jonak et al. (2004)
Zn	Effectively utilized the N content by wheat plants	Metal concentrations in plant tissue dropped to a minimum at the phosphorus saturation concentration of the plant	Seyhan and Erdincler (2003), Cakmak et al. (2010)

toxicity negatively affected the nitrogen metabolism and nitrate reductase in *Miscanthus sinensis* plant, while Cr toxicity altered the proteomics and oxidative mitochondrial phosphorylation in *Zea mays* plant (Wang et al. 2009). The high Cd concentration hindered nitrate uptake, its transportation, in addition to nitrate reductase, and significant variation occurred in trehalose-6-phosphate phosphatase (Lea and Miflin 2004). Whereas, N content increased when exposed to high arsenite in *P. vulgaris L.* plants. Similarly, P uptake also enhanced in tomato plant when exposed to low As level (Carbonell-Barrachina et al. 1997; Carbonell et al. 1998). Furthermore, maximum concentration of Cu hampered the root length and altered the auxin levels and disrupted the nitric oxide function. Likewise, phosphorylation process activated due to Cu metal stress, resulting in the stimulation of the four isoforms of mitogen-activated protein kinase (MAPK) in *Medicago sativa* plant (Zobel et al. 2007; Jonak et al. 2004). However, Zn at high concentration is efficiently employed the N by wheat plants, and phosphorus saturation in the plant led to drop and minimize the Zn concentration in plant tissues (Table 5.2) (Seyhan and Erdincler 2003; Cakmak et al. 2010).

5.6.2 *Influence of Toxic Elements on Carbohydrates in Plants*

High contamination of toxic elements strongly influences the carbohydrate contents in vegetable crops. Toxic elements alter the carbohydrate metabolism and their contents in different ways. As a result, certain enzymes could be damaged and inactivated that can take part in carbohydrate synthesis (Gawęda 2007; Nagor and Vyas 1997; Vikas et al. 2002). Moreover, it is important to mention that TEs including Ni and Cd have a stronger impact on carbohydrate transport by reducing photosynthesis (Moya et al. 1993). Elevated levels of Ni observed in shoots and particularly in the chloroplast were associated with a reduction in carbohydrate recirculation between leaves and roots (L'Huillier et al. 1996). Trace TEs at low concentration interfere with carbohydrate metabolism, enzyme and photosynthetic activities, and assimilation of important nutrients (Ashraf et al. 2010). Whereas, excessive toxic element accumulation may hinder the carbohydrate synthesis and destroy the ROS production and photosynthetic electron transport chain (Verma and Dubey 2003). Moreover, TEs also encourage ROS production within plants and are influenced by the intensity of the constant stress stages, plant age, and species (Verma and Dubey 2003). TEs may affect the carbohydrate metabolism by altering their contents, which could be attributed to damage and inactivation of some enzymes that usually take part in carbohydrate synthesis (Gawęda 2007). Elevated level of TEs such as Cd can induce to decrease the carbohydrate metabolism as well as upregulate the glycolytic pathway under metal stress (Rodríguez-Celma et al. 2010). Excessive concentration of Pb leads to decrease the sucrose content in vegetables as well as significantly affects the taste of rooted vegetables (Gawęda 2007). It has been stated in previous research on tomato plant species that carbohydrate metabolism reduced due to the stress of high Cd concentration, while stresses of TEs upregulate the glycolytic pathway (Rodríguez-Celma et al. 2010).

Similarly, low concentration of Cd can severely affect different enzyme activities to alter the carbohydrate and phosphorus metabolism, and Calvin cycle and carbon assimilation, thereby inhibit the plants' growth, cause chlorosis and variations in chloroplast, damage the antioxidant mechanism, and induce the peroxidation of lipids (Gill and Tuteja 2011). Toxicity of TE concentrations may hinder the plant physiological functions by inhibiting the germination, plants' growth, biomass accumulation, chlorosis, inhibition of growth, photosynthesis, alteration in water balance which ultimately cause the death of plant, and change the nutrient assimilation (Latef 2018; Singh et al. 2016), resulting to disrupt the carbohydrate and protein contents (Mondal et al. 2013). Moreover, Cd at high concentration may effectively alter and interrupt the plant physiology, resulting in low carbohydrate metabolism (Chaffei et al. 2004). Moreover, impact of toxic elements replaces the chlorophyll Mg and changes its function in plants (Farhat et al. 2016). Rodríguez-Celma et al. (2010) stated that total sugar contents increased in tomato plants grown in metal-contaminated soil. TEs have differet effects on the nutritional composition of the vegetables (Fig. 5.3). This variation might be due the antagonistic and synergetic effects of TEs with nutrient uptake. These variations also depend on specific metal, plant species, and metal concentrations (Legay et al. 2012; Ramos et al. 2002) as shown in Fig. 5.3. Existing literature revealed that toxic elements may strongly affect plant carbohydrates and other essential nutrients, and such contaminated food via consumption may cause malnutrition due to decrease in net carbohydrate contents.

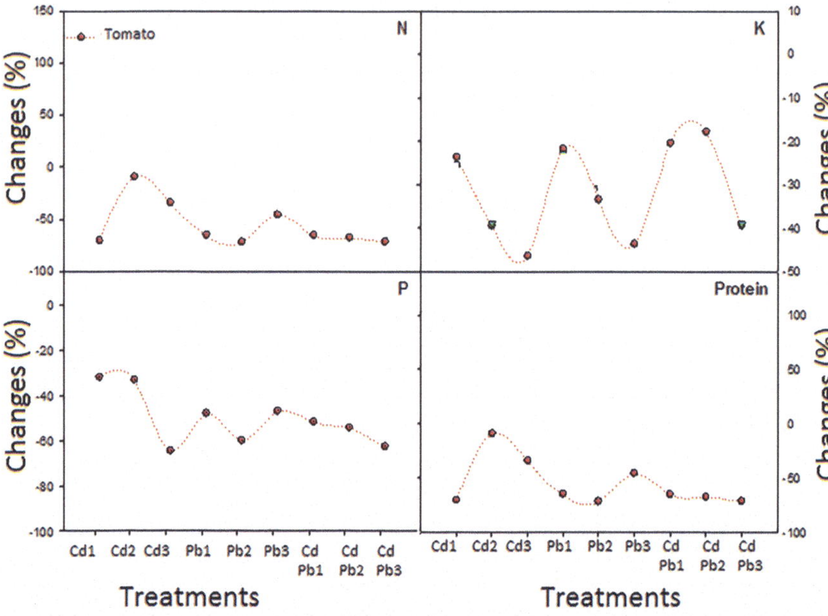

Fig. 5.3 Percentages changes in nutrients concentration of tomato grown under different treatments of toxic elements. The Cd1, Cd2, Cd3, Pb1, Pb2, Pb3, CdPb1 CdPb2 and CdPb3 represent the concentrations of Cd 1.0, 2.5 and 5.0 mg kg^{-1}, Pb 200, 300 and 400 mg kg^{-1} and Cd/Pb of 1.0/200, 2.5/300 and 5.0/400 mg kg^{-1}, respectively (Khan et al. 2016)

5.6.3 Impact of Toxic Elements on Plant Proteins and Amino Acids

Proteins are essential and useful dietary components of plants having high source of nutritional properties and basic requirement of beneficial nutrients for human beings. Protein contents are found varied from species to species in vegetable plants (Mosha and Gaga 1999; Odhav et al. 2007), the difference in protein contents could be attributed to irregular conditions of climate (Odhav et al. 2007) and alterations in structural characteristics. The richest and cheapest source of proteins are green leafy vegetables. The possible reason in the environment might be the abundant water, sunlight, nitrogen, and oxygen and their ability to create accumulative amino acids (Aletor et al. 2002). Leafy vegetables contain protein contents ranging from 1 to 7 g/100 g of fresh weight (Uusiku et al. 2010). Therefore, toxic elements are mostly accountable for reducing protein contents in plants cultivated on contaminated soil (Widowati 2012). High metal concentrations alter physiological functions as a result hindering protein synthetic activities and metabolism (Verma and Dubey 2003).

In addition, Mustafa and Komatsu (2016) demonstrated that TE stress damages the protein content associated with biosynthetic and nutrient metabolism with an increase in proteins connected with transcription and translational regulation and antioxidant pathways. TEs such as Cd, Pb, and Hg, may hinder and damage the nucleic acids as well as protein folding by interacting with proteins and DNA in plants. Hence, the metal stress makes the protein contents non-functional. In this condition, plant chaperones ease to improve, restore, and protect the content of proteins folding under TE toxicity (Bolin et al. 2006; Jain et al. 2018). Previous reports showed the contradictions associated to metal-induced harmful changes in protein contents (Sarry et al. 2006). Essential nutrients such as sulfur and nitrogen are very crucial for proteins synthesis and amino acids (Yu et al. 2018). The deficiency of sulfur and nitrogen contents may create harmful effects in plants' metabolic process (Yu et al. 2018). Furthermore, stress of high metal concentration could highly decompose the protein contents (Tangahu et al. 2011). Seneviratne et al. (2017) reported that TEs can adversely affect the protein profiles involved in plant's seed germination and enzymes such as proteases and acid phosphatases. Moreover, TEs reduce the starch content, and can affectively limit the level of essential nutrients, inhibit the chloroplast in plant, as well as induce the expression of heat shock proteins and proline. In addition, As has also been reported to decrease and alter the photosynthetic pigment, impair chloroplast membrane, alter the nutrient balance, hamper protein metabolism, and reduce the enzyme functions by interacting with the certain sulfhydryl group of proteins (Ahsan et al. 2010). Toxicity of TEs may adversely affect the plants in many ways, by altering to disrupt the biochemical and physiological processes including plant's seed germination, hinder photosynthesis and respiration process, reduce pigment synthesis, inactivate the enzymes and hormonal balance, thereby resulting in alteration of the nutrients and protein synthesis (Singh et al. 2013). Likewise, high toxic metal concentrations also hinder the protein synthesis by varying the complex accumulation of pigment lipoprotein in photosystems (Wang et al. 2009) and influence the ribulose-1, 5-bisphosphate carboxylase enzymes (Krantev et al. 2008). Toxic metals have encouraged instabilities of photosynthesis at dissimilar structural-functional stages: stains and light capture, photosynthetic electron transport and thylakoid ultrastructure, Calvin cycle enzymes activities, stomatal conductance, and CO_2 access (Clijsters and Van Assche 1985; Cuypers et al. 2001; Vassilev et al. 2011). Furthermore, TEs can effectively reduce protein synthesis and change the physiological characteristics in plants (Chaffei et al. 2004). The study of Leonard et al. (2004) showed that high concentrations of toxic elements interact with DNA and proteins, resulting in oxidative disruption in plant molecules.

5.6.4 Effects of Toxic Elements on Lipid Contents in Plants

The majority of studies in the previous literature showed the adverse impacts of toxic elements on food crops that are mostly focused upon the transfer, bioaccumulation, and their potential health effects. However, the information related to the impact of toxic element on the essential nutritional components such as lipids is found rarely (Upchurch 2008). Djebali et al. (2005) identified that effects of toxic

Table 5.3 Impact of TEs on proteins, carbohydrates, and lipids in plants

TEs	Effect on proteins, carbohydrates, and lipids	References
Cr, Ni, As, Pb, Cd, Co, Fe	Reduced proteins, carbohydrates, and photosynthetic pigments in plant	Galala et al. (2018)
Cr, Cd, Cu, Co	Decreased protein content and cell nitrogen	Jastaniah and Aburas (2016)
Cr	Reduced and inhibited the functions and regulations of various proteins	Dotaniya et al. (2014)
Zn, Fe, Cu, Al	Declined plasma membrane total lipids, glycolipids, phospholipids, and sterols in *Z. album* and *Z. coccineum*	Morsy et al. (2012)
Cu	Altered the protein profile and decreased antioxidant activities in *Withania somnifera*	Rout et al. (2013)
Fe, Cu, Zn	Inhibited several carbohydrates, namely, glucose, maltose, fructose, sucrose, and sorbitol at low and high concentrations	Prakash et al. (2013)
Fe	Caused oxidative stress. Increased peroxidation of lipids in *Eugenia uniflora L.* plant leaves	de Oliveira Jucoski et al. (2013)
Mo	Reduced total dry matter, seed protein and seed germination, protein and starch contents, nitrogen and sugars. Increased nitrate reductase activity in black gram (*Vigna mungo* L.).	Gopal et al. 2015

elements varied on different fatty acids. Another study revealed that fatty acids were negatively affected on C18:3 proportions due to TEs, while other fatty acids were positively affected (Ben et al. 2005). It has been reported that elevated Pb concentration resulted in lipid peroxidation in rice plants (Verma and Dubey 2003), due to activity of metal-induced lipoxygenase, which can displace the metal ions and change the enzyme processes (Bhaduri and Fulekar 2012). Effects of oxidative stress of metals damage the chloroplast and lipid peroxidation, while nutritional components and photosynthetic activities are also prominently affected by high concentration of toxic elements in contaminated plants (Dutta et al. 2018; Nagajyoti et al. 2010). While others reported that fatty acid contents reduced in plants grown in contaminated soil, they observed that the percentage of different levels of fatty acids increased to C18:0, C18:1, and C18:2. Therefore, it was concluded that fatty acids could be considerably used as bio indicator (Le Guédard et al. 2008).

Previous studies regarding impact of TEs on plant essential micronutrients have been contradictory and cannot be ignored. Table 5.3 shows the adverse impacts of TEs on proteins, carbohydrates, and lipids. Toxic effects of TEs are mainly responsible for low protein contents in plants grown on metal-contaminated soil (Widowati 2012). The toxicity stress of TEs such as Cr, Ni, As, Pb, Cd, Co, and Fe decreased the carbohydrates, proteins, and photosynthetic pigments. Proteins and carbohydrates were reduced by 53.1% and 11.5%, respectively, in the leaves. However, proteins decreased by 27.5% and carbohydrates increased by 29.7% in the roots (Galala et al. 2018). Previous research showed that all the tested TEs including Cr, Cd, Cu, and Co reduced significantly protein content and cell nitrogen (Jastaniah and Aburas 2016), as shown in Table 5.3. Several studies described that Cr toxic effects suppress the functions and regulations of various protein contents (Dotaniya et al. 2014). Morsy et al. 2012 reported the toxicity of metals Zn, Fe, Cu, and Al in *Z. album* and *Z. coccineum* plants, and efficient reduction was observed in plasma membrane total lipids, glycolipids phospholipids, and sterols.

High reduction and alteration in plant macronutrients may be attributed to the toxic effects of TEs. In previous literature, Rout et al. (2013) illustrated their study in context of Cu (200 μM) effect on *Withania somnifera* and identified reduction in root, shoot, and leaf length as well as decrease in growth and total number of leaves per plant. They also stated the outcome of study that antioxidant activities were decreased and alteration in protein content in plants. According to Prakash et al. (2013), Fe, Cu, Zn even at low concentration level strongly inhibited the activity of metabolism in plant. In contrast, several carbohydrates, namely, glucose, maltose, fructose, sucrose, and sorbitol were little inhibited at high elevated concentration. Another previous study showed that uptake of Fe and oxidative stress reduced the root and shoot growth in *Eugenia uniflora L.* and increased lipid peroxidation in plant leaves as shown in Table 5.3 (de Oliveira Jucoski et al. 2013). Mo at concentration of (2 μm) has also been reported in black gram (*Vigna mungo L.*), and study showed that total dry matter, seed protein and seed germination, protein and starch contents, nitrogen and sugars were decreased and increased nitrate reductase activity as reported by Gopal et al. (2015). From the existing literature, it has been found that toxicity of TEs has strong effect on proteins, carbohydrates, and lipid contents of vegetable plants. Such contaminated vegetables and lack of these essential nutrients may lead to malnutrition and serious health problems via consumption.

5.6.5 Influence of Metals on Plant Vitamins

Vitamins play an important role, particularly for living organisms as well as for plant microorganisms. Vitamins are mostly associated to significantly improve the plant growth in soil (Bergner 1997). Green leafy vegetables are nutritionally important and rich sources of vitamins such as carotene, riboflavin, ascorbic acid, carotene, folic acid, and proteins and minerals such as iron, phosphorous, and calcium contain important and rich sources of useful vitamins (Osler et al. 2001; Sheela et al. 2004). Plants usually contain a different kind of essential vitamins that are naturally important for human metabolism as well as for plants, due to their strong antioxidant potential with the key role of its cofactors and high redox chemistry between them. Essential nutrients such as vitamins play an important role to maintain good health, augment the functions of immune system, and provide protection against infectious diseases and other certain malignancies. However, high toxic elements may negatively affect the vitamin contents in addition to its functions (Jibril et al. 2017). In recent years, the effect of mineral fertilizers on the vitamin content of plants has received very little attention by scientists. The vitamin contents could be found varyingly in food crops, by addition of agriculture fertilizers to soil such as nitrogen fertilizers, and vitamin contents significantly reduced in plants grown with commercial fertilizers (Mozafar 1993), and their reduction could probably cause by TEs. It has been demonstrated that vitamin content could be enhanced when subjected to the concentration of TEs. Zengin (2013) reported that TEs such as Cr, Ni, Co, and Zn exposure to bean plants has increased the vitamin C content in the medium of growth. Previously, the content of vitamin E type of alpha-tocopherol has also been increased in *Arabidopsis thaliana* plants under TEs Cu and Cd stress (Collin et al. 2008). Vitamin contents are strongly influenced by environment (biological, physical and chemical), and due to the presence of high toxic element concentration in the extreme environment such as temperature and pH, as a result drastic reduction occurs in the vitamin contents (Vranova et al. 2013). Toxic elements and vitamins have a negative correlation; when metal concentrations increase then in contrary the vitamin contents decrease (Jaishankar et al. 2014; Munzuroglu et al. 2005). But some studies have also shown that plant growth-promoting bacteria such as the H3 strain of *Bacillus megaterium* can decrease the concentrations of Cd and Pb in green vegetables (i.e., Brassica sp.) and improve quality of soil, protein, and vitamin C content of the vegetables (Wang et al. 2018).

In the previous study, vitamin C level was effectively decreased under the toxicity and stress of Cd (12 mg/L) concentration, in comparison with the control lettuce plants. Significant reduction of 52%, and 67% of vitamin C was observed in lettuce plants when subjected to 9 mg/L, and 12 mg/L of Cd, concentrations, respectively, as compared to control (Jibril et al. 2017). Similarly, stress of toxic elements induces lipid peroxidation, resulting in reduction of vitamin contents in vegetables food crops cultivated in contaminated soil (Khan et al. 2019; Jibril et al. 2017; Seven et al. 2012). So far, the scientific interest of high-quality research has primarily been focused and devoted in the context of the functional molecules, including vitamins, polyphenols, and carotenoids that showed a tremendous variation trend depending upon various plant species and also depend on exposure to TE concentrations (Lajayer et al. 2017).

5.7 Conclusion

It is concluded that contamination of toxic elements has adverse effect on vegetable plants and can cause severe health problems via consumption. High toxicity, stress, and accumulation of elements may lead to high reduction in growth of vegetables grown in contaminated soil. Toxic elements may potentially affect the important parameters of plants such as plant morphology, photosynthesis process, and essential nutrients (carbohydrates, vitamins, proteins, lipids, nitrogen and phosphorus). Hence, essential nutrients are important source and required for plant growth, while these can be highly altered and affected due to toxicity of elements and thereby damage the food crops. Therefore, proper mitigation and implementation of environmentally friendly techniques are very necessary to avoid and reduce the health risks through consumption of vegetables.

References

Ahmad M, Ahmad M, El-Naggar AH, Usman AR, Abduljabbar A, Vithanage M, Elfaki J, Abdulelah AF, Al-Wabel MI (2018) Aging effects of organic and inorganic fertilizers on phosphorus fractionation in a calcareous sandy loam soil. Pedosphere 28(6):873–883

Ahsan N, Lee DG, Kim KH, Alam I, Lee SH, Lee KW, Lee H, Lee BH (2010) Analysis of arsenic stress-induced differentially expressed proteins in rice leaves by two-dimensional gel electrophoresis coupled with mass spectrometry. Chemosphere 78:224–231. https://doi.org/10.1016/j.chemosphere.2009.11.004

Aletor O, Oshodi A, Ipinmoroti K (2002) Chemical composition of common leafy vegetables and functional properties of their leaf protein concentrates. Food Chem 78(1):63–68

Alia N, Sardar K, Said M, Salma K, Sadia A, Sadaf S, Toqeer A, Miklas S (2015) Toxicity and bioaccumulation of toxic elements in spinach (Spinacia oleracea) grown in a controlled environment. Int J Environ Res Public Health 12(7):7400–7416

Angelis KJ, McGuffie M, Menke M, Schubert I (2000) Adaptation to alkylation damage in DNA measured by the comet assay. Environ Mol Mutagen 36:146–150

Arkhipchuk VV, Malinovskaya MV, Garanko NN (2000) Cytogenetic study of organic and inorganic toxic substances on Allium cepa, Lactuca sativa, and Hydra attenuata cells. Environ Toxicol 15:338–344

Ashraf M, Ozturk M, Ahmad MSA (eds) (2010) Plant adaptation and phytoremediation. Springer, Dordrecht, p 481

Babu TS, Akhtar TA, Lampi MA, Tripuranthakam S, Dixon DG, Greenberg BM (2003) Similar stress responses are elicited by copper and ultraviolet radiation in the aquatic plant Lemna gibba: implication of reactive oxygen species as common signals. Plant Cell Physiol 44(12):1320–1329

Bacaha N, Shamas R, Bakht J, Rafi A, Farhatullah, Gillani A (2015) Effect of heavy metal and EDTA application on plant growth and phyto-extraction potential of sorghum (Sorghum bicolor). Pak J Bot 47(5):1679–1684

Ben YN, Nouairi I, Ben TS, Taamalli W, Zarrouk M, Ghorbal MH, Ben MD (2005) Cadmium effects on lipid metabolism of rape (Brassica napus L.). C R Biol 328(8):745

Bergner P (1997) The healing power of minerals, special nutrients and trace elements. Prima Pub, Rocklin

Bhaduri AM, Fulekar MH (2012) Antioxidant enzyme responses of plants to heavy metal stress. Rev Environ Sci Biotechnol 11(1):55–69

Bilski JJ, Foy CD (1987) Differential tolerances of oat cultivars to aluminumin nutrient solutions and in acid soils of Poland. J Plant Nutr 10:129–141

Bolin CM, Basha R, Cox D, Zawia NH, Maloney B, Lahiri DK, Cardozo-Pelaez F (2006) Exposure to lead and the developmental origin of oxidative DNA damage in the aging brain. FASEB J 6:788–790. https://doi.org/10.1096/fj.05-5091fje

Cakmak I, Kalayci M, Kaya Y, Torun AA, Aydin N, Wang Y, Arisoy Z, Erdem HA, Yazici A, Gokmen O, Ozturk L (2010) Biofortification and localization of zinc in wheat grain. J Agric Food Chem 58(16):9092–9102

Carbonell AA, Aarabi MA, DeLaune RD, Gambrell RP, Patrick WH Jr (1998) Arsenic in wetland vegetation: availability, phytotoxicity, uptake and effects on plant growth and nutrition. Sci Total Environ 217:189–199. https://doi.org/10.1016/S0048-9697(98)00195-8

Carbonell-Barrachina AA, Burló-Carbonell F, Mataix-Beneyto J (1997) Effect of sodium arsenite and sodium chloride on bean plant nutrition (macronutrients). J Plant Nutr 20:1617–1633. https://doi.org/10.1080/01904169709365361

Castro E, Manas P, Heras JDL (2009) A comparison of the application of different waste products to a lettuce crop: effects on plant and soil properties. Sci Hortic 123:148–155

Chaffei C, Gouia H, Ghorbel MH (2003) Nitrogen metabolism in tomato plants under cadmium stress. J Plant Nutr 26:1617–1634. https://doi.org/10.1081/PLN-120022372

Chaffei C, Pageau K, Suzuki A, Gouia H, Ghorbel MH, Masclaux-Daubresse C (2004) Cadmium toxicity induced changes in nitrogen management in Lycopersicon esculentum leading to a metabolic safeguard through an amino acid storage strategy. Plant Cell Physiol 45:1681–1693

Chang LW, Meier JR, Smith MK (1997) Application of plant and earthworm bioassays to evaluate remediation of a lead-contaminated soil. Arch Environ Contam Toxicol 32:166–171

Chapman E, Dave G, Murimboh J (2010) Ecotoxicological risk assessment of undisturbed metal contaminated soil at two remote lighthouse sites. Ecotoxicol Environ Saf 73:961–969

Chapman EEV, Hedrei Helmer S, Dave G, Murimboh JD (2012) Utility of bioassays (lettuce, red clover, red fescue, Microtox, MetSTICK, Hyalella, bait lamina) in ecological risk screening of acid metal (Zn) contaminated soil. Ecotoxicol Environ Saf 80:161–171

Chibuike GU, Obiora SC (2014) Heavy metal polluted soils: effect on plants and bioremediation methods. Appl Environ Soil Sci 2014

Clijsters H, Van Assche F (1985) Inhibition of photosynthesis by toxic elements. Photosynth Res 7:31–40. https://doi.org/10.1007/BF0003292

Cobbett CS (2000) Phytochelatins and their roles in heavy metal detoxification. Plant Physiol 123(3):825–832

Collin VC, Eymery F, Genty B, Rey P, Havaux M (2008) Vitamin E is essential for the tolerance of Arabidopsis thaliana to metal-induced oxidative stress. Plant Cell Environ 31(2):244–257

Cuypers A, Vangronsve J, Clijsters H (2001) The redox status of plant cells (AsA and GSH) is sensitive to zinc imposed oxidative stress in roots and primary leaves of Phaseolus vulgaris. Plant Physiol Biochem 39:657–664. https://doi.org/10.1016/S0981-9428(01)01276-1

Czupyran G, Levy RD (1989) In situ immobilization of heavy metal contaminated soil, section II. Noyes Data Corp, Park Ridge

Dang YP, Chhabra R, Verma KS (1990) Effect of Cd, Ni, Pb and Zn on growth and chemical composition of onion and fenugreek. Commun Soil Sci Plant Anal 21:717–735

Datta JK, Kundu A, Hossein SD, Banerjee A, Mondal NK (2011) Studies on the impact of micronutrient (Molybdenum) on germination, seedling growth and physiology of bengal grain (Cicer arietinum) under laboratory condition. Asian J Crop Sci 3:55–67

de Dorlodot S, Lutts S, Bertin P (2005) Effects of ferrous iron toxicity on the growth and mineral composition of an inter specific rice. J Plant Nutr 28:1–20

de Oliveira Jucoski G, Cambraia J, Ribeiro C, de Oliveira JA, de Paula SO, Oliva MA (2013) Impact of iron toxicity on oxidative metabolism in young Eugenia uniflora L. plants. Acta Physiol Plant 35(5):1645–1657

DeWolf H, Blust R, Backeljau T (2004) The use of RAPD in ecotoxicology. Mutat Res Rev Mutat Res 566:249–262

Djebali W, Zarrouk M, Brouquisse R, El Kahoui S, Limam F, Ghorbel MH, Chaibi W (2005) Ultrastructure and lipid alterations induced by cadmium in tomato (Lycopersicon esculentum) chloroplast membranes. Plant Biol 7:358–368

Donma O, Donma MM (2005) Cadmium, lead and phytochemicals. Med Hypotheses 65:699–702

Dotaniya M, Das H, Meena V (2014) Assessment of chromium efficacy on germination, root elongation, and coleoptile growth of wheat (Triticum aestivum L.) at different growth periods. Environ Monit Assess 186:2957–2963

Dutta S, Mitra M, Agarwal P, Mahapatra K, De S, Sett U, Roy S (2018) Oxidative and genotoxic damages in plants in response to heavy metal stress and maintenance of genome stability. Plant Signal Behav 13(8):e1460048

Ebbs SD, Kochian LV (1997) Toxicity of zinc and copper to Brassica species: implications for phytoremediation. J Environ Qual 26(3):776–781

Emamverdian A, Ding Y, Mokhberdoran F, Xie Y (2015) Heavy metal stress and some mechanisms of plant defense response. Sci World J 2015

Environment Canada (2005) Biological test method: test for measuring emergence and growth of terrestrial plants exposed to contaminants in soil. Method development and applications section. Environmental Technology Centre, Ottawa

FAOSTAT (2007) FAOSTAT agriculture production database. http://faostat.fao.org/site/336/default.aspx

Farhat N, Elkhouni A, Zorrig W, Smaoui A, Abdelly C, Rabhi M (2016) Effects of magnesium deficiency on photosynthesis and carbohydrate partitioning. Acta Physiol Plant 38(6):145

Farinati S, DalCorso G, Varotto S, Furini A (2010) The Brassica juncea BjCdR15, an ortholog of Arabidopsis TGA3, is a regulator of cadmium uptake, transport and accumulation in shoots and confers cadmium tolerance in transgenic plants. New Phytol 185:964–978

Finglas PM, Faulks RM (1984) The HPLC analysis of thiamin and riboflavin in potatoes. Food Chem 15:37–44

Galal TM, Shehata HS (2015) Bioaccumulation and translocation of toxic elements by Plantago major L. grown in contaminated soils under the effect of traffic pollution. Ecol Indic 48:244–251

Galala TM, Khalafallah AA, Elawa OE, Hassan LM (2018) Human health risks from consuming cabbage (Brassica oleracea L. var. capitata) grown on wastewater irrigated soil. Int J Phytoremediation 20(10):1007–1016

Garate A, Ramos I, Manzanares M, Lucena JJ (1993) Cadmium uptake and distribution in three cultivars of Lactuca sp. Bull Environ Contam Toxicol 50:709–716

Gawęda M (2007) Changes in the contents of some carbohydrates in vegetables cumulating lead. Pol J Environ Stud 16(1):57–62

Gil C, Boluda R, Ramos J (2004) Determination and evaluation of cadmium lead and nickel in greenhouse soils of Almeria (Spain). Chemosphere 55:1027–1034

Gill SS, Tuteja N (2010) Reactive oxygen species and antioxidant machinery in abiotic stress tolerance in crop plants. Plant Physiol Biochem 48:909–930

Gill SS, Tuteja N (2011) Cadmium stress tolerance in crop plants: probing the role of sulfur. Plant Signal Behav 6:215–222. https://doi.org/10.4161/psb.6.2.14880

Giovanelli G, Paradise A (2002) Stability of dried and intermediate moisture tomato pulp during storage. J Agric Food Chem 50:7277–7281

Gopal R, Sharma YK, Shukla AK (2015) Effect of molybdenum stress on growth, yield and seed quality in black gram. J Plant Nutr 39:463–469. https://doi.org/10.1080/01904167.2015.1016176

Guala SD, Vega FA, Covelo EF (2010) The dynamics of toxic elements in plant–soil interactions. Ecol Model 221:1148–1152

Hagemeyer J (2004) Ecophysiology of plant growth under heavy metal stress. In: Heavy metal stress in plants. Springer, Berlin/Heidelberg, pp 201–222

He WY, Yang XE, Yang JY, He ZL (2018) Effect of lead on plant availability of phosphorus and potassium in a vegetable–soil system. Environ Sci Pollut Res 25(34):34793–34797

Hernández LE, Lozano E, Gárate A, Carpena R (1998) Influence of cadmium on the uptake, tissue accumulation and subcellular distribution of manganese in pea seedlings. Plant Sci 132:139–151

Hernández-Allica J, Becerril JM, Garbisu C (2008) Assessment of the phytoextraction potential of high biomass crop plants. Environ Pollut 152(1):32–40

Hossain Z, Huq F (2002) Studies on the interaction between Cd2+ ions and nucleobases and nucleotides. J Inorg Biochem 90:97–105

Hossain MA, Piyatida P, da Silva JA, Fujita M (2012) Molecular mechanism of heavy metal toxicity and tolerance in plants: central role of glutathione in detoxification of reactive oxygen species and methylglyoxal and in heavy metal chelation. J Bot 2012

Intawongse M, Dean JR (2006) Uptake of toxic elements by vegetable plants grown on contaminated soil and their bioavailability in the human gastrointestinal tract. Food Addit Contam 23(1):36–48

Iqbal MZ, Nayab S, Shafiq M (2018) Effects of copper on seed germination and seedling growth performance of lens culinaris medik. J Plant Develop 25:85

Irfan M, Hayat S, Ahmad A, Alyemeni MN (2013) Soil cadmium enrichment: allocation and plant physiological manifestations. Saudi J Biol Sci 20(1):1–10

Jain S, Muneer S, Guerriero G, Liu S, Vishwakarma K, Chauhan DK, Dubey NK, Tripathi DK, Sharma S (2018) Tracing the role of plant proteins in the response to metal toxicity: a comprehensive review. Plant Signal Behav 13(9):e1507401

Jaishankar M, Tseten T, Anbalagan N, Mathew BB, Beeregowda KN (2014) Toxicity, mechanism and health effects of some toxic elements. Interdiscip Toxicol 7(2):60–72

Jastaniah SD, Aburas MM (2016) Effects of some toxic elements on growth, protein content and pigment production by Streptomyces Coelicolor SM1. Biosci Biotechnol Res Asia 13(4):1975–1981

Jibril SA, Hassan SA, Ishak CF, Megat Wahab PE (2017) Cadmium toxicity affects phytochemicals and nutrient elements composition of lettuce (Lactuca sativa L.). Adv Agric 3:1–7

John MK, Van Laerhoven CJ (1976) Differential effects of cadmium on lettuce varieties. Environ Pollut 4:7–15

John R, Ahmad P, Gadgil K, Sharma S (2009) Heavy metal toxicity: effect on plant growth, biochemical parameters and metal accumulation by Brassica juncea L. Int J Plant Prod 3:65–76

Jonak C, Nakagami H, Hirt H (2004) Heavy metal stress. Activation of distinct mitogen-activated protein kinase pathways by copper and cadmium. Plant Physiol 136(2):3276–3283

Karami N, Clemente R, Moreno-Jiménez E, Lepp NW, Beesley L (2011) Efficiency of green waste compost and biochar soil amendments for reducing lead and copper mobility and uptake to ryegrass. J Hazard Mater 191:41–48

Khan MR, Khan MM (2010) Effect of varying concentration of nickel and cobalt on the plant growth and yield of chickpea. Aust J Basic Appl Sci 4(6):1036–1046

Khan S, Cao Q, Zheng YM, Huang YZ, Zhu YG (2008) Health risks of toxic elements in contaminated soils and food crops irrigated with wastewater in Beijing, China. Environ Pollut 152:686–692

Khan S, Reid BJ, Li G, Zhu Y-G (2014) Application of biochar to soil reduces cancer risk via rice consumption: a case study in Miaoqian village, Longyan, China. Environ Int 68:154–161

Khan S, Waqas M, Ding F, Shamshad I, Arpd HPH, Li G (2015) The influence of various biochars on the bioaccessibility and bioaccumulation of PAHs and potentially toxic elements to turnips (Brassica rapa L.). J Hazard Mater 300:1–11. https://doi.org/10.1016/j.jhazmat.2015.06.050

Khan MA, Ding X, Khan S, Brusseau ML, Khan A, Nawab J (2018) The influence of various organic amendments on the bioavailability and plant uptake of cadmium present in mine-degraded soil. Sci Total Environ 636:810–817

Khan S, Khan A, Khan MA, Aamir M, Li G (2019) Arsenic interaction and bioaccumulation in food crops grown on degraded soil: effect on plant nutritional components and other dietary qualities. Land Degrad Dev 30(16):1954–1967

Khan AZ, Khan S, Ayaz T, Brusseau ML, Khan MA, Nawab J, Muhammad S (2020) Popular wood and sugarcane bagasse biochars reduced uptake of chromium and lead by lettuce from mine-contaminated soil. Environ Pollut 263:114446

Kim D-Y, Bovet L, Kushnir S, Noh EW, Martinoia E, Lee Y (2006) AtATM3 is involved in heavy metal resistance in Arabidopsis. Plant Physiol 140:922–932

Krantev A, Yordanova R, Janda T, Szalai G, Popova L (2008) Treatment with salicylic acid decreases the effect of cadmium on photosynthesis in maize plants. J Plant Physiol 165:920–993

Lajayer BA, Ghorbanpour M, Nikabadi S (2017) Toxic elements in contaminated environment: destiny of secondary metabolite biosynthesis, oxidative status and phytoextraction in medicinal plants. Ecotoxicol Environ Saf 145:377–390. https://doi.org/10.1016/j.ecoenv.2017.07.035

Latef AAA (2018) Growth and some physiological activities of pepper (Capsicum annuum L.) in response to cadmium stress and mycorrhizal symbiosis. J Agric Sci Technol 15:1437–1448

Le Guédard M, Schraauwers B, Larrieu I, Bessoule J-J (2008) Development of a biomarker for metal bioavailability: the lettuce fatty acid composition. Environ Toxicol Chem 27:1147–1151

Lea PJ, Miflin BJ (2004) Glutamate synthase and the synthesis of glutamate in plants. Plant Physiol Biochem 41:555–564. https://doi.org/10.1016/S0981-9428(03)00060-3

Lee CW, Choi JM, Pak CH (1996) Micronutrient toxicity in seed geranium (*Pelargonium x Hortorum Bailey*). J Am Soc Hortic Sci 121:77–82
Legay S, Guignard C, Ziebel J, Evers D (2012) Iron uptake and homeostasis related genes in potato cultivated in vitro under iron deficiency and overload. Plant Physiol Biochem 60:180–189
Leonard SS, Harris GK, Shi XL (2004) Metal-induced oxidative stress and signal transduction. Free Radic Biol Med 37:1921–1942
Lewis JS, Sharp TL, Laforest R, Fujibayashi Y, Welch MJ (2001) Tumor uptake of copper-diacetyl-bis (N4-methylthiosemicarbazone): effect of changes in tissue oxygenation. J Nucl Med 42(4):655–661
L'Huillier L, d'Auzac J, Durand M, Michaud-Ferrière N (1996) Nickel effects on two maize (Zea mays) cultivars: growth, structure, Ni concentration, and localization. Can J Bot 74(10):1547–1554
Lim H-S, Lee J-S, Chon H-T, Sager M (2008) Heavy metal contamination and health risk assessment in the vicinity of the abandoned Songcheon Au–Ag mine in Korea. J Geochem Explor 96:223–230
Lin C, Zhu T, Liu L, Wang D (2010) Influences of major nutrient elements on Pb accumulation of two crops from a Pb-contaminated soil. J Hazard Mater 174(1–3):202–208
Lin T, Zhu X, Zhang F, Wan X (2011) The detoxification effect of nitrogen on cadmium stress in Populus yunnanensis. Bot Res J 4(1):13–19
Lindsey PA, Lineberger RD (1981) Toxicity, cadmium accumulation and ultrastructural alterations induced by exposure of Phaseolus seedlings to cadmium. Hortic Sci 16:434
Liu H, Probst A, Liao B (2005) Metal contamination of soils and crops affected by the Chenzhou lead/zinc mine spill (Hunan, China). Sci Total Environ 339:153–166
López-Millán A-F, Sagardoy R, Solanas M, Abadía A, Abadía J (2009) Cadmium toxicity in tomato (Lycopersicon esculentum) plants grown in hydroponics. Environ Exp Bot 65:376–385
Loureiro S, Santos C, Pinto G, Costa A, Monteiro M, Nogueira AJ, Soares A (2006) Toxicity assessment of two soils from Jales mine (Portugal) using plants: growth and biochemical parameters. Arch Environ Contam Toxicol 50:182–190
Lutz JAJ, Genter CF, Hawkins GW (1972) Effect of soil pH on element concentration and uptake by maize. II. Cu, B, Zn, Mn, Mo, Ai & Fe. Agron J 64:583–585
Mackenzie EL, Iwasaki K, Tsuji Y (2008) Intracellular iron transport and storage: from molecular mechanisms to health implications. Antioxid Redox Signal 10(6):997–1030
Mahmood T, Islam KR (2006) Response of rice seedlings to copper toxicity and acidity. J Plant Nutr 29(5):943–957
Mahmood Q, Ahmad R, Kwak SS, Rashid A, Anjum NA (2010) Ascorbate and glutathione: protectors of plants in oxidative stress. In: Ascorbate–glutathione pathway and stress tolerance in plants. Springer, New York, pp 209–229
Marangoni F, Cetin I, Verduci E, Canzone G, Giovannini M, Scollo P, Corsello G, Poli A (2016) Maternal diet and nutrient requirements in pregnancy and breastfeeding. An Italian consensus document. Nutrients 8(10):629
Martinez-Valvercle I, Periage MJ, Provan G, Chesson A (2002) Phenolic compounds, lycopene and antioxidant activities in commercial varieties of tomato (Lycopersicon esculentum). J Sci Food Agric 82:323–330
Mellem JJ, Baijnath H, Odhav B (2009) Translocation and accumulation of Cr, Hg, As, Pb, Cu and Ni by Amaranthus dubius (Amaranthaceae) from contaminated sites. J Environ Sci Health A Tox Hazard Subst Environ Eng 44:568–575
Mendoza RE, García IV, de Cabo L, Weigandt CF, de Iorio AF (2015) The interaction of toxic elements and nutrients present in soil and native plants with arbuscular mycorrhizae on the riverside in the Matanza-Riachuelo River Basin (Argentina). Sci Total Environ 505:555–564
Mondal NK, Das C, Roy S, Datta JK, Banerjee A (2013) Effect of varying cadmium stress on chickpea (Cicer arietinum l) seedlings: an ultrastructural study. Ann Environ Sci 7:59–70

Monni S, Salemaa M, White C, Tuittila E, Huopalainen M (2000) Copper resistance of Calluna vulgaris originating from the pollution gradient of a Cu-Ni smelter in Southwest Finland. Environ Pollut 109:211–219
Morsy AA, Salama KH, Kamel HA, Mansour MM (2012) Effect of toxic elements on plasma membrane lipids and antioxidant enzymes of Zygophyl I urn species. Eurasian J Biosci 6:1–10
Mosha TC, Gaga HE (1999) Nutritive value and effect of blanching on the trypsin and chymotrypsin inhibitor activities of selected leafy vegetables. Plant Foods Hum Nutr 54:271–283
Moya JL, Ros R, Picazo I (1993) Influence of cadmium and nickel on growth, net photosynthesis and carbohydrate distribution in rice plants. Photosynth Res 36(2):75–80
Mozafar A (1993) Nitrogen fertilizers and the amount of vitamins in plants: a review. J Plant Nutr 16(12):2479–2506
Mukhopadhyay M, Das A, Subba P, Bantawa P, Sarkar B, Ghosh P, Mondal TK (2013) Structural, physiological, and biochemical profiling of tea plantlets under zinc stress. Biol Plant 57(3):474–480
Munzuroglu O, Obek E, Karatas F, Tatar SY (2005) Effects of simulated acid rain on vitamins A, E, and C in strawberry (Fragaria vesca). Pak J Nutr 4(6):402–406
Mustafa G, Komatsu S (2016) Toxicity of toxic elements and metal-containing nanoparticles on plants. Biochim Biophys Acta Proteins Proteom 1864(8):932–944. https://doi.org/10.1016/j.bbapap.2016.02.020
Nagajyoti PC, Lee KD, Sreekanth TVM (2010) Toxic elements, occurrence and toxicity for plants: a review. Environ Chem Lett 8:199–216
Nagor S, Vyas AV (1997) Heavy metal induced changes in growth and carbohydrate metabolism in wheat seedlings. Indian J Environ Toxicol 7(2):98
Nawab J, Khan S, Shah MT, Qamar Z, Din I, Mahmood Q, Gul N, Huang Q (2015) Contamination of soil, medicinal, and fodder plants with lead and cadmium present in mine-affected areas, Northern Pakistan. Environ Monit Assess 187(9):605
Nawab J, Khan S, Shah MT, Gul N, Ali A, Khan K, Huang Q (2016a) Heavy metal bioaccumulation in native plants in chromite impacted sites: a search for effective remediating plant species. Clean (Weinh) 44(1):37–46
Nawab J, Khan S, Aamir M, Shamshad I, Qamar Z, Din I, Huang Q (2016b) Organic amendments impact the availability of TMs(loid)s in mine-impacted soil and their phytoremediation by Penisitum americanum and Sorghum bicolor. Environ Sci Pollut Res 23:2381–2390
Nawab J, Li G, Khan S (2016c) Health risk assessment from contaminated foodstuffs: a field study in chromite mining-affected areas northern Pakistan. Environ Sci Pollut Res 23, 12227–12236 https://doi.org/10.1007/s11356-016-6379-9
Nawab J, Farooqi S, Xiaoping W, Khan S, Khan A (2018a) Levels, dietary intake, and health risk of potentially toxic metals in vegetables, fruits, and cereal crops in Pakistan. Environ Sci Pollut Res 25(6):5558–5571
Nawab J, Ghani J, Khan S, Xiaoping W (2018b) Minimizing the risk to human health due to the ingestion of arsenic and toxic metals in vegetables by the application of biochar, farmyard manure and peat moss. J Environ Manag 214:172–183
Nawab J, Khan N, Ahmed R, Khan S, Ghani J, Rahman Z, Khan F, Wang X, Muhammad J, Sher H (2019) Influence of different organic geo-sorbents on Spinacia oleracea grown in chromite mine-degraded soil: a greenhouse study. J Soils Sediments 19(5):2417–2432
Odhav B, Beekrum S, Akula US, Baijnath H (2007) Preliminary assessment of nutritional value of traditional leafy vegetables in KwaZulu- Natal, South Africa. J Food Compos Anal 20:430–435
Onasanya RO, Aiyelari OP, Onasanya A, Nwilene FE, Oyelakin OO (2009) Effect of different levels of nitrogen and phosphorus fertilizers on the growth and yield of maize (Zea mays L.) in Southwest Nigeria. Int J Agric Res 6:193–203
Opeolu BO, Adenuga OO, Ndakidemi PA, Olujimi OO (2010) Assessment of phyto-toxicity potential of lead on tomato (Lycopersicon esculentum L) planted on contaminated soils. Int J Phys Sci 5(2):68–73

Osler M, Heitmann BL, Gerdes LU, Jorgensen LM, Schroll M (2001) Dietary patterns and mortality in Danish men and women: a prospective observational study. Br J Nutr 85(2):219–225

Parrotta L, Guerriero G, Sergeant K, Cai G, Hausman JF (2015) Target or barrier? The cell wall of early-and later-diverging plants vs cadmium toxicity: differences in the response mechanisms. Front Plant Sci 6:133

Peralta-Videa JR, Gardea-Torresdey JL, Tiemann KJ, Gomez E, Arteaga S, Rascon E, Parsons JG (2001) Uptake and effects of five toxic elements on seed germination and plant growth in alfalfa (Medicago sativa L.). Bull Environ Contam Toxicol 66(6):727–734

Pourrut B, Shahid M, Douay F, Dumat C, Pinelli E (2013) Molecular mechanisms involved in lead uptake, toxicity and detoxification in higher plants. In: Heavy metal stress plants. Springer, Berlin, pp 121–147

Prakash O, Rai AK, Singh J, Singh PM (2013) Effect of heavy metal ions and carbohydrates on the activity of cauliflower (Brassica oleracea Var. botrytis) Myrosinase. J Stress Phys Biochem 9(2):108–117

Pruvot C, Douay F, Herve F, Waterlot C (2006) Toxic elements in soil, crops and grass as a source of human exposure in the former mining areas. J Soils Sediments 6:215–220

Radwan MA, Salama AK (2006) Market basket survey for some toxic elements in Egyptian fruits and vegetables. Food Chem Toxicol 44(8):1273–1278

Ramos I, Esteban E, Lucena JJ, Gárate A (2002) Cadmium uptake and sub cellular distribution in plants of Lactuca sp. . Cd-Mn interaction. Plant Sci 162:761–767

Reeves RD, Baker AJM (2000) Metal accumulating plants. In: Raskin I, Finsley BD (eds) Phytoremediation of toxic metals: using plants to Clean up the environment. Wiley, New York, pp 193–229

Rodríguez-Celma J, Rellán-Álvarez R, Abadía A, Abadía J, López- Millán A-F (2010) Changes induced by two levels of cadmium toxicity in the 2-DE protein profile of tomato roots. J Proteome 73:1694–1706

Rout GR, Das P (2009) Effect of metal toxicity on plant growth and metabolism: I. zinc. In: Sustainable agriculture. Springer, Dordrecht, pp 873–884

Rout JR, Ram SS, Das R, Chakraborty A, Sudarshan M, Sahoo SL (2013) Copper-stress induced alterations in protein profile and antioxidant enzymes activities in the in vitro grown Withania somnifera L. Physiol Mol Biol Plants 19:353–361. https://doi.org/10.1007/s12298-013-0167-5

Sarry JE, Kuhn L, Ducruix C, Lafaye A, Junot C, Hugouvieux V et al (2006) The early responses of Arabidopsis thaliana cells to cadmium exposure explored by protein and metabolite profiling analyses. J Proteome 6:2180–2198

Sarwar N, Saifullah, Malhi SS, Zia MH, Naeem A, Bibia S, Farid G (2010) Role of mineral nutrition in minimizing cadmium accumulation by plants. J Sci Food Agric 90:925–929

Seneviratne M, Rajakaruna N, Rizwan M, Madawala TESP, Ok YS, Vithanage M (2017) Heavy metal-induced oxidative stress on seed germination and seedling development: a critical review. Environ Geochem Health 41(4):1813–1831. https://doi.org/10.1007/s10653-017-0005-8

Seven PT, Yilmaz S, Seven I, Tuna Kelestemur G (2012) The effects of propolis in animals exposed oxidative stress. Oxidative stress—environmental induction and dietary antioxidants. In: Volodymyr Lushchak (ed) (Chapter 13)/ InTECH BOOK (ISBN 978–953–51-0553-4). https://doi:10.5772/2536

Seyhan D, Erdincler A (2003) Plant availability of toxic elements and phosphorus in the agricultural reuse of biological sludges. J Environ Sci Health A Tox Hazard Subst Environ Eng 38(10):2413–2423

Shahid M, Pinelli E, Dumat C (2012) Review of Pb availability and toxicity to plants in relation with metal speciation; role of synthetic and natural organic ligands. J Hazard Mater 219:1–2

Sharma SS, Dietz KJ (2006) The significance of amino acids and amino acid-derived molecules in plant responses and adaptation to heavy metal stress. J Exp Bot 57(4):711–726

Sharma P, Jha AB, Dubey RS, Pessarakli M (2012) Reactive oxygen species, oxidative damage, and antioxidative defense mechanism in plants under stressful conditions. J Bot 2012

Sharmin SA, Alam I, Kim KH, Kim YG, Kim PJ, Bahk JD, Lee BH (2012) Chromium-induced physiological and proteomic alterations in roots of Miscanthus sinensis. Plant Sci 187:113–126

Sheela K, Nath KG, Vijayalakshmi D, Yankanchi GM, Patil RB (2004) Proximate composition of underutilized green leafy vegetables in Southern Karnataka. J Hum Ecol 15(3):227–229

Sheppard SC, Evenden WG, Abboud SA, Stephenson M (1993) A plant life-cycle bioassay for contaminated soil, with comparison to other bioassays: mercury and zinc. Arch Environ Contam Toxicol 25:27–35

Shi HP, Zhu YF, Wang YL, Tsang PK (2014) Effect of cadmium on cytogenetic toxicity in hairy roots of Wedelia trilobata L. and their alleviation by exogenous CaCl 2. Environ Sci Pollut Res 21(2):1436–1443

Siedlecka A (1995) Some aspects of interactions between toxic elements and plant mineral nutrients. Acta Soc Bot Pol 64(3):265–272

Singh VP, Srivatava PK, Prasad SM (2013) Nitric oxide alleviates arsenic-induced toxic effects in ridged Luffa seedlings. Plant Physiol Biochem 71:155–163. https://doi.org/10.1016/j.plaphy.2013.07.003

Singh S, Parihar P, Singh R, Singh VP, Prasad SM (2016) Heavy metal tolerance in plants: role of transcriptomics, proteomics, metabolomics, and ionomics. Front Plant Sci 6:1143. https://doi.org/10.3389/fpls.2015.01143

Slavin JL, Lloyd B (2012) Health benefits of fruits and vegetables. Adv Nutr 3(4):506–516

Small J, Jackson T (1949) Buffer index values in relation to soil—pH tolerances. Plant Physiol 24:75–83

Smith SR (2009) A critical review of the bioavailability and impacts of toxic elements in municipal solid waste composts compared to sewage sludge. Environ Int 35:142–156

Srikumar TS, Ockerman PA (1990) The effects of fertilization and manuring on the content of some nutrients in potato (var. Provita). Food Chem 37:47–60

Stangeland T, Remberg SF, Lye KA (2009) Total antioxidant activity in 35 Ugandan fruits and vegetables. Food Chem 113:85–91

Suzuki N, Yamaguchi Y, Koizumi N, Sano H (2002) Functional characterization of heavy metal binding protein Cdl19 from Arabidopsis. Plant J 32:165–173

Tan J, He S, Yan S, Li Y, Li H, Zhang H, Zhao L, Li L (2014) Exogenous EDDS modifies copper-induced various toxic responses in rice. Protoplasma 251(5):1213–1221

Tangahu BV, Sheikh ASR, Basri H, Idris M, Anuar N, Mukhlisin M (2011) A review on toxic elements (As, Pb, and Hg) uptake by plants through phytoremediation. Int J Chem Eng 2011

Tchounwou PB, Yedjou CG, Patlolla AK, Sutton DJ (2012) Heavy metal toxicity and the environment. In: Molecular, clinical and environmental toxicology. Springer, Basel, pp 133–164

Thys C, Vanthomme P, Schrevens E, De Proft M (1991) Interactions of cadmium with zinc, copper, manganese, and iron in lettuce (Lactuca sativa L.) in hydroponic culture. Plant Cell Environ 14:713–717

Upchurch RG (2008) Fatty acid unsaturation, mobilization, and regulation in the response of plants to stress. Biotechnol Lett 30:967–977

Uusiku NP, Oelofse A, Duodu KG, Bester MJ, Faber M (2010) Nutritional value of leafy vegetables of sub-Saharan Africa and their potential contribution to human health: a review. J Food Compos Anal 23:499–509

Vahter M, Åkesson A, Lidén C, Ceccatelli S, Berglund M (2007) Gender differences in the disposition and toxicity of metals. Environ Res 104(1):85–95

Vanassche F, Clijsters H (1990) Effects of metals on enzyme-activity in plants. Plant Cell Environ 13:195–206

Vassilev A, Nikolova A, Koleva L, Lidon F (2011) Effects of excess Zn on growth and photosynthetic performance of young bean plants. J Phytology 3:58–62

Verma S, Dubey R (2003) Lead toxicity induces lipid peroxidation and alters the activities of antioxidant enzymes in growing rice plants. Plant Sci 164:645–655

Vikas D, Kaushalya G, Sawhney SK (2002) Effect of lead on starch mobilization in germinating chickpea seeds. J Plant Biol 29(1):85

Vranova V, Rejsek K, Formanek P (2013) Aliphatic, cyclic, and aromatic organic acids, vitamins, and carbohydrates in soil: a review. Sci World J 2013:524239

Wang H, Zhao SC, Liu RL, Zhou W, Jin JY (2009) Changes of photosynthetic activities of maize (Zea mays L.) seedlings in response to cadmium stress. Photosynthetica 47(2):277–283

Wang Q, Zhang WJ, He LY, Sheng XF (2018) Increased biomass and quality and reduced heavy metal accumulation of edible tissues of vegetables in the presence of Cd-tolerant and immobilizing Bacillus megaterium H3. Ecotoxicol Environ Saf 148:269–274

Widowati H (2012) The influence of cadmium heavy metal on vitamins in aquatic vegetables. Makara J Sci 16(1):33–38

Wong SC, Li XD, Zhang G, Qi SH, Min YS (2002) Toxic elements in agricultural soils of the Pearl River Delta, South China. Environ Pollut 119:33–44

Younis U, Malik SA, Rizwan M, Qayyum MF, Ok YS, Shah MHR, Rehman RA, Ahmad N (2016) Biochar enhances the cadmium tolerance in spinach (Spinacia oleracea) through modification of Cd uptake and physiological and biochemical attributes. Environ Sci Pollut Res. https://doi.org/10.1007/s11356-016-7344-3

Yu Z, Juhasz A, Islam S, Diepeveen D, Zhang J, Wang P, Ma W (2018) Impact of mid-season sulphur deficiency on wheat nitrogen metabolism and biosynthesis of grain protein. Sci Rep 8(1):1–13

Zechmann B, Müller M (2010) Subcellular compartmentation of glutathione in dicotyledonous plants. Protoplasma 246:15–24

Zengin F (2013) Physiological behavior of bean (phaseolus vulgaris l.) seedlings under metal stress. Biol Res 46(1):79–85

Zhang X, Zhang X, Gao B, Li Z, Xia H, Li H, Li J (2014) Effect of cadmium on growth, photosynthesis, mineral nutrition and metal accumulation of an energy crop, king grass (Pennisetum americanum 3 P. purpureum). Biomass Bioenergy 67:179–187

Zhao H, Wu L, Chai T, Zhang Y, Tan J, Ma S (2012) The effects of copper, manganese and zinc on plant growth and elemental accumulation in the manganese-hyperaccumulator Phytolacca americana. J Plant Physiol 169(13):1243–1252

Zobel RW, Kinraide TB, Baligar VC (2007) Fine root diameters can change in response to changes in nutrient concentrations. Plant Soil 297:243–254. https://doi.org/10.1007/s11104-007-9341-2

Chapter 6
Plant Nitrogen Nutrition, Environmental Issues, and Crop Productivity

Moddassir Ahmad and Nasir Ahmad Saeed

6.1 Introduction

Nitrogen (N) is the essential macro-nutrient required for plant growth and development and an important component of crop production. However, its excessive application may result in N loss that could have serious environmental concerns (Gao et al. 2020). Although N is abundantly available in the atmosphere, plants cannot take it directly from the environment. Plants could take it only if it exists in the form of nitrate or ammonium. Nitrogen fertilization applied in the form of ammonium or nitrate meets plant's need. Nitrate is the major form of nitrogen readily acquired by most of plant species to be utilized for metabolism, assimilation, and its conversion to yield (Colla et al. 2011). Nonetheless, if the rate of nitrate acquisition is higher than the rate of metabolism or assimilation, then it has the tendency to get accumulated in tuber, bulb, roots, stem, fruits, seeds, and leaves. However, predominantly it accumulates in non-leguminous crop plants, roots, and leafy vegetables (Santamaria 2006).

Nitrate-contaminated drinking water and consumption of raw seeds, processed food, and leafy vegetables are the main source of human exposure to nitrogenous fertilizers (Rathod et al. 2016). Though, nitrate by itself does not pose a risk to human health but upon its conversion to nitrite, nitric oxide and N-nitroso compounds arbitrated in the acidic stomach conditions pose associated risk to develop methemoglobinemia syndrome and gastric bladder cancer (Ahmed et al. 2017; Parks et al. 2008). Nitrite is cytotoxic to plant cells; hence, plants tend to maintain it to a minimum level to meet ionic homeostasis balance (Riens and Heldt 1992). However, nitrate gets accumulated in various tissues and edible parts of crop plants depending upon plant species, genotype, environmental conditions, agronomic factors, crop variety, pre-harvest aspects, harvest stage, and post-harvest storage

M. Ahmad · N. A. Saeed (✉)
National Institute for Biotechnology and Genetic Engineering (NIBGE), Faisalabad, Pakistan

Q. Mahmood (ed.), *Sustainable Plant Nutrition under Contaminated Environments*, Sustainable Plant Nutrition in a Changing World,
https://doi.org/10.1007/978-3-030-91499-8_6

(Andrews et al. 2013; Colonna et al. 2016). Post-harvest aspects, e.g., storage conditions of plant biomass may aggravate or hamper conversion of nitrate to nitrite endogenously. Accumulation of nitrate in crop plants depends upon agronomic, genetic background and environmental factors (Santamaria et al. 2001).

6.2 Agronomic Procedures to Maintain Nitrate Concentration to Minimal Level

With the dramatic increase of nitrogenous fertilizer application since the last four decades, scientists are adapting various agronomic strategies to maintain available soil nitrate content to a minimal level. These strategies include the following:

1. Adaptation of nitrate free fluid N fertilizer (Marsic and Osvald 2002)
2. Replacement of nitrate with ammonium-based fertilizer
3. Nitrate starvation 5 days before harvest (Borgognone et al. 2013)
4. Utilization of low nitrate accumulating plant genotypes (Burns et al. 2011)
5. Enhancement of nitrate reductase activity through optimization of light spectral fluxes (Gaudreau et al. 1995)
6. Substitution of nitrate with chloride, i.e., calcium nitrate with calcium chloride (Borgognone et al. 2016)

In this chapter, we explained various agronomic, environmental, and genetic factors for the mitigation of excessive nitrate accumulation in various crop plants and tissues, e.g., seeds, fruit, tuber, root, and stem. Variations with respect to nitrate accumulation do exist among different plant species and even in cultivars of same plant species (Anjana et al. 2006). Nitrate absorbed by plant roots are transported to various tissues by xylem vessels through transpiration stream. The nutrient absorption through root tissues might be accomplished by passive mode, i.e., nutrients enter in root tissue with water stream or through active mode, i.e., entry of nutrient molecule with the involvement of carrier molecule. The plant tissues with largely scattered laminae are prone to accumulate more nitrates in vacuole than the tissues with compact laminae, e.g., hypogeal storage organs, fruit, stem, and petiole (Chen et al. 2004).

Nitrate tends to accumulate differentially in various plant tissues in the order lea f > stem > root > tuber > fruit > seeds (Santamaria et al. 1999), depending upon the age and type of plant tissue. Nevertheless, younger leaves accumulate relatively less amount of nitrate than older or mature leaves (Konstantopoulo et al. 2010). Additionally, lately harvested plant tissues accumulate high nitrate concentration than earlier harvested tissues (Anjana et al. 2006). Similarly, some families (*Amaranthaceae, Asteraceae, Apiaceae,* and *Brassicaceae*) are known for hyper nitrate accumulations compared to *Lamiaceacae*, i.e., low nitrate accumulator (Santamaria 2006). Fruits, among other farm produce, are observed to accumulate less amount of nitrate, i.e., <15 mg kg^{-1} FW (Susin et al. 2006).

6.3 Genotypic and Nutritional Factors Associated with Nitrate Accumulation

Efficient nitrogen conversion process has been observed in low nitrate accumulating cultivars than high nitrate accommodating plant species. Low nitrate accumulating plant cultivars are found to possess lesser transcripts level of nitrate transporter (*LsNRT*) gene than high nitrate accommodating species. However, nitrate accumulation depends upon nitrate uptake and differences in its assimilation during reduction process (Luo et al. 2006). Plant nitrate and water content exist in a positive linear relationship to maintain plant homoeostasis (Qiu et al. 2014).

Excessive nitrogenous fertilizer applications lead to higher nitrate acquisition and nitrate accumulation in different plant tissues (Soleymani and Shahrajabian 2012). Unnecessary application of N fertilizer at a wrong plant growth stage/time when plant does not need it, enhances crop production cost, contaminate groundwater through leaching of nitrate, and elevate the nitrate content in various plant tissues (Ahmed et al. 2017). The demand of N decreases as the plant grows toward maturity; therefore, application of N fertilizer at maturity stage results in nitrate accumulation. To reduce nitrate level in farm produce, it is recommended to deprive crop plants with N fertilization few days or weeks before harvest (Borgognone et al. 2016).

Nitrate accumulation in edible parts of crop plant is directly related to the type of N fertilizer (Pavlou et al. 2007). Usually, nitrogen exists in the form of urea, nitrate, ammonium, or as organic compounds. Nitrate-based fertilizer dose increases nitrate uptake in crop plants, while ammonium-based fertilizer leads to reduction in plant nitrate content (Wang and Li 2004). Hence, type and choice of N fertilizer application is important to keep nitrate level to a minimum level in different plant species (Zhou et al. 2000). The accumulation of nitrate with reference to various N fertilizer types in petioles exists in order of: urea > ammonium nitrate > ammonium sulfate > ammonium carbonate (Renseigné et al. 2007). The release of N from mineralization of organic matter is much slower than that from inorganic sources (Herencia et al. 2011). Therefore, significantly less nitrate content is found in organic farm produce than in crop plants conventionally applied with inorganic N fertilizers (Nunez de Gonzalez et al. 2015).

The supply of other macro- and micro-nutrients, e.g., phosphorus, potassium, magnesium, calcium, sulfur, iron, and molybdenum can influence accumulation of nitrate in crop plants. In phosphorus-rich soil, nitrate uptake in plant is positively associated (Buwalda and Warmenhoven 1999), while it is linked negatively with sulfate supply in soil mix (Blom-Zandstra and Lampe 1983). However, potassium promotes nitrate acquisition, remobilization, and metabolism (assimilation) in plants (Ahmed et al. 2000). Magnesium is usually involved in metabolic processes and for chloroplast development (Kessler 1964); nonetheless, no evidence supports its role in nitrate accumulation. Calcium is required for root development, and its deficiency may influence nitrate uptake. Iron can affect the reduction in nitrate content in root and leaf (Borlotti et al. 2012). The role of molybdenum in nitrate reductase activity has been well demonstrated as it reduces nitrate to nitrite during assimilation process (Hewitt and Smith 1975).

6.4 Nitrate Accumulation and Environmental Factors

A number of environmental factors such as temperature, light, salt, drought, and carbon dioxide concentration affect the nitrate accumulation in crop plants. Intensive nitrogen fertilizer dose and low light intensity result in high accumulation of nitrate (Fu et al. 2017); however, in case of high light spectrum, nitrate accumulation in plants is tending to decrease (Liu et al. 2016). As nitrate and chloride ions are interchangeable during osmoregulation process, low to medium salt stress may lead to the avoidance of nitrate accumulation in soilless or soiled cultivation method. Nevertheless, under drought stress, plants are activated for hyper nitrate uptake through roots and speed up its mobilization in shoot (Talouizite and Champigny 1988). Energy requirements for nitrate acquisition and metabolism are met through biochemical process by involving low- and high-affinity nitrate transport system (Bose and Srivastava 2001). Photosynthesis triggers nitrate assimilation in plant system, providing carbon skeleton essentially required for ammonium integration and releasing electrons for reduction of nitrate to nitrite (Cavaiuolo and Ferrante 2014). Therefore, air temperature, intensity of carbon dioxide in the atmosphere and availability of suitable light conditions influence accumulation of nitrate in various crop plants (Santamaria 2006).

6.5 Indexing of Nitrogen Nutrition and Dilution Curve

Since after green revolution, crop yield enhancement reflects corresponding increase in N fertilizer usage; however, after reaching to maximum fertilizer dose, no further improvement could be achieved (Scharf and Lory 2009). Consequently, N fertilization beyond threshold may increase N concentration in surrounding environment, which creates serious environmental issues (Ju et al. 2009). Before applying N fertilizer to the crop plants, certain diagnostic tools are required to check existing (available) N status in a field. The soil-plant testing to measure N content could provide important direction to achieve sustainable crop productivity (He et al. 2009; Cui et al. 2008; Alivelu et al. 2006). In addition to soil analytical testing, N supply from soil has also been depicted from plant growth parameters. Therefore, plant physiological or morphological based testing seems crucial for determination of soil N status, which offers to adapt suitable N management actions for optimal crop productivity and to minimize the risk of N accumulation in terrestrial or aquatic environment (Lemaire et al. 2008; Ziadi et al. 2008; Naud et al. 2009; Voillot and Barret 1999). Correlation of shoot biomass and plant nitrogen content (Nc) is an important parameter to achieve optimal plant growth under suitable N fertilizer dose (Ulrich 1952). However, in general, N concentration in plants usually decreases with the advancement in plant growth (Lemaire and Salette 1984a, b). The decrease in plant N concentration with the enhancement of crop growth is called as dilution curve and empirically represented as:

$$Nc = aW^{-b} \tag{6.1}$$

Various crop cultivars demonstrate different dilution curve, which is affected by morphological traits and histological factors. The nitrogen dilution curve for a number of crop plant species has been constructed, e.g., *Brassica napus* L. (Colnenne et al. 1998), *Gossypium* spp. (Xue et al. 2007), *Linum usitatissimum* L. (Flenet et al. 2006), *Lolium multiflorum* L. (Marino et al. 2004), *Lycopersicon esculentum* Mill. (Tei et al. 2002), *Oryza sativa* L. (Sheehy et al. 1998), *Pisum sativum* L. (Ney et al. 1997), and for *Triticum aestivum* L. (Greenwood et al. 1990). Ziadi et al. (2008) associated the value of N dilution curve with the succeeding estimation of Nitrogen Nutrition Index (NNI), which is a reliable parameter to manage appropriate N nutrition for a crop. However, various experimental sites and different crop species require different validations in said climatic condition for specific plant genotypes. Usually the estimation in this respect involves calculation of critical N dilution curve, which is a comparison between existing and newly established N curves for different plant genotypes and probability of utilizing the dilution curve to determine N nutrition of crop plants. The assessment of critical N concentration is believed to be a reliable approach to measure existing N nutrition status at different growth stages of a crop plant. Therefore, N dilution curve is an important indicator to detect N nutrition status of a growing crop plant to achieve optimum crop yield.

6.6 Application of N Fertilizers and Environmental Consequences

Enhancing crop production by various fertilizer management strategies is crucial for environmental pollution and fertilizer economy (Schulze 1989). The application of N fertilizer (ammonium or urea) in alkaline calcareous soil leads to volatilization of ammonia. Although N losses through ammonia volatilization were reported up to 32–35% (Zhang et al. 1992; Gandhi and Paliwal 2004), N fertilizer losses during early days of application were observed up to 80% of the total loss (Akhtar et al. 2012). The pH of soil strongly linked to the extent of ammonia volatilization. The acidic (pH 5.2) soil requires a twice longer time to hydrolyze urea as compared to highly alkaline (pH 8.2 or higher) soil (Fenn and Richards 1989). However, N losses vary differently depending upon soil textures; higher losses were observed in light texture soil due to low capacity of cation exchange (Bernard et al. 2009). In the soil, urea is hydrolyzed to carbamic acid by the action of urease enzyme, but carbamic acid is an unstable entity, which is ultimately converted to carbon dioxide and ammonia. Ammonia keeps on escaping from the soil to the atmosphere until it chemically reacts with water to be transformed into ammonium (NH_4^+) (Benini et al. 1999). Nonetheless, hydrolysis process by itself tends to enhance the soil pH to 9.0, which is sufficient for ammonia volatilization (Brady and Weil 2001).

In the soil, nitrification is the key aerobic reaction in the N-cycle, which transforms ammonium/ ammonia to nitrate. The reduction of nitrate is affected by different microbial activities, which lead to the formation of nitrous oxide (N_2O) and other products (Carlson and Ingraham 1983). However, nitrous oxide has 300 times more global warming potential than CO_2, which is present in higher concentration in the atmosphere. N_2O contributes~6% in the global warming, and it is very reactive with ozone in the stratosphere (Dalsgaard et al. 2005). Its emission from arable lands causes huge losses of applied nitrogenous fertilizers, which may reach up to 65% of total greenhouse gas emission (Ali et al. 2008; Prather et al. 1995). The major processes are nitrification and de-nitrification, liable for gaseous emission from soil (Sahrawat and Keeney 1986). Soil physiochemical characteristics, organic manures, inorganic N fertilizer, and mode of fertilizer application affect N_2O emission (Smith 1997). Aerobic soil results higher N_2O emission (Ciarlo et al. 2008) as compared to submerged soil condition (Sahrawat and Keeney 1986).

Globally urea is the major N fertilizer contributing around 50% of total nitrogenous fertilizer consumption (IFIA 2007) leading to a series of biochemical reactions influencing the availability of N to plants. Urea undergoes a slow diffusion process and hydrolyzes in microcites of its granules. The reduced NUE of urea is basically due to volatilization of ammonia from agricultural lands upon its application in dry, hot, and high pH environment (Cabrera et al. 2001; Latifah et al. 2011; Gioacchini et al. 2002; Kissel et al. 2004; Pacholski et al. 2006; Rochette et al. 2009).

6.7 N Utilization and Sustainable Crop Productivity

Crop productivity in cereal cropping system, i.e., wheat-rice, maize-wheat can be sustained by crop rotation, growing legumes in fallow period especially prior to rice cultivation and instigating biodiversity (Garrity and Becker 1994; Lauren et al. 2001). However, the main issue in legume introduction is limited available time period and unavailability of soil moisture required for legume cultivation especially in pre-rice niche (Garrity and Becker 1994). In spite of having several benefits in growing legumes in fallow period and as pre-rice niche, it has not been practiced due to the involvement of additional production cost and no cash returns (Ali and Narciso 1994). Nevertheless, in rice-wheat cropping system, cultivation of pulses or forage is more popular as compared to sole manure crops due to resource sustainability and farmers' financial benefits (Ali and Narciso 1994; Lauren et al. 2001). However, the green manure of legume residue significantly enhances the crop productivity by increasing soil fertility due to fixation of atmospheric N. Thus, reducing the need of inorganic N fertilizer application restores soil for cultivation of different crops (Yaqub et al. 2010). Therefore, it is very important to keep the record of nitrogen application and quantification of nutrient budget for sustainable nutrient management. To address the challenge, there is need to improve knowledge of inputs and outputs of a given system and complex nutrient cycles and technological advancements of specific observations, modes, and bookkeeping of known activity data (Zhang et al. 2020).

Sustainable crop productivity can be achieved by involving restorative crops, i.e., exhaustive and legume crops. Additionally, intercropping system is believed to be environment friendly as legumes are proved to fix N from the atmosphere (Vankessel et al. 1985). Groundnut and cowpea could gather around 80–250 kg N/ha for proceeding crops, which become available by mineralization of their residues (Liebman and Dyck 1993; Mohr et al. 1999; Przednowek 2003). The cultivation of cereal after leguminous crop is crucial for biological nitrogen fixation as it enhances the capabilities of legumes to fix atmospheric N and by efficient utilization of excessive nitrate (Fujita et al. 1992). However, uninterrupted legume cultivation leads to accumulation of nitrate, which decreases the capacity of legumes to fix N (Anil et al. 1998). Contrarily, mono cropping of cereal depletes other nutrient and N resulting in stagnant or lesser yield. The use of legumes green manure before cereal cultivation improves soil fertility and enhances crop productivity (Andrews 1979; Gill et al. 2009). Intercropping system is known for enhancing utilization of natural resources and crop productivity (Hauggaard-Nielsen et al. 2001). Many reports highlighted the importance of mix cropping system to sustain soil fertility and N assimilation by crop plants for higher yield output (Karpenstein-Machan and Stuelpnagel 2000; Li et al. 2001; Fujita et al. 1992; Izaurralde et al. 1992; Ofosu-Budu et al. 1995; Shahzad et al. 2019). Wheat crop yield enhancement to 34, 27, and 19% has been reported by intercropping of cowpea, soybean, and groundnut, respectively (Nair et al. 1981). Subsequent or associated cereal crop consumes the nutrient supplies obtained from the decomposition of legume residues. Therefore, crop diversification, balanced and sustainable crop nutrition management, irrigation management, induction of legumes in crop rotation, precision in season management, and use of enhanced efficiency fertilizers are suggested to be a better strategy for efficient utilization of natural resources and lemmatizing dependence on inorganic nitrogenous fertilizers to achieve sustainable crop productivity.

References

Ahmed AHH, Khalil MK, Farrag AM (2000) Nitrate accumulation, growth, yield and chemical composition of rocket (*Eruca vesicaria* subsp. *sativa*) plant as affected by NPK fertilization, kinetin and salicylic acid. In: Proceedings of ICEHM 2000. Cairo University, Egypt, pp 495–508

Ahmed M, Rauf M, Mukhtar Z, Saeed NA (2017) Excessive use of nitrogenous fertilizers: an unawareness causing serious threats to environment and human health. Environ Sci Pollut Res Int 24:26983–26987

Akhtar M, Hussain F, Ashraf MY, Mahmood TQ, Akhter J, Rasool AA (2012) Influence of salinity on nitrogen transformations in soil. Commun Soil Sci Plant Anal 43(12):1674–1683

Ali M, Narciso JH (1994) Economic evaluation and farmers' perception of green manure use in rice-based farming systems. In: Ladha JK, Garrity DP (eds) Green manure production systems for Asian rice lands. International Rice Research Institute, Manila, pp 173–195

Ali R, Iqbal J, Tahir GR, Mahmood T (2008) Effect of 3, 5-dimethylpyrazole and nitrapyrin on nitrification under high soil temperature. Pak J Bot 40:1053–1062

Alivelu K, Rao SA, Sanjay S, Raju NS, Madhuri P (2006) Prediction of optimal nitrogen application rate of rice based on soil test values. Eur J Agron 25(1):71–73

Andrews RW (1979) Intercropping, its importance and research needs. Field Crops Res 32:1–10

Andrews M, Raven JA, Lea PJ (2013) Do plants need nitrate? The mechanisms by which nitrogen form affects plants. Ann Appl Biol 163:174–199

Anil L, Park J, Phillip RH, Miller FA (1998) Temperate intercropping of cereals for forage: a review of the potential for growth and utilization with particular reference to the UK. Grass Forage Sci 53:301–317

Anjana, Umar S, Iqbal M, Abrol YP (2006) Are nitrate concentrations in leafy vegetables within safe limits? In: Proceedings of the workshop on nitrogen in environment. Industry and Agriculture, New Delhi, pp 81–84

Benini S, Rypniewski WR, Wilson KS, Miletti S, Ciurli S, Mangani S (1999) A new proposal for urease mechanism based on the crystal structures of the native and inhibited enzyme from *Bacillus pasteurii*: why urea hydrolysis costs two nickels. Structure 7:205–216

Bernard UM, Dudoignon P, Pons Y (2009) Characterization of structural profiles in clay-rich marsh soils by cone resistance and resistivity measurements. Soil Sci Soc Am J 73:46. https://doi.org/10.2136/sssaj2007.0347

Blom-Zandstra M, Lampe JEM (1983) The effect of chloride and sulphate salts on the nitrate content in lettuce plants. J Plant Nutr 6:611–628

Borgognone D, Colla G, Rouphael Y, Cardarelli M, Rea E, Schwarz D (2013) Effect of nitrogen form and nutrient solution pH on growth and mineral composition of self-grafted and grafted tomatoes. Sci Hortic 149:61–69

Borgognone D, Rouphael Y, Cardarelli M, Lucini L, Colla G (2016) Changes in biomass, mineral composition, and quality of cardoon in response to NO_3^-:Cl^- ratio and nitrate deprivation from the nutrient solution. Front Plant Sci 7:978

Borlotti A, Vigani G, Zocchi G (2012) Iron deficiency affects nitrogen metabolism in cucumber (*Cucumis sativus* L.) plants. BMC Plant Biol 12:189. https://doi.org/10.1186/1471-2229-12-189

Bose B, Srivastava HS (2001) Absorption and accumulation of nitrate in plants: influence of environmental factors. Indian J Exp Biol 39:101–110

Brady NC, Weil RR (2001) The nature and properties of soils, 13th edn. Prentice Hall, Upper Saddle River. ISBN: 0130167630

Burns IG, Zhang KF, Turner MK, Meacham M, Al-Redhiman K, Lynn J, Broadley MR, Hand P, Pink D (2011) Screening for genotype and environment effects on nitrate accumulation in 24 species of young lettuce. J Sci Food Agric 91(3):553–562

Buwalda F, Warmenhoven M (1999) Growth-limiting phosphate nutrition suppresses nitrate accumulation in greenhouse lettuce. J Exp Bot 50:813–821

Cabrera ML, Kissel DE, Davis RC, Qafoku NP, Segars WI (2001) Design and ammonia-recovery evaluation of a wind speed sensitive chamber system. Soil Sci Soc Am J 65:1302–1306

Carlson CA, Ingraham JL (1983) Comparison of denitrification by *Pseudomonas stutzeri*, *Pseudomonas aeruginosa*, and *Paracoccus denitrificans*. Appl Environ Microbiol 45:1247–1253

Cavaiuolo M, Ferrante A (2014) Nitrates and glucosinolates as strong determinants of the nutritional quality in rocket leafy salads. Nutrients 6:1519–1538

Chen BM, Wang ZH, Li SX, Wang GX, Song HX, Wang XN (2004) Effects of nitrate supply on plant growth, nitrate accumulation, metabolic nitrate concentration and nitrate reductase activity in three leafy vegetables. Plant Sci 167:635–643

Ciarlo E, Conti M, Bartoloni N, Rubio G (2008) Soil N_2O emissions and $N_2O/(N_2O+N_2)$ ratio as affected by different fertilization practices and soil moisture. Biol Fertil Soils 44:991–995

Colla G, Rouphael Y, Mirabelli C, Cardarelli M (2011) Nitrogen-use efficiency traits of mini-watermelon in response to grafting and nitrogen-fertilization doses. J Plant Nutr Soil Sci 174:933–941

Colnenne C, Meynard M, Reau R, Justes E, Merrien A (1998) Determination of a critical nitrogen dilution curve for winter oilseed rape. Ann Bot 81:311–317
Colonna E, Rouphael Y, Barbieri G, De Pascale S (2016) Nutritional quality of ten leafy vegetables harvested at two light intensities. Food Chem 199:702–710
Cui ZL et al (2008) On-farm evaluation of an in-season nitrogen management strategy based on soil N_{min} test. Field Crops Res 105(1–2):48–55
Dalsgaard T, Thamdrup B, Canfield DE (2005) Anaerobic ammonium oxidation (anammox) in the marine environment. Res Microbiol 156:457–464
Fenn LB, Richards J (1989) Ammonia loss from surface applied urea-acid products. Fertil Res 9:265–275
Flenet F, Guerif M, Boiffin J, Dorvillez D, Champolivier L (2006) The critical N dilution curve for linseed (*Linum usitatissimum* L.) is different from other C3 species. Eur J Agron 24(4):367–373
Fu Y, Li HY, Yu J, Liu H, Cao ZY, Manukovsky NS, Liu H (2017) Interaction effects of light intensity and nitrogen concentration on growth, photosynthetic characteristics and quality on lettuce (*Lactuca sativa* L. Var. *youmaicai*). Sci Hortic 214:51–57
Fujita K, Ofosu-Budu KG, Ogata S (1992) Biological nitrogen fixation in mixed legume cereal cropping systems. Plant Soil 141:155–175
Gandhi AP, Paliwal KV (2004) Mineralization and gaseous losses of nitrogen from urea and ammonium sulphate in salt-affected soils. Plant Soil 45:247–255
Gao L, Li W, Ashraf U, Lu W, Li Y, Li C, Li G, Li G, Hu J (2020) Nitrogen fertilizer management and maize straw return modulate yield and nitrogen balance in sweet corn. Agronomy 10:362. https://doi.org/10.3390/agronomy10030362
Garrity DP, Becker M (1994) Where do green manures fit in Asian rice farming systems? In: Ladha JK, Garrity DP (eds) Green manure production systems for Asian ricelands. International Rice Research Institute, Manila, pp 1–10
Gaudreau L, Charbonneau J, Vezina LP, Gosselin A (1995) Effects of photoperiod and photosynthetic photon flux on nitrate content and nitrate reeducates activity in greenhouse-grown lettuce. J Plant Nutr 18:437–453
Gill S, Abid M, Azam F (2009) Mixed cropping effects on growth of wheat (*Triticum aestivum* L.). Pak J Bot 41:1029–1036
Gioacchini P, Nastri A, Marzadori C, Giovannini C, Vittori Antisari L, Gessa C (2002) Influence of urease and nitrification inhibitors on N losses from soils fertilized with urea. Biol Fertil Soils 36:129–135
Greenwood DJ, Lemaire G, Gosse G, Cruz P, Draycott A, Neeteson JJ (1990) Decline in percentage N of C3 and C4 crops with increasing plant mass. Ann Bot 67:181–190
Hauggaard-Nielsen H, Ambus P, Jensen ES (2001) Inter-specific competition, N use and interference with weeds in pea–barley intercropping. Field Crop Res 70:101–109
He P, Li ST, Jin JY, Wang HT, Li CJ, Wang YL, Cui RZ (2009) Performance of an optimized nutrient management system for double-cropped wheat–maize rotations in North-Central China. Agron J 101:1489–1496
Herencia JF, García-Galavís PA, Dorado JAR, Maqueda C (2011) Comparison of nutritional quality of the crops grown in an organic and conventional fertilized soil. Sci Hortic 129:882–888
Hewitt EJ, Smith TA (1975) Plant mineral nutrition. English Universities Press, London
International Fertilizer Industry Association (IFIA) (2007) A survey of the anticipated impact of biofuels development on short-, medium-and long-term fertilizer demands. Report prepared by the IFA Task Force on Bio energy. 75th IFA Annual Conference, International Fertilizer Industry Association Agriculture Committee. Istanbul, Turkey, May 2007
Izaurralde RC, McGill WB, Juma NG (1992) Nitrogen fixation efficiency, interspecies N transfer, and root growth in barley-field pea intercrop on a Black Chernozemic soil. Biol Fertil Soils 13:11–16
Ju XT, Xing GX, Chen XP, Zhang XL, Zhang LJ, Liu XJ, Cui ZL, Yin B, Christie P, Zhu ZL, Zhang FS (2009) Reducing environmental risk by improving N management in intensive Chinese agricultural systems. PNAS 106(9):3041–3046

Karpenstein-Machan M, Stuelpnagel R (2000) Biomass yield and nitrogen fixation of legumes monocropped and intercropped with rye and rotation effects on subsequent maize crop. Plant Soil 218:215–232

Kessler E (1964) Nitrate assimilation by plants. Annu Rev Plant Physiol 15:57–72

Kissel DM, Cabrera N, Vaio J, Craig J, Rema A, Morris L (2004) Rainfall timing and ammonia loss from urea in a loblolly pine plantation. Soil Sci Soc Am J 68:1744–1750

Konstantopoulo E, Kapotis E, Salachas G, Petropoulos SA, Karapanos IC, Passam HC (2010) Nutritional quality of greenhouse lettuce at harvest and after storage in relation to N application and cultivation season. Sci Hortic 125:93–95

Latifah MN, Zaulia O, Aida MPN, Fauziah O, Hairiyah M, Talib Y (2011) Effect of citric acid treatment on the quality of fresh-cut pineapple. Acta Hortic 902:467–476

Lauren JG, Shreshta R, Sattar MA, Yadev RL (2001) Legumes and diversification of the rice-wheat cropping system. J Crop Prod 3:67–102

Lemaire G, Salette J (1984a) Relationship between growth and nitrogen uptake in a pure grass stand: environmental effects. Agronomie 4:423–430

Lemaire G, Salette J (1984b) Relationship between growth and nitrogen uptake in a pure grass stand: II. Study on genotype variation. Agronomie 4:431–436

Lemaire G, Jeuffroy MH, Gastal F (2008) Diagnostic tools for crop and plant N status in vegetative stage. Theory and practices for crop N management. Eur J Agron 28:614–624

Li L, Sun JH, Zhang F, Li XL, Yang SC, Rengel Z (2001) Wheat/maize or wheat/soybean strip intercropping: I. yield advantage and interspecific interactions on nutrients. Field Crops Res 71:123–137

Liebman M, Dyck E (1993) Crop rotation and intercropping strategies for weed management. Ecol Appl 3(1):92–122

Liu H, Fu Y, Yu J, Liu H (2016) Accumulation and primary metabolism of nitrate in lettuce (*Lactuca sativa* L. var. *youmaicai*) grown under three different light sources. Commun Soil Sci Plant Anal 17:1994–2002

Luo J, Sun S, Jia L, Chen W, Shen Q (2006) The mechanism of nitrate accumulation in pakchoi [*Brassica campestris* L. ssp. Chinensis (L.)]. Plant Soil 282:291–300

Marino MA, Mazzanti A, Assuero SG, Gastal F, Echeverria HE, Andrade F (2004) Nitrogen dilution curves and nitrogen use efficiency during winter-spring growth of annual ryegrass. Agron J 96:601–607

Marsic NK, Osvald J (2002) The influence of different concentration of nitrogen in nutrient solution on plant growth and nitrate accumulation in aeroponically grown lettuce (*Lactuca sativa* L.). Agrochimica -Pisa 46(1):56–65

Mohr RM, Entz MH, Janzen HH, Bullied WJ (1999) Plant-available nitrogen supply as affected by method and timing of alfalfa termination. Agron J 91:622–630

Nair KPP, Patel UK, Singh RP, Kaushik MK (1981) Evaluation of legume intercropping in conservation of fertilizer nitrogen in maize culture. Exp Agric 17:312–314

Naud C, Makowski D, Jeuffroy MH (2009) Leaf transmittance measurements can improve predictions of the nitrogen status for winter wheat crop. Field Crops Res 110:27–34

Ney B, Doré T, Sagan M (1997) The nitrogen requirement of major agricultural crops: grain legumes. In: Lemaire G (ed) Diagnosis of the nitrogen status in crops. Springer-Verlag, Heigelberg, pp 107–118

Nunez de Gonzalez MT, Osburn WN, Hardin MD, Longnecker M, Garg HK, Bryan NS, Keeton JT (2015) A survey of nitrate and nitrite concentrations in conventional and organic-labeled raw vegetables at retail. J Food Sci 80:C942–C949

Ofosu-Budu KG, Noumura K, Fujita K (1995) Nitrogen fixation, N transfer and biomass production of soybean cv. bragg or its super-nodulating nts1007 and sorghum mixing cropping at two rates of N fertilizer. Soil Biol Biochem 27:311–317

Pacholski A, Cai G, Nieder R, Richter J, Fan X, Zhu Z, Roelcke M (2006) Calibration of a simple method for determining ammonia volatilization in the field-comparative measurements in Henan Province, China. Nutr Cycl Agroecosyst 74:259–273

Parks SE, Huett DO, Campbell LC, Spohr LJ (2008) Nitrate and nitrite in Australian leafy vegetables. Aust J Agric Res 59:632–638

Pavlou GC, Ehaliotis CD, Kavvadias VA (2007) Effect of organic and inorganic fertilizers applied during successive crop seasons on growth and nitrate accumulation in lettuce. Sci Hortic 111:319–325

Prather M, Derwent R, Ehhalt D, Fraser P, Sanhueza E, Zhou X (1995) Other trace gases and atmospheric chemistry. In: Houghton JT et al (eds) Climate change 1994: radiative forcing of climate change and an evaluation of the IPCC IS92 emission scenarios. Cambridge University Press, Cambridge, pp 73–126

Przednowek DWA (2003) The effect of pulse crop rotation and controlled-release urea on the nitrogen accumulation and end-use quality of Canada Western Red Spring Wheat, M.S. thesis. University of Manitoba, Winnipeg

Qiu W, Wang Z, Huang C, Chen B, Yang R (2014) Nitrate accumulation in leafy vegetables and its relationship with water. J Soil Sci Plant Nutr 14:761–768

Rathod KS, Velmurugan S, Ahluwalia A (2016) A 'green' diet-based approach to cardiovascular health? Is inorganic nitrate the answer? Mol Nutr Food Res 60:185–202

Renseigné N, Umar S, Iqbal M (2007) Nitrate accumulation in plants, factors affecting the process, and human health implications. A review. Agron Sustain Dev 27:45–57

Riens B, Heldt HW (1992) Decrease of nitrate reductase activity in spinach leaves during a light-dark transition. Plant Physiol 98:573–577

Rochette P, Angers DA, Chantigny MH, MacDonald JD, Gasser MO, Bertrand N (2009) Reducing ammonia volatilization in a no till soil by incorporating urea and pig slurry in shallow bands. Nutr Cycl Agroecosyst 84:71–80

Sahrawat K, Keeney DR (1986) Nitrous oxide emission from soils. Adv Soil Sci 4:103–148

Santamaria P (2006) Nitrate in vegetables: toxicity, content, intake and EC regulation. J Sci Food Agric 86:10–17

Santamaria P, Elia A, Serio F, Todaro E (1999) A survey of nitrate and oxalate content in fresh vegetables. J Sci Food Agric 79:1882–1888

Santamaria P, Elia A, Gonnella M, Parente A, Serio F (2001) Ways of reducing rocket salad nitrate content. Acta Hortic 548:529–537

Scharf P, Lory JA (2009) Calibrating reflectance measurements to predict optimal sidedress nitrogen rate for corn. Agron J 101(3). https://doi.org/10.2134/agronj2008.0111

Schulze ED (1989) Air pollution and forest decline in a spruce (*Piceaabies*) forest. Science 244:776–783

Shahzad A, Qureshi M, Wakeel A, Misselbrook T (2019) Crop production in Pakistan and low nitrogen use efficiencies. Nat Sustain 2:1106–1114

Sheehy JE, Dionora MJA, Mitchell PL, Peng S, Cassman KG, Lemaire G, Williams RL (1998) Critical nitrogen concentrations: implications for high-yielding rice (Oryza sativa L.) cultivars in the tropics. Field Crops Res 59:31–41

Smith KA (1997) The potential for feedback effects induced by global warming on emissions of nitrous oxide by soils. Glob Chang Biol 3:327–338

Soleymani A, Shahrajabian MH (2012) Effects of different levels of nitrogen on yield and nitrate content of four spring onion genotypes. Int J Agric Crop Sci 4:179–182

Susin J, Kmcel V, Gregorcic A (2006) A survey of nitrate and nitrite content of fruit and vegetables grown in Slovenia during 1996–2002. Food Addit Contam 23:385–390

Talouizite A, Champigny ML (1988) Response of wheat seedlings to short-term drought with particular respect to nitrate utilization. Plant Cell Environ 11:149–155

Tei F, Benincasa P, Guiducci M (2002) Critical nitrogen concentration in processing tomato. Eur J Agron 18(1):45–55

Ulrich A (1952) Physiological bases for assessing the nutritional requirements of plants. Annu Rev Plant Physiol 3:207–228

Vankessel C, Singleton PW, Hoben HJ (1985) Enhanced N-transfer from a soybean to maize by vesicular arbuscular mycorrhizal (VAM) fungi. Plant Physiol 79:562–563

Voillot MO, Barret FD (1999) Accumulation and remobilization of nitrogen in a vegetative winter wheat crop during or following nitrogen deficiency. Ann Bot 83(5):569–575

Wang Z, Li S (2004) Effects of nitrogen and phosphorus fertilization on plant growth and nitrate accumulation in vegetables. J Plant Nutr 27:539–556

Xue X, Wang J, Wang Z, Guo W, Zhou Z (2007) Determination of a critical dilution curve for nitrogen concentration in cotton. J Plant Nutr Soil Sci 170(6):811–817

Yaqub M, Mahmood T, Akhtar M, Iqbal MM, Ali S (2010) Induction of mungbean [(VignaRadiata L.) Wilczek] as a grain legume in the annual rice-wheat double cropping system. Pak J Bot 42(5):3125–3135

Zhang SL, Cai GX, Wang XZ, Xu YH, Zhu ZL, Freney JR (1992) Losses of urea-nitrogen applied to maize grown on a calcareous Fluvo-Aquic soil in North China plain. Pedosphere 2:171178

Zhang X, Davidson EA, Zou T, Lassaletta L, Quan Z, Li T, Zhang W (2020) Quantifying nutrient budgets for sustainable nutrient management. Glob Biogeochem Cycles 34. https://doi.org/10.1029/2018GB006060

Zhou ZY, Wang MJ, Wang JS (2000) Nitrate and nitrite contamination in vegetables in China. Food Rev Intl 16:61–76

Ziadi N, Brassard M, Bélanger G, Cambouris AN, Tremblay N, Nolin MC, Claessens A, Parent LE (2008) Critical nitrogen curve and nitrogen nutrition index for corn in eastern Canada. Agron J 100:271–276

Chapter 7
Zn Biofortification in Crops Through Zn-Solubilizing Plant Growth-Promoting Rhizobacteria

Allah Ditta, Naseer Ullah, Muhammad Imtiaz, Xiaomin Li, Amin Ullah Jan, Sajid Mehmood, Muhammad Shahid Rizwan, and Muhammad Rizwan

7.1 Introduction

In the quest to enhance food production for an ever-increasing population of the world, research in agriculture has mainly focused on planting the crops with higher yields and crop intensification through the utilization of chemical fertilizers in

A. Ditta (✉)
Department of Environmental Sciences, Shaheed Benazir Bhutto University Sheringal, Upper Dir, Khyber Pakhtunkhwa, Pakistan

School of Biological Sciences, The University of Western Australia, Perth, WA, Australia
e-mail: allah.ditta@sbbu.edu.pk

N. Ullah · X. Li
Environmental Chemistry Laboratory, Department of Environmental Science and Engineering, School of Space and Environment, Beihang University, Beijing, P. R. China

M. Imtiaz
Soil and Environmental Biotechnology Division, National Institute for Biotechnology and Genetic Engineering, Faisalabad, Pakistan

A. U. Jan
Department of Biotechnology, Shaheed Benazir Bhutto University Sheringal, Upper Dir, Khyber Pakhtunkhwa, Pakistan

S. Mehmood
Key Laboratory of Agro-Forestry Environmental Processes and Ecological Regulation of Hainan Province, Hainan University, Haikou, China

M. S. Rizwan
Cholistan Institute of Desert Studies, The Islamia University of Bahawalpur, Bahawalpur, Pakistan

M. Rizwan
Institute of Soil Science, PMAS-Arid Agriculture University, Rawalpindi, Pakistan

Q. Mahmood (ed.), *Sustainable Plant Nutrition under Contaminated Environments*, Sustainable Plant Nutrition in a Changing World,
https://doi.org/10.1007/978-3-030-91499-8_7

unbalanced amounts (Singh 2000; Foley et al. 2005; Elkoca et al. 2010; Gliessman 2014). Detrimental effects such as nutrient equilibrium disturbance, deterioration of soil structure, and function have been reported in many studies due to the overuse of chemical fertilizers (Ali et al. 2018). The unbalanced use of chemical fertilizers also disrupts the ecology of soils leading to adverse impacts on the community's propagation of macro/microfauna and flora along with the health of human beings (Hafeez et al. 2013; Lockhart et al. 2013; Wu and Ma 2015). Application of chemical fertilizers in higher than the recommended fertilizer dose also results in micronutrient deficiency to the crop plants due to low soil organic matter and phosphorus, soil erosion, etc., and these effects are extremely serious crops grown in salt-affected, calcareous, arid, waterlogged and sandy soils (Alloway 2008).

Zn is among the 17 basic elements (micro- and macro-nutrients) required for optimum growth and production of crop plants and the most important among eight plant-based micronutrients. Zn plays a critical role in various biochemical reactions occurring in plants such as it is the structural component or regulatory cofactor of a large variety of different enzymes and proteins, which are primarily involved in carbohydrate, protein, and auxin metabolism (King 2006; Broadley et al. 2007). It also plays an important role in the production of pollen, ensuring biological membrane stability and resistance to invasion by some pathogenic substances. Zn deficiency disturbs photosynthesis and metabolism of nitrogen, decreases the development of fruit and flowering, protracts cycles of growth leading to maturity postponement, which in turn results in lower crop yield and quality and ultimately in suboptimal efficiency of nutrient usage in crop plants.

The idea of hidden hunger (insufficiency of some vitamins and micronutrients) has been widely known over the last two decades (Nilson and Piza 1998). The shortage of micronutrients such as Zn in soil not just decreases agricultural productivity but also reduces the food quality of the edible portion of the crop plants that leads to malnutrition among humans (Kumssa et al. 2015). Zn deficiency is widespread, and in humans, it may result in underdevelopment in neonatal, dysfunction in the immune, cognitive disability, and hypogonadism (Sauer et al. 2016). About one-third of the world's population especially pre-school children and women are affected by Zn deficiency and around 433,000 children die from Zn malnutrition annually (Hotz and Brown 2004; World Health Organization 2009; Stein 2010; Wessells and Brown 2012; Zhang et al. 2012). It is required in very small quantities for human beings and other living organisms throughout their lives to sustain their proper physiological functions. Zn deficiency, on the other hand, is hard to estimate because there is currently no accurate biochemical predictor suitable to indicate the status of Zn (Wieringa et al. 2015). Due to the detrimental effect of the unbalanced use of chemical fertilizers, the mobility of micronutrients such as Zn and iron (Fe) is reduced in soil (Alloway 2008). The need of the hour is to devise some economical and sustainable strategies, which have the potential to cope with Zn malnutrition in humans as well as in other living organisms.

Among different strategies, biofortification has been an effective strategy to mitigate Zn deficiency in foodstuffs and boosting the nutritional properties of crops. The main objective of biofortification is to grow crops that have increased the contents of bioavailable nutrients in their edible parts (Cakmak et al. 2010; Abaid-Ullah et al. 2011). Despite of various biofortification strategies such as breeding strategies through conventional and latest genetic engineering approaches, agronomic practices have been proposed to ameliorate the Zn deficiency in crop plants (Di Tomaso 1995; Rengel 2001; Cakmak et al. 2010; Garcia et al. 2016). The Zn bioavailability could be enhanced with certain amendments such as chelators (EDTA), organic acids (citrate), and amino acids (histidine and methionine) (Lonnerdal 2000). Different organic acids such as ascorbic acid, lactic acid, citric acid, and malic acid, may improve Zn absorption. Insoluble phytate-Zn and EDTA are a constituent part of Zn-EDTA. Harvest Plus (2012) showed an increase in the content of Zn in wheat by 25 ppm in Pakistan by fortification. Genetic and agronomic biofortification are two approaches that comprise them. Despite the effectiveness of these strategies in amelioration of Zn deficiency, these strategies are time-consuming, costly, and may not have up-to-the-mark satisfaction in combating Zn deficiency. The main disadvantages of biofortification are the barriers of root or shoot and the grain filling process.

Deficiency in Zn is usually found in arid and semiarid regions. Another sustainable and promising strategy is the use of Zn-solubilizing rhizobacteria that can enhance the growth and productivity of crop plants through various mechanisms such as macro- and micronutrient solubilization (such as phosphorus, potassium, Zn), mobilization, and translocation to various edible portions of the crop plants, fixation of atmospheric nitrogen, production of phytohormones (gibberellic acid, IAA, and kinetin), organic acids, abiotic stress tolerance through 1-aminocycloprop ane-1-carboxylate (ACC) deaminase activity, inducing systemic resistance among crops plants, etc. The interactions of plants and microbes in the alleviation of micronutrient deficiency are considered to play a pivotal role in enhancing the soil's nutritional status and empowering micronutrients by metal solubilization, mobilization, and translocation (Chen et al. 2014). Various researchers have reported microbial strategy as an effective way to enhance Zn biofortification in cereals such as rice and wheat (Mader et al. 2011; Rana et al. 2012; Zhang et al. 2012; Vaid et al. 2014; Prasanna et al. 2016; Singh et al. 2017a, b, 2018).

A soluble form of Zn is created by the conversion of an insoluble form of Zn by Zn-solubilizing bacteria (Zn-SB), which can be taken up by the plants. Microorganisms capable to solubilize Zn generate numerous organic acids by acidifying the soil, sequestrating cations of Zn, and subsequently reducing the soil pH in surrounding areas (Alexander 1997). Additionally, the solubility of Zn can also be increased by the application of anionic chelate Zn (Jones and Darrah 1994). The solubilization of Zn consists of the production of siderophores (Saravanan et al. 2011) and the cell membrane's oxidoreductase systems, chelating ligands, and proton (Chang et al. 2005). A variety of advantageous strains of bacteria, such as *Bacillus* spp., *Bacillus thuringiensis*, *Burkholderia cenocepacia*, *Gluconacetobacter diazotrophicus*, *Pseudomonas aeruginosa*, *Pseudomonas striata*, *Pseudomonas fluorescens*, *Serratia marcescens*, *S. liquefaciens* (Fasim et al. 2002; Saravanan et al. 2007c; Abaid-Ullah et al. 2011; Pawar et al. 2015), have been reported to induce solubilization of Zn at laboratory scale. Soil pH and moisture contents affect

Zn solubility in the soil. The type and quantity of various biochemical acids formed by diverse soil microbes are mostly contingent on the pH of the medium, source of carbon, and capacity of buffering (Mattey 1992).

This chapter outlines the role of Zn-solubilizing bacteria in enhancing Zn biofortification under Zn-deficient soils, mechanisms of Zn solubilization, and their application to alleviate Zn deficiency in crops especially in cereals under controlled and field condition.

7.2 Zn Deficiency in Agricultural Soils and Causes behind

Naturally, Zn occurs in soils as hopeite [$Zn_3(PO_4)_2 \cdot 4H_2O$], smithsonite ($ZnCO_3$), zinkosite ($ZnSO_4$), zincite (ZnO), franklinite ($ZnFe_2O_4$), and sphalerite (ZnS). A natural process such as (i) parent rock's physical and chemical weathering (Alloway 1995) and (ii) Zn addition to soils via atmospheric activities such as volcanoes, forest fires, and dust from the surface are among the main sources of Zn in soil. The initial phase of Zn's rhizosphere absorption is its deposition in plants until it is transferred to the seeds of crops (Giehl et al. 2009). Zn is an integral element for crops due to its involvement in various essential physiological activities required for the normal growth of plants. Zn deficiency in agricultural soils is widespread and has been reported in many countries. Around 50% of arable land worldwide is Zn deficient. More specifically, approximately 50% of agricultural fields in China, India, and Turkey are Zn deficient (FAO WHO 2002). However, approximately 70% of agricultural land in Pakistan is Zn deficient (Kauser et al. 2001; Hamid and Ahmad 2001). In India, the soils are mostly Zn deficient particularly under wheat-rice cropping systems, and the concentration of Zn in a grain of cereals is expected to decline (Gupta 2005; Prasad 2005). Owing to the existence of an anti-nutrition factor, that is, phytic acid that reduces the bioavailability of mineral, a reduced amount of Zn is present in cereal grains (Pahlvan-Rad and Pressaraki 2009).

Widespread Zn deficiency in crop plants due to Zn-impoverished soils results in a significant decline in crop productivity. Numerous factors affect Zn bioavailability in soils. For example, the use of Zn-containing fertilizers is very rare among the farmers around the world who generally apply chemical fertilizers mainly containing nitrogen, phosphorus, and potassium. Moreover, the farmers using Zn-containing fertilizers use certain forms, which are not easily accessible for the crop plants after their application in soil due to solubility issues. Zn deficiency may also emerge in soils with high levels of salinity, sodium, phosphorus, and silicon; soils with acidic or calcareous nature, poorly drained, sandy, and coarse-textured soils as well as heavily weathered acidic soils (Sillanpaa 1982; Hacisalihoglu and Kochian 2003; Alloway 2008).

The critical limit of DTPA-extractable Zn in the soil is 0.6 mg kg^{-1} (Sillanpaa 1982; Alloway 2009). The critical limit is defined as the minimum soil test value of Zn that is required for the maximum productivity of crops. It reflects the intensity where the deficiencies arise, as the lower end of the sufficiency spectrum is specified. Plants uptake Zn through the roots as Zn^{2+} (Havlin et al. 2005). The amount of Zn available to plants depends on various factors such as the concentration of total Zn, pH, organic matter, calcium carbonate, clay, rhizosphere microbial activity, redox conditions, moisture contents, macronutrient concentrations, in particular climate and phosphorus and other trace element concentrations (Alloway 2008; Ibrahim et al. 2011; Ramzan et al. 2014). The pH of soils had a great influence on the Zn stream, as it is readily adsorbed in sites of cation exchange at a neutral pH and accessible at low pH values (Havlin et al. 2005; Broadley et al. 2007). Unlike animal-based foods or pulses, cereal grains contain much less concentration of Zn. Chirwa and Yerokun (2012) and Harter (1983) reported a decline in soil bioavailability of Zn, with a rise in soil pH owing to precipitation or adsorption of Zn on the $CaCO_3$ and Fe oxide surfaces. A significant negative relationship exists between the available Zn or Fe and soil cation exchange capability (Yoo and James 2002). Sidhu and Sharma (2010) reported a lower abundance of Zn with increased clay content in the soils. The usable Zn associated negatively with electrical conductivity (Chattopadhyay et al. 1996).

7.3 Plant Growth Promotion Characteristics of Zn-Solubilizing PGPR

Traditional exogenous application of Zn amendments partly meets the plant requirements as most of the available Zn is converted into insoluble forms depending on the soil type and reactions involved (Saravanan et al. 2004). Plant growth-promoting rhizobacteria (PGPR) have the potential to accomplish the necessity of plants with Zn via solubilization of fixed Zn in the soil through various mechanisms. In this regard, various PGPR genera such as *Bacillus* and *Pseudomonas* are known to solubilize the fixed Zn by secreting chelating ligands, protons, and performing redox reactions on cell membranes (Mumtaz et al. 2017; Zeb et al. 2018; Hussain et al. 2020). PGPR have several other advantageous properties toward plants, such as the synthesis of cyanide hydrogen, siderophores, antifungal substances, enzymes, antibiotics, and phytohormone (Rodriguez and Fraga 1999; Sarfraz et al. 2019).

Three decades earlier, the concept of plant growth-promoting rhizobacteria (PGPR) was developed. These bacteria were nonpathogenic, isolated from the rhizosphere of crop plants, colonized plant roots rapidly, and enhanced crop yield through various direct and indirect mechanisms (Akhtar et al. 2012; Agbodjato et al. 2016). Several species of bacteria are present both on the surface of the plant roots in the rhizosphere and within plant roots as endophytes (Desai et al. 2012). These can migrate from the soil's bulk into the rhizosphere of living plants and colonize belligerently in the zone of the plant's rhizosphere (Islam et al. 2014). According to research, numerous strains of bacteria in soil could survive within the plant rhizosphere but that can flourish throughout in tissues of the plant and promote the growth of the plant through a variety of mechanisms that are generally referred to as PGPR (Usha Rani and Reddy 2012).

Research findings earlier revealed that the associations of PGPR differ in the degree of bacterial specificity to the association's source and familiarity. Overall, these PGPR may be intracellular (iPGPR), present in the cells of the root, particularly in the areas of nodular and extracellular (ePGPR), occurring on the rhizoplane, in the rhizosphere, or the spaces between the cortex of cells in the root (Gopalakrishnan et al. 2014). PGPR are naturally occurring beneficial soil microbes that provide essential micronutrients to the plants through atmospheric nitrogen fixation; solubilization of soil-fixed nutrients; production of phytohormones such as kinetin, IAA, and GA; and the production of ACC deaminase that improves to regulate the production of ethylene or enzymes such as chitinase and cellulase produced under abiotic stress conditions (Siddiqui and Shaukat 2004; Saleem et al. 2007; Ahmad et al. 2020; Sabir et al. 2020; Ullah et al. 2020a, b). PGPR induce stimulatory effects on plant growth via nutrient antibiosis, competition, parasitism, induced systemic resistance (ISR), and development of metabolite (hydrogen cyanide, siderophore) suppressing detrimental microbial populations. Such processes are essentially helpful in plant growth.

PGPR also play a major function in the dissolution of phosphates and the bioavailability of soil phosphorus, potassium, iron, and silicate to the roots of the plant (Abaid-Ullah et al. 2011; Ditta et al. 2015, 2018a, b; Ditta and Khalid 2016). Several studies have confirmed that the inoculation of an active strain of rhizobacteria capable of mobilizing Zn improved yields of field crops such as maize, barley, corn, and wheat. Moreover, Zn-mobilizing PGPR significantly alleviated the Zn deficiency symptoms and caused a significant increase in overall biomass and grain yield (Tariq et al. 2007).

Earlier, it was revealed that *Bacillus* sp. (Zn-solubilizing bacteria) can be used as a Zn biofertilizer in soils where natural Zn is elevated or mixed with insoluble cheaper Zn compounds such as Zn sulfide (ZnS), Zn oxide (ZnO), and Zn carbonate ($ZnCO_3$) as an alternative for expensive sulfates of Zn (Mahdi et al. 2010a, b). Consistent research on PGPR has suggested that some effective strains are tethering as well as characteristics of PGPR are continuously distributed among numerous genera and species of microorganisms, several of which are indigenous species of the bacterial communities in the soil. Inclusively, the performance of specific strains varies greatly. Indigenous PGPR may have a comparable effect on the efficiency of introduced inoculants of PGPR. Consequently, awareness and details about the context of PGPR and its role are crucial otherwise; the reaction of inoculants in soil with various PGPR is difficult to estimate.

PGPR in a wide range also take part in the solubilization of nutrients such as iron, silicate, Zn, and phosphorus and generate auxins that promote the growth of root and generate siderophores and antibiotics, which might mitigate infections in the root. Bacteria capable of solubilizing Zn, soil-borne, colonize into the area of the rhizosphere, reproduce, and interact with other rhizobacteria and thus increase the growth and yield of the plant (Kloepper and Okon 1994). Glick (2012) described the application of PGPR stimulates plant growth by releasing phytohormones, solubilization and nutrient acquisition assistance, and biocontrol agents to defend crops against various pathogens. It has been found that the various PGPR are very efficient in Zn solubilization. These powerful PGPR facilitate the advancement of plant growth by colonizing the rhizosphere and by solubilizing the insoluble complex Zn compounds into simpler ones, making the crops very easy to access. In the situation of environmental stress, reactive oxygen species, hydrogen cyanide, and ethylene are produced by plants, which may also be minimized compounds (enzymes) secreted by PGPR in the soil environment.

7.4 Mechanisms of Zn Solubilization Employed by Zn-Solubilizing PGPR

Microbial comparative and functional genomics analysis has opened new avenues for genetic and biochemical approaches to these fundamental processes. Numerous experiments were undertaken to investigate the pathways of Zn-solubilizing PGPR. PGPR have potential mechanisms for nutrient solubilization in the soil, including organic acid generation, chelation, acidification, and exchange reactions (Chung et al. 2005; Hafeez et al. 2005). PGPR employ certain mechanisms to mobilize Zn from soil to the crop plants, which include the secretion of siderophores, gluconate or the derivatives of gluconic acids (5-keto-gluconic acid and 2-keto-gluconic acid), and several others (Di Simine et al. 1998; Fasim et al. 2002; Saravanan et al. 2007a, b, 2011; Tariq et al. 2007; Wani et al. 2007). Acidification is the most favored mechanism employed by the microbes for Zn solubilization. The mechanisms employed by Zn-solubilizing bacteria for Zn solubilization/mobilization are summarized in Fig. 7.1.

7.5 Role of PGPR as Zn Mobilizers

PGPR play a significant role in sustainable agriculture by promoting plant growth. These are a unique community of microorganisms found on the root surface and in conjunction with roots in the rhizosphere (Maheshwari et al. 2012; Zeb et al. 2018; Hussain et al. 2020). The impact of different Zn-solubilizing bacteria on Zn biofortification in different crop plants is summarized in Table 7.1. Such microbes migrate from the surrounding soil to a rhizosphere of plant roots and colonize insensitively toward the region of rhizosphere and the crop roots (Hafeez et al. 2005; Khan 2005). To resolve Zn deficiency, soil microorganisms which can mobilize unavailable Zn, improve the assimilation of Zn, and foster growth in the plants, can be utilized (Singh et al. 2017a, b).

Tariq et al. (2007) found symptoms of Zn deficiency were significantly reduced by the application of PGPR capable of mobilizing Zn along with continuous

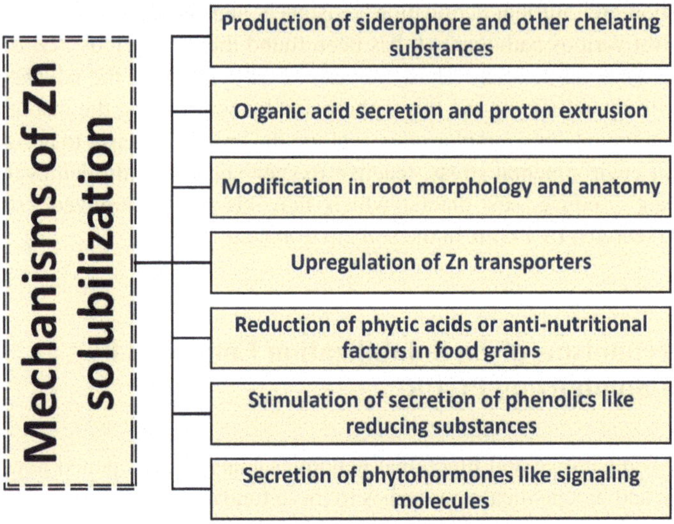

Fig. 7.1 The mechanisms employed by Zn-solubilizing bacteria for Zn solubilization/mobilization

improvement in yield of grain, total biomass, and harvest index as well as an increase in the concentration of Zn in grains of rice. Ahmad (2007) isolated 50 strains of PGPR capable of mobilizing the Zn in the rhizosphere of maize and were very successful strains based on simple, transparent zone forming on the respective petri-plates. Similarly, Yasmin (2011) also confirmed the potential of *Pseudomonas* sp. Z5 in solubilizing Zn, isolated from the rhizosphere of rice. Abaid-Ullah et al. (2011) qualitatively and quantitatively picked nine out of 50 PGPR capable of mobilizing Zn on different insoluble Zn ores such as $Zn(PO_4)_3$, $Zn(CO_3)_2$, ZnS, and ZnO. A strong association between the qualitative and quantitative research in *Serratia* sp. was found in Zn solubilization. Likewise, greater Zn solubilization was identified with ZnO comparison with other insoluble ores. *Serratia* sp. capable of mobilizing Zn was verified for in vivo beneficial effects that maximized the production and quality qualities of wheat crops.

PGPR play an important role in the solubilization of many essential elements, such as phosphorus, magnesium, Zn, and potassium, while increasing the bioavailability of these critical nutrients to crops (Glick 1995). Penrose and Glick (2003) reported that PGPR improved plant growth by enhancing the solubilization of nutrient and triggering hormones and siderophore production, contributing to increased uptake of nutrients by crops (Fig. 7.1). Several studies have reported the potential of various PGPR strains such as *Bacillus aryabhattai*, *Bacillus* sp., *Pseudomonas aeruginosa*, *Pseudomonas fluorescens*, *Pseudomonas striata*, *Rhizobium*, *Gluconacetobacter diazotrophicus*, *Serratia liquefaciens*, *Burkholderia cenocepacia*, *S. marcescens,* and *Bacillus thuringiensis* in improving growth and Zn content of various crops (Saravanan et al. 2007c; Joshi et al. 2013; Ramesh et al. 2014; Hussain et al. 2015; Naz et al. 2016, b). Gadd (2007) reported that microorganisms

Table 7.1 Impact of different Zn-solubilizing bacteria on Zn biofortification in different crop plants

Zn-solubilizing bacteria	Crop	Experimental conditions	Effects	References
Bacillus megaterium	*Capsicum annuum* L.	Pot experiment	Improved plant growth and Zn contents	Bhatt and Maheshwari (2020)
Bacillus sp. AZ6	Maize	Field experiment	Increased the Zn contents in grain and shoot by 46 and 52%, respectively, while reduced the phytate contents by 73% as compared to control	Hussain et al. (2020)
Bacillus strains	Rice	Axenic conditions	Improved rice growth, hydrolyzed starch, and can produce siderophores	Naseer et al. (2020)
Enterobacter sp. MN17	Desi Chickpea	Field experiment	Zn soil application with PGPB proved economical	Ullah et al. (2020a, b)
Pseudomonas sp., *P. putida* and *P. fluorescens*	Maize	Pot experiments	Increased Zn contents of roots and shoots	Shafgh et al. (2019)
Bacillus sp. strain AZ17 and *Pseudomonas* sp. strain AZ5	Chickpea	Field experiment	Strain AZ5 improved growth, yield, and Zn uptake compared to control	Zaheer et al. (2019)
B. megaterium, (ZnSB2)	Turmeric	Greenhouse experiment	ZnSB2 enhanced Zn dissolution in soil	Dinesh et al. (2018)
Bacillus sp. ZM20 and *Bacillus aryabhattai* ZM31	Okra (*Abelmoschus esculentus* L.)	Pot experiment	Improved growth, yield, and biochemical parameters	Fatima et al. (2018)
Bacillus strains	Soybean and wheat	Microcosm experiment	Modulated growth, yield, and Zn biofortification	Khande et al. (2017)
Pseudomonas fragi, *Pantoea dispersa*, and *Pantoea agglomerans*	Wheat	Pot experiment	Increased plant growth and Zn contents	Kamran et al. (2017)
Bacillus spp.	Maize	Jar trial	Promoted growth and Zn biofortification in maize	Mumtaz et al. (2017)
Acinetobacter sp. and *Serratia* sp.	Rice	Growth chamber under sand culture conditions	Inoculation significantly improved growth and Zn nutrition	Othman et al. (2017)

(continued)

Table 7.1 (continued)

Zn-solubilizing bacteria	Crop	Experimental conditions	Effects	References
Exiguobacterium aurantiacum	*Triticum aestivum*	Field experiment	Improved micronutrient (Zn and Fe) accumulation	Shaikh and Saraf (2017)
Acinetobacter sp.	Rice	Pot experiment	Improved available Zn compared to control	Gandhi and Muralidharan (2016)
Enterobacter cloacae	Rice	Culture media	Upregulated OsZIP1 and OsZIP5 expressions	Krithika and Balachandar (2016)
Rhizobium, Azospirillum, and *Pseudomonas*	Wheat	Field experiment	Increased Zn contents in wheat	Naz et al. (2016, b)
Bacillus sp. AZ6	Maize	Growth room	Improved growth and physiology of plants	Hussain et al. (2015)
Bacillus sp. and *Bacillus cereus*	Rice	Pot experiment	Acted as biocontrol agents, enhanced Zn translocation in grains, and improved yield	Shakeel et al. (2015, b)
Bradyrhizobium japonicum	Cowpea	Field and screen house experiment	Phosphorus application improved micronutrients uptake but a decrease in Zn content was observed in few organs	Nyoki and Ndakidemi (2014)
Bacillus aryabhattai	Soybean and wheat	Microcosm	Increased growth, yield, and Zn contents in seeds	Ramesh et al. (2014)
Burkholderia and *Acinetobacter*	Rice	Greenhouse	Enhanced growth, yield, and total Zn uptake	Vaid et al. (2014)
Sphingomonas sp., Enterobacter sp.	Rice	Greenhouse	Enhanced Zn bioavailability and improved yield	Wang et al. (2014)
Pseudomonas strains	Wheat	Pot experiment	Increased grain Zn concentration (31%)	Joshi et al. (2013)
Pseudomonas spp. and *Bacillus* spp.	Maize	Pot experiment	Increased the total dry mass and uptake of N, K, Mn, and Zn	Goteti et al. (2013)
Providencia sp., Anabaena sp., Calothrix sp., and *Anabaena sp.*	Wheat	Field	Improved protein, Fe, Cu, Zn, and Mn contents	Rana et al. (2012)
Bacillus isolates	Soybean	Microcosm experiment	Increase Zn accumulation in seeds	Sharma et al. (2012)
Zn-solubilizing bacterial isolates	Mung bean	Pot experiment	Improved growth and yield	Iqbal et al. (2010)

(continued)

Table 7.1 (continued)

Zn-solubilizing bacteria	Crop	Experimental conditions	Effects	References
Zn-mobilizing PGPR	Wheat	Greenhouse experiment	Enhanced shoot and root biomass	Kutman et al. (2010)
Pseudomonas strain and *Bacillus* strain	Strawberry	Field experiment	Increased fruit yield per plant	Esitken et al. (2010)
P. aeruginosa	Rye	Pot experiment	Increase in root and shoot growth	Shahab et al. (2009)
Azotobacter, *Azospirillum*	Corn	Greenhouse experiment	Increased Zn contents of grain	Biari et al. (2008)
A. lipoferum, *Pseudomonas sp.*, *Agrobacterium sp.*	Rice	Field experiment	Improved growth, physiology, yield, and Zn biofortification in rice	Hafeez et al. (2002) and Tariq et al. (2007)

such as *Pseudomonas*, *Bacillus*, *Acinetobacter*, and *Gluconacetobacter* possess the potential to solubilize fixed Zn in soil and could be used to increase Zn bioavailability to the crop plants.

Inoculants of PGPR having the ability to mobilize Zn are applied as biofertilizers that can restore degraded land and increase soil fertility status. Such PGPR increase survival and growth in the plant, optimize production of grain, diminish rates of malnutrition, and reduce chemical fertilizer dependency (Hafeez et al. 2001). The use of Zn-SB along with some other chemical fertilizers would be of crucial benefit in formulating efficient biofertilizers (Gull et al. 2004). According to Ramesh et al. (2014), the application of bacteria capable of solubilizing Zn affiliated with genera of *Bacillus* have the potential to improve plant growth. The efficacy of Zn-mobilizing PGPR has a positive effect on fresh and dry biomass of root and shoot, surface area, and width of the root, and index of panicle emergence. Some PGPR work in the symbiotic relationship of plant-microbes. Similarly, in another study, He et al. (2010) noted that genera of *Bacillus*, Zn-mobilizing bacteria, when employed as inoculant enhances growth parameters (Zhao et al. 2011). Comparable rises are reported through PGPR inoculation in Zn absorption and dry matter production (Rana et al. 2012). Supportive utilization in high-quality microbial interactions is required and it would ultimately improve the promising result of their application under the field conditions (Usha Rani and Reddy 2012).

7.6 Conclusions and Future Perspectives

The existence of micronutrients at low levels is commonly referred to as "hidden hunger" and attracts less attention than people's visible hunger. It is obvious that the implementation of pesticides, chemical fertilizers, farming techniques, and transgenic plant production possess the capacity to improve the Zn biofortification in agricultural production, but these strategies have exceptionally high prices, cause degradation of the atmosphere, and pose various socio-political and economic challenges. In this regard, Zn-solubilizing rhizobacteria being economical and eco-friendly nature have the potential to alleviate Zn deficiency and ultimately lead us toward sustainable agriculture. It is also inferred that integrated nutrient management (INM) involving effective biofertilizers with chemical fertilizers is a safer option for farmers to minimize the use of chemical fertilizers and to grow healthier crops. With the identification of new strains of bacteria, which are abundantly successful in plant growth and yield, the use of various microbial technologies in agriculture currently grows very quickly and popularly.

For sustainable agriculture, using powerful Zn-SB strains as an inoculant would be beneficial for the promotion of plant growth, soil health, and soil fertility. Prospective research thus needs to be conducted on boosting micronutrient bioavailability via inoculating seeds with microbes or their products to diminish phytic acid in grains, sustaining and examining existing native wild cereal germplasm to determine the innovative plant-microbiome combos liable to augmentation of micronutrients. More research to understand the interactions among plants and microbes is essential. New challenges would require technological convergence to provide an increasing population with healthy nutrition utilizing safe and environmentally friendly innovations. PGPR that serve multipurpose tasks such as solubilization of P, K, and Zn tend to be efficient biofertilizers. Genotypic analysis of the strains of Zn-SB and plant's molecular characterization is important to understand the processes in plants toward the absorption of Zn and their requirements in plants. Quest for new effective Zn-SB strains as biofertilizers to enhance microbial diversity for either area is required in the future. Co-inoculation is being developed for other synergistically advantageous bacterial species, and several recent studies indicate a positive development in inoculation technology. This approach can also be employed for the Zn biofortification of crops. The researchers have to answer how to monitor factors of nutritional and root exudation in ways that benefit even more from the application of co-inoculation.

References

Abaid-Ullah M, Hassan MN, Nawaz MK, Hafeez FY (2011) Biofortification of wheat (*Triticum aestivum* L.) through Zn mobilizing PGPR. In: Proceedings of international science conference "prospects and challenges to sustainable agriculture". Azad Jammu and Kashmir University, Muzaffarabad, p 298

Agbodjato NA, Noumavo PA, Adjanohoun A, Agbessi L, Baba-moussa L (2016) Synergistic effects of plant growth promoting rhizobacteria and chitosan on in vitro seeds germination, greenhouse growth, and nutrient uptake of maize (*Zea mays* L). Hindawi Publishing Corporation. Biotechnol Res Int 11:7830182

Ahmad M (2007) Biochemical and molecular basis of phosphate and zinc mobilization by PGPR in rice. M. Phil Dissertation, NIBGE, Faisalabad

Ahmad HT, Hussain A, Aimen A, Jamshaid MU, Ditta A, Asghar HN, Zahir ZA (2020) Improving resilience against drought stress among crop plants through inoculation of plant growth-promoting rhizobacteria. In: Husen A, Jawaid M (eds) Resilient environment and plant potential. Springer, New York. (In Press)

Akhtar A, Hisamuddin RMI, Abbasi SR (2012) Plant growth promoting rhizobacteria: an overview. J Nat Prod Plant Resour 2(1):19–31

Alexander M (1997) Introduction to soil microbiology. Wiley, New York

Ali A, Bilal K, Ahmad AD (2018) How to sustain agricultural land for safe food production? Acta Sci Agric 2(10):131–133

Alloway BJ (1995) Heavy metals in soils, 2nd edn. Blackie Academic & Professional, London

Alloway BJ (2008) Zinc in soils and crop nutrition. International Fertilizer Association, Paris and International Zinc Association, Brussels, p 135

Alloway BJ (2009) Soil factors associated with zinc deficiency in crops and humans. Environ Geochem Health 31:537–548

Bhatt K, Maheshwari DK (2020) Zinc solubilizing bacteria (*Bacillus megaterium*) with multifarious plant growth promoting activities alleviates growth in *Capsicum annuum* L. 3 Biotech 10:36. https://doi.org/10.1007/s13205-019-2033-9

Biari A, Gholami A, Rahmani HA (2008) Growth promotion and enhanced nutrient uptake of maize (*Zea mays* L.) by application of plant growth promoting rhizobacteria in arid region of Iran. J Biol Sci 8:1015–1020

Broadley MR, White PJ, Hammond JP, Zelko I, Lux A (2007) Zinc in plants. New Phytol 173(4):677–702

Cakmak I, Pfeiffer WH, McClafferty B (2010) Biofortification of durum wheat with zinc and iron. Cereal Chem 87:10–20

Chang HB, Lin CW, Huang HJ (2005) Zinc-induced cell death in rice (*Oryza sativa* L.) roots. Plant Growth Regul 46:261–266

Chattopadhyay T, Sahoo AK, Singh RS, Shyampura RL (1996) Available micronutrient status in the soils of Vindhyan scarplands of Rajasthan in relation to soil characteristics. J Indian Soc Soil Sci 44(4):678–681

Chen B, Shen J, Zhang X, Pan F, Yang X, Feng Y (2014) The endophytic bacterium, *Sphingomonas* SaMR12, improves the potential for zinc phytoremediation by its host, *Sedum alfredii*. PLoS One 9(9):e106826. https://doi.org/10.1371/journal.pone.0106826

Chirwa M, Yerokun OA (2012) The distribution of zinc fractions in surface samples of selected agricultural soils of Zambia. Int J Soil Sci 7:51–60. https://doi.org/10.3923/ijss.2012.51.60

Chung H, Park M, Madhaiyan M, Seshadri S, Song J, Cho H (2005) Isolation and characterization of phosphate solubilizing bacteria from the rhizosphere of crop plants of Korea. Soil Biol Biochem 37:1970–1974

Desai S, Praveen Kumar G, Sultana U (2012) Potential microbial candidate strains for management of nutrient requirements of crops. Afr J Microbiol Res 6(17):3924–3931

Di Simine CD, Sayer JA, Gadd GM (1998) Solubilization of zinc phosphate by a strain of Pseudomonas fluorescens isolated from forest soil. Biol Fertil Soils 28(1):87–94

Di Tomaso JM (1995) Approaches for improving crop competitiveness through the manipulation of fertilization strategies. Weed Sci 43:491–497. https://doi.org/10.1017/S0043174500081522

Dinesh R, Srinivasan V, Hamza S, Sarathambal C, Gowda SJA, Ganeshamurthy AN, Gupta SB, Nair VA, Subila KP, Lijina A, Divya VC (2018) Isolation and characterization of potential Zn solubilizing bacteria from soil and its effects on soil Zn release rates, soil available Zn and plant Zn content. Geoderma 321:173–186. https://doi.org/10.1016/j.geoderma.2018.02.013

Ditta A, Khalid A (2016) Bio-organo-phos: a sustainable approach for managing phosphorus deficiency in agricultural soils. In: Larramendy M, Soloneski S (eds) Organic fertilizers - from basic concepts to applied outcomes. InTech, Rijeka, pp 109–136. https://doi.org/10.5772/62473

Ditta A, Arshad M, Zahir ZA, Jamil A (2015) Comparative efficacy of rock phosphate enriched organic fertilizer vs. mineral phosphatic fertilizer for nodulation, growth and yield of lentil. Int J Agric Biol 17:589–595

Ditta A, Imtiaz M, Mehmood S, Rizwan MS, Mubeen F, Aziz O, Qian Z, Ijaz R, Tu S (2018a) Rock phosphate enriched organic fertilizer with phosphate solubilizing microorganisms improves nodulation, growth and yield of legumes. Commun Soil Sci Plant Anal 49(21):2715–2725. https://doi.org/10.1080/00103624.2018.1538374

Ditta A, Muhammad J, Imtiaz M, Mehmood S, Qian Z, Tu S (2018b) Application of rock phosphate enriched composts increases nodulation, growth and yield of chickpea. Int J Recycl Org Waste Agric 7(1):33–40. https://doi.org/10.1007/s40093-017-0187-1

Elkoca E, Turan M, Donmez MF (2010) Effects of single, dual and triple inoculation with *Bacillus subtilis*, *Bacillus megaterium* and *Rhizobium leguminosarum* bv. phaseoil on nodulation, nutrient uptake, yield and yield parameters of common bean (*Phaseolus vulgaris* L. cv. 'Elkoca-05'). J Plant Nutr 33:2104–2119. https://doi.org/10.1080/01904167.2010.519084

Esitken A, Yildiz HE, Ercisli S, Donmez MF, Turan M, Gunes A (2010) Effects of plant growth promoting bacteria (PGPB) on yield, growth and nutrient contents of organically grown strawberry. Sci Hortic 124(1):62–66

FAO WHO (2002) Human vitamin and mineral requirements. Food and Agriculture Organization of the United Nations, Bangkok. ISBN 1014–9228

Fasim F, Ahmed N, Parsons R, Gadd GM (2002) Solubilization of zinc salts by a bacterium isolated from the air environment of a tannery. FEMS Microbiol Lett 213:1–6

Fatima I, Jamil M, Hussain A, Mumtaz MZ, Luqman M, Hussain S, Kashif SUR, Ahmad M (2018) Zinc solubilizing *Bacillus* sp. ZM20 and *Bacillus aryabhattai* ZM31 promoted the productivity in Okra (*Abelmoschus esculentus* L.). Biologia (Pakistan) 64(II):179–185

Foley JA, DeFries R, Asner GP, Barford C, Bonan G, Carpenter SR, Helkowski JH (2005) Global consequences of land use. Science 309(5734):570–574

Gadd GM (2007) Geomycology: biogeochemical transformations of rocks, minerals, metals and radionuclides by fungi, bioweathering and bioremediation. Mycol Res 111:3–49

Gandhi A, Muralidharan G (2016) Assessment of zinc solubilizing potentiality of Acinetobacter sp. isolated from rice rhizosphere. Eur J Soil Biol 76:1–8. https://doi.org/10.1016/j.ejsobi.2016.06.006

Garcia C, Maria N, Juan PPR, Helena P, Luz MDR, Elizabeth CT, Monica CFU (2016) Staple crops biofortified with increased micronutrient content: effects on vitamin and mineral status, as well as health and cognitive function in the general population. Cochrane Database Syst Rev 8. https://doi.org/10.1002/14651858.CD012311

Giehl RFH, Meda AR, Wiren N (2009) Moving up, down, and everywhere: signaling of micronutrients in plants. Curr Opin Plant Biol 12:320–327

Glick BR (1995) The enhancement of plant growth by free-living bacteria. Can J Microbiol 41(2):109–117

Glick B (2012) Plant growth-promoting bacteria: mechanisms and applications. Hindawi Publishing Corporation, New York

Gliessman SR (2014) Agroecology: the ecology of sustainable food systems. CRC Press, Boca Raton

Gopalakrishnan S, Sathya A, Vijayabharathi R, Varshney RK, Gowda CLL (2014) Plant growth promoting rhizobia: challenges and opportunities. Biotech 5(4):355–377

Goteti PK, Daniel L, Emmanuel A, Desai S, Hassan M (2013) Prospective zinc solubilizing bacteria for enhanced nutrient uptake and growth promotion in maize (*Zea mays* L.). Int J Microbiol 2013:1

Gull M, Hafeez FY, Saleem M, Malik KA (2004) Phosphate-uptake and growth promotion of chickpea (*Cicer arietinum* L.) by co-inoculation of mineral phosphate solubilizing bacteria and mixed rhizobial culture. Aust J Exp Agric 44(6):623–628

Gupta AP (2005) Micronutrient status and fertilizer use scenario in India. J Trace Elem Med Biol 18:325–331

Hacisalihoglu G, Kochian LV (2003) How do some plants tolerate low levels of soil zinc? Mechanisms of zinc efficiency in crop plants. New Phytol 159:341–350

Hafeez FY, Hameed S, Ahmad T, Malik KA (2001) Competition between effective and less effective strains of *Bradyrhizobium* sp. for nodulation on *Vigna radiata*. Biol Fertil Soils 33:382–386

Hafeez FY, Hameed S, Zaidi AH, Malik KA (2002) Biofertilizer for sustainable agriculture. In: Azam F, Iqbal MM, Inayatullah C, Malik KA (eds) Techniques for sustainable agriculture. ISBN/NIAB, Faisalabad, pp 67–73

Hafeez FY, Naeem FI, Naeem R, Zaidi AH, Malik KA (2005) Symbiotic effectiveness and bacteriocin production by *Rhizobium leguminosarum* bv. viciae isolated from agriculture soils in Faisalabad. Environ Exp Bot 54:142–147

Hafeez FY, Abaid-Ullah M, Hassan MN (2013) Plant growth-promoting rhizobacteria as zinc mobilizers: a promising approach for cereals biofortification. In: Bacteria in agrobiology: crop productivity. Springer, Berlin, pp 217–235

Hamid A, Ahmad N (2001) Paper at a regional workshop on integrated plant nutrition system (IPNS): development and rural poverty alleviation, Bangkok, pp 18–21

Harter RD (1983) Effect of soil pH on adsorption of lead, copper, zinc, and nickel. Soil Sci Soc Am J 47:47–51. https://doi.org/10.2136/sssaj1983.03615995004700010009x

HarvestPlus (2012) Breeding crops for better nutrition. Web page of HarvestPlus. International Food Policy Research Institute, Washington, DC. http://www.harvestplus.org/content/zinc-wheat

Havlin JL, Beaton JD, Tisdale SL, Nelson WL (2005) Soil fertility and fertilizers. Upper Saddle River, Pearson, p 515

He CQ, Tan GE, Liang X, Du W, Chen YL, Zhi GY (2010) Effect of Zn-tolerant bacterial strains on growth and Zn accumulation in *Orychophragmus violaceus*. Appl Soil Ecol 44:1–5

Hotz C, Brown KH (2004) Assessment of the risk of zinc deficiency in populations and options for its control. Food Nutr Bull 25:S91–S204

Hussain A, Arshad M, Zahir ZA, Asghar M (2015) Prospects of zinc solubilizing bacteria for enhancing the growth of maize. Pak J Agric Sci 52:915–922

Hussain A, Zahir ZA, Ditta A, Tahir MU, Ahmad M, Mumtaz MZ, Hayat K, Hussain S (2020) Production and implication of bio-activated organic fertilizer enriched with zinc-solubilizing bacteria to boost up maize (Zea mays L.) production and biofortification under two cropping seasons. Agron J 10(39). https://doi.org/10.3390/agronomy10010039

Ibrahim AK, Usman A, Abubakar B, Aminu UH (2011) Extractable micronutrients status in relation to other soil properties in Billiri local government area. J Soil Sci Environ Manag 3(10):282–285

Iqbal U, Jamil N, Ali I, Hasnain S (2010) Effect of zinc-phosphate-solubilizing bacterial isolates on the growth of Vigna radiata. Ann Microbiol 60:243. https://doi.org/10.1007/s13213-010-0033-4

Islam F, Yasmeen T, Ali Q, Ali S, Arif MS (2014) Influence of *Pseudomonas aeruginosa* as PGPR on oxidative stress tolerance in wheat under Zn stress. Ecotoxicol Environ Saf 104(1):285–293

Jones DL, Darrah PR (1994) Role of root-derived organic acids in the mobilization of nutrients from the rhizosphere. Plant Soil 166:247–257

Joshi D, Geeta N, Sachin V, Anita S (2013) Enhancement of wheat growth and Zn content in grains by zinc solubilizing bacteria. Int J Agric Environ Biotechnol 6:363–370

Kamran S, Shahid I, Baig DN, Rizwan M, Malik KA, Mehnaz S (2017) Contribution of zinc solubilizing bacteria in growth promotion and zinc content of wheat. Front Microbiol 8:2593

Kauser MA, Hussain F, Ali S, Iqbal MM (2001) Zinc and Cu nutrition of two wheat varieties on calcareous soil. Pak J Soil Sci 20:21–26

Khan AG (2005) Role of soil microbes in the rhizospheres of plants growing on trace metal contaminated soils in phytoremediation. J Trace Elem Med Biol 18:355–364

Khande R, Sushil KS, Ramesh A, Mahaveer PS (2017) Zinc solubilizing Bacillus strains that modulate growth, yield and zinc biofortification of soybean and wheat. Rhizosphere 4:126–138

King JC (2006) Zinc. In: Shils ME, Shike M (eds) Modern nutrition in health and disease, 10th edn. Lippincott Williams & Wilkins, Philadelphia, pp 271–285

Kloepper JW, Okon Y (1994) Plant growth-promoting rhizobacteria (other systems). In: Okon Y (ed) *Azospirillum*/plant associations. CRC Press, Boca Raton, pp 111–118

Krithika S, Balachandar D (2016) Expression of zinc transporter genes in rice as influenced by zinc-solubilizing Enterobacter cloacae strain ZSB14. Front Plant Sci 7:446. https://doi.org/10.3389/fpls.2016.00446

Kumssa DB, Joy EJ, Ander EL, Watts MJ, Young SD, Walker S, Broadley MR (2015) Dietary calcium and zinc deficiency risks are decreasing but remain prevalent. Sci Rep 5:10974

Kutman UB, Yildiz B, Ozturk L, Cakmak I (2010) Biofortification of durum wheat with zinc through the soil and foliar applications of nitrogen. Cereal Chem 87(1):1–9

Lockhart K, King A, Harter T (2013) Identifying sources of groundwater nitrate contamination in a large alluvial groundwater basin with highly diversified intensive agricultural production. J Contam Hydrol 151:140–154. https://doi.org/10.1016/j.jconhyd.2013.05.008

Lonnerdal IB (2000) Dietary factors influencing zinc absorption. J Nutr 130:1378–1387

Mader P, Kaiser F, Adholeya A, Singh R, Uppal HS, Sharma AK, Srivastava R, Sahai V, Aragno M, Wiemken A (2011) Inoculation of root microorganisms for sustainable wheat–rice and wheat–black gram rotations in India. Soil Biol Biochem 43(3):609–619. https://doi.org/10.1016/j.soilbio.2010.11.031

Mahdi SS, Hassan GI, Samoon SA, Rather HA, Dar SA, Zehra B (2010a) Bio-fertilizers in organic agriculture. J Phytopathol 2(10):42–54

Mahdi SS, Dar SA, Ahmad S, Hassan GI (2010b) Zinc availability – a major issue in agriculture. Res J Agric Sci 3(3):78–79

Maheshwari DK, Kumar S, Maheshwari NK, Patel D, Saraf M (2012) Nutrient availability and management in the rhizosphere by microorganisms. In: Maheshwari DK (ed) Bacteria in agrobiology: stress management. Springer, Berlin, pp 301–326

Mattey M (1992) The production of organic acids. Crit Rev Biotechnol 12:87–132

Mumtaz MZ, Ahmad M, Jamil M, Hussain T (2017) Zinc solubilizing *Bacillus* spp. potential candidates for biofortification in maize. Microbiol Res 202:51–60. https://doi.org/10.1016/j.micres.2017.06.001

Naseer I, Ahmad M, Hussain A, Jamil M (2020) Potential of zinc solubilizing bacillus strains to improve rice growth under axenic conditions. Pak J Agric Sci 57(4):1057–1071

Naz I, Ahmad H, Khokhar SN, Khan K, Shah AH (2016) Impact of zinc solubilizing bacteria on zinc contents of wheat. Am Euras J Agric Environ Sci 16:449–454

Nilson A, Piza J (1998) Food fortification: a tool for fighting hidden hunger. Food Nutr Bull 19(1):49–60. https://doi.org/10.1177/156482659801900109

Nyoki D, Ndakidemi PA (2014) Effects of phosphorus and Bradyrhizobium japonicum on growth and chlorophyll content of cowpea (*Vigna unguiculata* (L) Walp). Am J Exp Agric 4(10):1120

Othman NMI, Othman R, Saud HM, Wahab PEM (2017) Effects of root colonization by zinc-solubilizing bacteria on rice plant (Oryza sativa MR219) growth. Agric Nat Resour 51:532–537. https://doi.org/10.1016/j.anres.2018.05.004

Pahlvan-Rad MR, Pressaraki M (2009) Response of wheat plant to zinc, iron and manganese applications and uptake and concentration of zinc, iron, and manganese in wheat grains. Commun Soil Sci Plant Anal 40:1322–1332

Pawar A, Ismail S, Mundhe S, Patil VD (2015) Solubilization of insoluble zinc compounds by different microbial isolates in vitro condition. Int J Trop Agric 33:865–869

Penrose DM, Glick BR (2003) Methods for isolating and characterizing ACC deaminase containing plant growth-promoting rhizobacteria. Physiol Plant 118(1):10–15

Prasad R (2005) Rice-wheat cropping system. Adv Agron 86:255–339

Prasanna R, Nain L, Rana A, Shivay YS (2016) Biofortification of crop plants using microorganisms: present status and challenges. In: Singh U et al (eds) Biofortification of food crops. Springer, New Delhi, pp 249–262. https://doi.org/10.1007/978-81-322-2716-8_19

Ramesh A, Sharma SK, Sharma MP, Yadav N, Joshi OP (2014) Inoculation of zinc solubilizing *Bacillus aryabhattai* strains for improved growth, mobilization, and biofortification of zinc in soybean and wheat cultivated in Vertisols of Central India. Appl Soil Ecol 73:87–96

Ramzan S, Bhat MA, Kirmani NA, Rasool R (2014) Fractionation of zinc and their association with soil properties in soils of Kashmir Himalayas. Ind J Agric Soil Sci 2:132–142

Rana A, Joshi M, Prasanna R, Shivay YS, Nain L (2012) Biofortification of wheat through inoculation of plant growth promoting rhizobacteria and cyanobacteria. Eur J Soil Biol 50:118–126. https://doi.org/10.1016/j.ejsobi.2012.01.005

Rengel Z (2001) Genotypic differences in micronutrient use efficiency in crops. Commun Soil Sci Plant Anal 32(7–8):1163–1186. https://doi.org/10.1081/CSS-100104107

Rodriguez H, Fraga R (1999) Phosphate solubilizing bacteria and their role in plant growth promotion. Biotechnol Adv 17:319–339

Sabir A, Naveed M, Bashir MA, Hussain A, Mustafa A, Zahir ZA, Kamran M, Ditta A, Núñez-Delgado A, Saeed Q, Qadeer A (2020) Cadmium mediated phytotoxic impacts in Brassica napus: managing growth, physiological and oxidative disturbances through combined use of biochar and Enterobacter sp. MN17. J Environ Manag 265(110522). https://doi.org/10.1016/j.jenvman.2020.110522

Saleem M, Arshad M, Hussain S, Saeed A (2007) Perspective of plant growth promoting rhizobacteria (PGPR) containing ACC deaminase in stress agriculture. J Ind Microbiol Biotechnol 34(10):635–648

Saravanan VS, Subramoniam SR, Raj SA (2004) Assessing in vitro solubilization potential of different zinc solubilizing bacterial (ZSB) isolates. Braz J Microbiol 35(1–2):121–125

Saravanan VS, Madhaiyan M, Thangaraju M (2007a) Solubilization of zinc compounds by the diazotrophic, plant growth promoting bacterium *Gluconacetobacter diazotrophicus*. Chemosphere 66:1794–1798

Saravanan VS, Osborne J, Madhaiyan M, Mathew L, Chung J, Ahn K, Sa T (2007b) Zinc metal solubilization by *Gluconacetobacter diazotrophicus* and induction of pleomorphic cells. J Microbiol Biotechnol 17(9):1477–1482

Saravanan VS, Kalaiarasan P, Madhaiyan M, Thangaraju M (2007c) Solubilization of insoluble zinc compounds by *Gluconacetobacter diazotrophicus* and the detrimental action of zinc ion (Zn^{2+}) and zinc chelates on root-knot nematode *Meloidogyne incognita*. Lett Appl Microbiol 44(3):235–241

Saravanan VS, Kumar MR, Sa TM (2011) Microbial zinc solubilization and their role on plants. In: Maheshwari DK (ed) Bacteria in agrobiology: plant nutrient management. Springer, Berlin, pp 47–63

Sarfraz R, Hussain A, Sabir A, Fekih IB, Ditta A, Xing S (2019) Role of biochar and plant growth-promoting rhizobacteria to enhance soil carbon sequestration– a review. Environ Monit Assess 191:251. https://doi.org/10.1007/s10661-019-7400-9

Sauer AK, Hagmeyer S, Grabrucker AM (2016) Zinc deficiency. In: Nutritional deficiency. InTech Open, London

Shafgh M, Hamidpour M, Abbaszadeh-Dahaji P, Mozafari V, Furrer G (2019) Bioavailability of Zn from layered double hydroxides: the effects of plant growth-promoting rhizobacteria (PGPR). Appl Clay Sci 182:105283. https://doi.org/10.1016/j.clay.2019.105283

Shahab S, Ahmed N, Khan NS (2009) Indole acetic acid production and enhanced plant growth promotion by indigenous PSBs. Afr J Agric Res 4(11):1312–1316

Shaikh S, Saraf M (2017) Biofortification of Triticum aestivum through the inoculation of zinc solubilizing plant growth promoting rhizobacteria in field experiment. Biocatal Agric Biotechnol 9:120–126. https://doi.org/10.1016/j.bcab.2016.12.008

Shakeel M, RaisA HMN, Hafeez FY (2015) Root associated Bacillus sp. improves growth, yield and zinc translocation for basmati rice (*Oryza sativa*) varieties. Front Microbiol 6:1–7

Sharma SK, Sharma MP, Ramesh A, Joshi OP (2012) Characterization of zinc-solubilizing bacillus isolates and their potential to influence zinc assimilation in soybean seeds. J Microbiol Biotechnol 22(3):352–359. https://doi.org/10.4014/jmb.1106.05063

Siddiqui IA, Shaukat SS (2004) *Trichoderma harzianum* enhances the production of nematicidal compounds in vitro and improves biocontrol of *Meloidogyne javanica* by *Pseudomonas fluorescens* in tomato. Lett Appl Microbiol 38(2):169–175

Sidhu GS, Sharma BD (2010) Diethylene triamine penta acetic acid– extractable micronutrients status in soil under a rice–wheat system and their relationship with soil properties in different agro-climatic zones of indo-Gangetic plains of India. Commun Soil Sci Plant Anal 41:29–51. https://doi.org/10.1080/00103620903360262

Sillanpaa M (1982) Micronutrients and the nutrient status of soils. A global study, FAO soil bulletin no. 48. FAO, Rome

Singh RB (2000) Environmental consequences of agricultural development: a case study from the green revolution state of Haryana, India. Agric Ecosyst Environ 82(1):97–103. https://doi.org/10.1016/S0167-8809(00)00219-X

Singh D, Rajawat MVS, Kaushik R, Prasanna R, Saxena AK (2017a) Beneficial role of endophytes in biofortification of Zn in wheat genotypes varying in nutrient use efficiency grown in soils sufficient and deficient in Zn. Plant Soil 416(1–2):107–116. https://doi.org/10.1007/s11104-017-3189-x

Singh D, Geat N, Rajawat MVS, Mahajan MM, Prasanna R, Singh S, Kaushik R, Singh RN, Kumar K, Saxena AK (2017b) Deciphering the mechanisms of endophyte-mediated biofortification of Fe and Zn in wheat. J Plant Growth Regul 37(1):174–182. https://doi.org/10.1007/s00344-017-9716-4

Singh D, Geat N, Rajawat MVS, Prasanna R, Kar A, Singh AM, Saxena AK (2018) Prospecting endophytes from different Fe or Zn accumulating wheat genotypes for their influence as inoculants on plant growth, yield, and micronutrient content. Ann Microbiol 68(12):815–833. https://doi.org/10.1007/s13213-018-1388-1

Stein AJ (2010) Global impacts of human mineral malnutrition. Plant Soil 335:133–154

Tariq M, Hameed S, Malik KA, Hafeez FY (2007) Plant root-associated bacteria for zinc mobilization in rice. Pak J Bot 39(1):245–253

Ullah A, Farooq M, Nadeem F, Rehman A, Hussain M, Nawaz A, Naveed M (2020a) Zinc application in combination with zinc solubilizing Enterobacter sp. MN17 improved productivity, profitability, zinc efficiency, and quality of desi chickpea. J Soil Sci Plant Nutr. https://doi.org/10.1007/s42729-020-00281-3

Ullah N, Ditta A, Khalid A, Mehmood S, Rizwan MS, Mubeen F, Imtiaz M (2020b) Integrated effect of algal biochar and plant growth promoting rhizobacteria on physiology and growth of maize under deficit irrigations. J Soil Sci Plant Nutr 20:346–356. https://doi.org/10.1007/s42729-019-00112-0

Usha Rani M, Reddy G (2012) Screening of rhizobacteria containing plant growth promoting (PGPR) traits in rhizosphere soils and their role in enhancing the growth of pigeon pea. Afr J Biotechnol 11(32):8085–8091

Vaid SK, Kumar B, Sharma A, Shukla AK, Srivastava PC (2014) Effect of zinc solubilizing bacteria on growth promotion and zinc nutrition of rice. J Soil Sci Plant Nutr 14(4):889–910

Wang T, Liu M-Q, Li H-X (2014) Inoculation of phosphate-solubilizing bacteria *Bacillus thuringiensis* B1 increases available phosphorus and growth of peanut in acidic soil. Acta Agric Scand B Soil Plant Sci 64(3):252–259

Wani PA, Khan MS, Zaidi A (2007) Impact of zinc tolerant plant growth promoting rhizobacteria on lentil grown in zinc-amended soil. Agron Sustain Dev 28:449–455

Wessells KR, Brown KH (2012) Estimating the global prevalence of zinc deficiency: results based on zinc availability in national food supplies and the prevalence of stunting. PLoS One 7(11):e50568. https://doi.org/10.1371/journal.pone.0050568

Wieringa F, Dijkhuizen M, Fiorentino M, Laillou A, Berger J (2015) Determination of zinc status in humans: which indicator should we use? Nutrients 7(5):3252–3263. https://doi.org/10.3390/nu7053252

World Health Organization (2009) Global health risks: mortality and burden of disease attributable to selected major risks. World Health Organization, Geneva

Wu W, Ma B (2015) Integrated nutrient management (INM) for sustaining crop productivity and reducing environmental impact: a review. Sci Total Environ 512:415–427. https://doi.org/10.1016/j.scitotenv.2014.12.101

Yasmin S (2011) Characterization of growth promoting and antagonistic bacteria associated with rhizosphere of cotton and rice. NIBGE, Faisalabad

Yoo SM, James BR (2002) Zinc extractability as a function of pH in organic waste amended soils. Soil Sci 167:246–259. https://doi.org/10.1097/00010694-200204000-00002

Zaheer A, Malik A, Sher A, Qaisrani MM, Mehmood A, Khan SU, Ashraf M, Mirza Z, Karim S, Rasool M (2019) Isolation, characterization, and effect of phosphate-zinc-solubilizing bacterial strains on chickpea (Cicer arietinum L.) growth. Saudi J Biol Sci 26:1061–1067. https://doi.org/10.1016/j.sjbs.2019.04.004

Zeb H, Hussain A, Naveed M, Ditta A, Ahmad S, Jamshaid MU, Ahmad HT, Hussain B, Aziz R, Haider MS (2018) Compost enriched with ZnO and Zn-solubilizing bacteria improves yield and Zn-fortification in flooded rice. Ital J Agron 13(4):310–316. https://doi.org/10.4081/ija.2018.1295

Zhang YQ, Sun YX, Ye YL, Karim MR, Xue YF, Yan P, Meng QF, Cui ZL, Cakmak I, Zhang FS (2012) Zinc biofortification of wheat through fertilizer applications in different locations of China. Field Crops Res 125:1–7

Zhao A, Lu X, Chen Z, Tian X, Yang X (2011) Zinc fertilization methods on zinc absorption and translocation in wheat. J Agric Sci 3:28–35

Chapter 8
Elemental Composition of Medicinal Plants Under Changing Environmental and Edaphic Conditions

Shaista Anjum, Zahoor Ahmed Bazai, Cinzia Benincasa, Sabeena Rizwan, and Ashif Sajjad

8.1 Introduction

Medicinal plants being curative are valuable in human civilization. A great variety of these medicinal plants are utilized in crude form in various remedial processes, viz., herbal tea, powders, tinctures, and decoction. These natural herbal products are used in several traditional medicinal systems such as Unani, Chinese, and Ayurveda (Mathew and Abraham 2006). More than 50,000 medicinal plants are believed to possess therapeutic potentials to cure diseases around the globe (Hamilton 2004). Several modern allopathic medicines such as morphine, aspirin, atropine, tubocurarine, and digoxin may have originated from indigenous medicinal plants locally used in tribal communities (Ghani et al. 2014; Gilani 2005). It was further discovered in the near past that the actual medicinal properties of these valuable plants are because of certain active chemical compounds, mostly the secondary metabolites. Many researchers have carried out different studies for the qualitative screening and quantitative isolation of these active compounds. A part from these Phyto-chemicals it was

S. Anjum · Z. A. Bazai
Department of Botany, University of Balochistan, Quetta, Pakistan

C. Benincasa
CREA Research Centre for Olive, Citrus and Tree Fruit, Rende, Italy

S. Rizwan
Department of Chemistry, Sardar Bahadur Khan, Women's University, Quetta, Pakistan

A. Sajjad (✉)
Institute of Biochemistry, University of Balochistan, Quetta, Pakistan
e-mail: ashif.sajjad@um.uob.edu.pk

Q. Mahmood (ed.), *Sustainable Plant Nutrition under Contaminated Environments*, Sustainable Plant Nutrition in a Changing World,
https://doi.org/10.1007/978-3-030-91499-8_8

also documented that elements present in different concentration may also affect remedial efficacy of herbal and/or medicinal plants products. Plants being sensitive to all environmental condition have been found to show alterations in their elemental composition in response to these changing factors (Kabata-Pendias and Pendias 1984; Vtorova 1987). Alongside environmental factors, edaphic factors such as soil pH and soil organic matter content are also important as soil serves as a reservoir of nutrients for plants, and their availability to plant roots is dependent on these edaphic factors (Kloke et al. 1984). Several anthropogenic pressures such as uncontrolled urbanization, industrialization, and mining practices around the natural ecosystems resulted in unpredictable toxic concentrations of metals and rare earth elements (REEs) in soil (Ahmed and Ishiga 2006), and their intake and composition in plants are directly related to the concentration of these elements in the nutrient solutions of soils (Fifield and Haines 2000). Plants are the major source of nutritional elements to human beings; therefore, excess of these elements may affect the normal functioning of different organs in human bodies (Ahmed and Ishiga 2006). Although medicinal plants with variety of therapeutic properties are considered harmless, it was established that prolonged use of medicinal plants in raw forms resulted in noxious accumulation of different elements causing lethal health issues in human being (Sharma et al. 2009; Ernst et al. 1992). All such findings have raised questions about safety of these wild flora for consumption as food and utilization in medicinal products, thus; made it a crucial step to quantify elemental composition of medicinal plants to ensure their quality as consumable products in food and medicines (Arceusz et al. 2010; Liang and Tao 2004; Schroeder 1973; Somers 1974). Many studies have been conducted to investigate the intake levels of elemental constituents, and all the findings reached one conclusion that essential elements when consumed in high concentrations may cause hazardous heath issues, while non-essential metals are lethal even in low concentration for human health (Al Moaruf Olukayode et al. 2004; Başgel and Erdemoğlu 2006; Kanias and Loukis 1987; Koc and Sari 2009; Sharma et al. 2009; Sheded et al. 2006; Wong et al. 1993). Most of the elements were detected in aerial parts of medicinal plants which are directly utilized in raw form by indigenous people (Bahadur et al. 2011). It is, therefore, an important task to quantify essential as well as toxic inorganic elements in medicinal plants in order to determine their safety and effectiveness against different diseases. Soil associated with medicinal plants must also be examined as it may serve as a main source of these contaminations caused by various biotic and abiotic stresses.

According to the World Health Organization (WHO), over 80% of local communities living in rural area of developing countries rely on traditional remedies for their primary healthcare (Wilkinson 2013). Indigenous people use these herbal plants as an alternative to modern medicine (Robinson and Zhang 2011). With the increasing popularity of these traditional herbs, concern about their safety, effectiveness, and viability has also increased, especially in today's changing climate where plants being static entities are most vulnerable.

8.2 Medicinal Plants as a Source of Mineral Composition

Traditional medicinal plants have a long history of utilization by indigenous communities around the globe. Ethnobotanical information played an immense role in conservation of biodiversity, and traditional cultural ethics are followed for the development of a drug or a cosmetic. The knowledge regarding the use of these medicinal plants is passed from one generation to other as a traditional heritage. Medicinal plants although are always used in raw form without any high technical processing. However, it is documented that for each particular ailment, a different drug preparation method is commonly opted. The traditional recipes include either whole plant or some specific plant organs such as buds, flowers, leaves, stems, barks, or plant products such as latex, resins, gum, etc. (Khan et al. 2011). The reason to use a specific part might be the chemical constituents present in that particular tissue against the target disease. Chemical constituents present in medicinal plant when enter in blood stream may interact with body chemistry and may influence metabolic activities inside human body (Kolasani et al. 2011). Medicinal plants being one of the major sources of health and nutrition actually supply minerals, protein, carbohydrates, and vitamins. Simultaneously, elements from these medicinal plants are also absorbed and enter into the body's organs or tissues. Therefore, it is indispensable to characterize the elements in various plant parts to estimate the proper therapeutic dose for intake by consumers. Several efforts have been made to quantify the elemental composition of numerous medicinal plants. Findings exposed encroaching data that demonstrate the value of elemental composition in various parts of plants, viz., leaves, seeds, flowers, stem/bark, and rhizomes/roots. Considering the value of elemental composition of medicinal plants, Raju et al. (2016) collected leaves of two medicinal plants (*Sphaeranthus indicus* and *Cassia fistula*) to evaluate their elemental composition like, aluminum, bromide, calcium, iron, lanthanum, magnesium, manganese, sodium, scandium, vanadium, and zinc. A different concentration of each element was detected in each selected medicinal plant depending on their location, climatic condition, and edaphic factors. Further research on the elemental concentration of medicinal plants like coriander, lavender, chamomile, mint, dill, and plantain cultivated in unpolluted area was done by Haidu et al. (2017), and their findings revealed that the decoction prepared from these plants was a good source of essential nutrients, while the level of other toxic elements was either low or below the tolerable daily intake values. These findings of the above studies suggested that the cultivation of medicinal plants in unpolluted areas support protection and conservation of these valuable plants from environmental pollution and other factors that may cause contamination. Another study was conducted by Adazabra et al. (2012), in which 15 elements, viz, sodium, potassium, calcium, copper, iron, arsenic, chlorine, chromium, magnesium, manganese, phosphorous, zinc, antimony, selenium, and vanadium were analyzed in six medicinal plants of Ghana and all were found to be a good source of mineral elements and also were considered safe according to the international safety standards.

In numerous parts of the world, medicinal plants are the only therapeutic source of basic healthcare system, thus serving as a major source of mineral nutrition. High dependency on these natural resources imposed deleterious effects on their regeneration potential and made it important to analyze the elemental composition alongside other chemical analysis. These medicinal plants are also used as food by local communities beside their therapeutic uses, which is why it is very important to evaluate the mineral and heavy metal composition in order to rule out their suitability for human health and nutritional requirements (Fernandes et al. 2019). The human body requires different elements in different quantities based on their function in metabolic processes.

8.3 Elemental Composition of Medicinal Plants in Relation to Changing Environmental and Edaphic Factors

Plants in general are considered as major sources of elements for the living organisms consuming these as food. Environmental conditions dramatically affect human health, especially in the areas where local communities depend on vegetation growing in their natural habitat as major part of their diet. Medicinal plants in most of the communities around the world are one such source of nutrition. Although medicinal plants are genetically programmed to produce active ingredients with therapeutic properties, the impact of various environmental and edaphic factors on their quality and quantity cannot be neglected. In most of the cases, amounts of active ingredients especially secondary metabolites in medicinal plants are influenced by growing environmental condition (Valls et al. 2007). Therefore, monitoring and evaluation of environmental factors are crucial for quality control of medicinal plants as growth, yield, biomass, active organic constituents, and inorganic elemental composition may get altered under the influence of these factors.

Elements occur naturally in different forms and play essential roles in body function at different cellular levels. Despite their crucial nature in stabilizing life on earth, knowledge regarding their exact mode of action inside living tissues is still limited and yet to be explored. Medicinal plants while coping with different climatic catastrophes are at risk in terms of their quality indices. For better chances of survival, competition in limited resources and harsh environmental stresses has raised a question mark on safety and sustainability of wild medicinal flora. Undesirable environmental factors (temperature, rainfall, snow, or wind patterns lasting for decades or longer) not only impose detrimental effects on medicinal plant quality by disrupting their therapeutic chemical composition but also affect their availability and production. These changes in medicinal plants then ultimately may impose a stress on life forms including human being which specifically depends on these plants for normal and balanced life processes. It is noteworthy to understand that stress is an altered physiological condition caused by elements that tend to disrupt the equilibrium of life or any unfavorable environmental parameter that confines plant expansion and production (Claeys et al. 2014). Any physical and chemical change produced by a stress is known as strain (Gaspar et al. 2002), and it is also a noticeable fact that every deviation of a factor from its optimum does not necessarily result in stress. Stress being a constraint or highly unpredictable fluctuations imposed on regular metabolism causes injury, disease, or aberrant physiology. The natural environment for plants is composed of a complex set of biotic and abiotic stresses. Plant responses to these stresses are equally complex, especially when it comes to elemental composition of wild flora. Medicinal plants in natural habitat are prone to all environmental stresses; thus, it is compulsory to understand and assess their elemental composition as these are presumed to have a vital therapeutic action (Serfor-Armah et al. 2003).

It was further documented that elemental profile of medicinal plant is conditional, depending upon multiple factors including soil as a prime source of trace and REEs followed by air and water serving as the major source of some macro elements. Any contamination/pollution by anthropogenic contributions or natural processes altering elemental composition of soil, air, or water ultimately influences the elemental profile of plants, which may attribute to the differences in elemental profile of same plant species of different sites. Differences in the elemental profile of different species of same site further indicated plant genotype and absorption affinity toward a particular element and group of elements as another important factor that may influence the elemental profile of medicinal plants. Thus, external factors such as soil composition and environmental dispersion of elements and internal factors such as plant genotype and selective absorption ability may be considered as determining factors of elemental profile of medicinal plants (Anjum et al. 2019).

8.3.1 Potential Influence of Abiotic Environmental Factors on Elemental Composition of Medicinal Plants

There are so many abiotic environmental stress factors such as elevated temperature, heavy metal stress, elevated CO_2, salinity, and drought posing a threat to biodiversity, especially plant growth and biomass productivity. All these abiotic stresses are usually inter-related and individually or in combination cause morphological, physiological, biochemical, and molecular changes, which badly affect plant growth and productivity. Medicinal plants are categorized as a nutritional class of pharmaceuticals and are also considered risk-free by indigenous people of rural communities. Accumulation of undesirable elements in medicinal plants and procedures to eliminate and control hazardous effects of these elements must be considered as a criterion for their utilization (Aksuner et al. 2012). It was established that the raw material prepared from these medicinal plants is greatly influenced by surrounding environment and soil condition in which these plants are flourishing. The relation between the geographical location of a plant and its elemental contents is complex. There are many plant species which show affinity toward particular elements. This accumulation and affinity may be considered as a characteristic feature of a given plant species; for example, *Equisetum arvense* L. accumulates a silicon compound (Kohlmünzer 2003). In contrast to this, it is also well examined that when a plant is grown on a soil rich in certain elements easily absorbable by roots, it accumulates those elements (Brooks 1994). Considering the value of ambient environmental factors on elemental composition of medicinal plants, Abdykhalikova et al. (2018) collected samples such as hawthorn, sea buckthorn, and yarrow from an industrial zone of Rudny City, Kazakhstan (Table 8.1). Natural habitat in particular

Table 8.1 Potential influence of environmental and edaphic factors on elemental composition of medicinal plants

Factors	Medicinal plants	Elemental composition			References
		Increased/ accumulated	Decreased/ suppressed	Conditional variation	
Elevated temperature	*Zingiber officinale*	Na, K, Mg, Ca, Fe	–		Ajagun et al. (2017)
Elevated CO_2	*Ocimum basilicum* L. *Mentha piperita* L.	K, Ca, Mg, Na Cu, Fe, Mn, Cd, Zn	–		Al Jaouni et al. (2018)
	Quercus acutissima Carr. *Fraxinus rhynchophylla* Roxburgh, Fl.	C	Ca, K, N	P	Cha et al. (2017)
Salinity	*Portulaca oleracea* L.	Na, Mg Cl	Ca, K, Zn		Kamal Uddin et al. (2012)
	Amaranthus tricolor L.	Ca, Mg, Fe, Mn, Cu, Zn, Na	K		Sarker et al. (2018)
	Origanum majorana L.	Na, Cl	K		Baatour et al. (2018)
	Andrographis paniculata Nees.	Na, Fe, Zn, Cu	N, P, K, Ca, Mg, Mn		Talei et al. (2012)
	Trachyspermum ammi [L.] Sprague	Na, Cl	K, Ca		Ashraf and Orooj (2006)
	Foeniculum vulgare Mill subsp. *vulgare* var. *vulgare*	–	N		Abd El-Wahab (2006)
	Peppermint and lemon verbena	–	K, N		Tabatabaie and Nazari (2007)
	Achillea fragrantissima Forssk	Na	K		Abd EL-Azim and Ahmed (2009)
	Simmondsia chinensis (Link) Schneider	Na, Cl	N, P, K, Ca		Ali et al. (2013)
	Satureja hortensis L.	Na	K		Mohammadi et al. (2019)
	Catharanthus roseus (Linn.) G.	Na	K, N, P, Fe		Jaleel et al. (2008)

Factors	Medicinal plants	Elemental composition			References
		Increased/ accumulated	Decreased/ suppressed	Conditional variation	
Drought	*Ricinus communis* L.	K, Ca, Na	Fe, Cu, Mg Zn		Tadayyon et al. (2018)
	Matricaria chamomilla L.		K, P		Salehi et al. (2016)
	Lavandula latifolia Med. *Mentha piperita* L. *Salvia lavandulifolia* Vahl. *Thymus capitatus* (L.) Hoff. et Link. *Thymus mastichina* L.		N, P		García-Caparrós et al. (2019)
	Gongronema latifolium (Benth)	N	Ca, K		Osuagwu and Edeoga (2012)
Heavy metal stress	*Petroselinum crispum*, *Ocimum basilicum*, *Salvia officinalis*, *Origanum vulgare*, *Mentha spicata*, *Thymus vulgaris*, *Matricaria chamomilla*	Cd, Pb	Zn	Cu, Fe	Dghaim et al. (2015)
Soil pH	*Brassica parachinensis* *Brassica chinensis*	Mo, Mn	Fe, Zn		Wong and Wong (1990)
	Spiraea alba Du Roi *Spiraea tomentosa* L.	N, K, Zn, Mn	Mg		Mickelbart et al. (2012)
	Valeriana officinalis L.	Cu, Mn	Zn		Adamczyk-Szabela et al. (2015)

(continued)

Table 8.1 (continued)

Factors	Medicinal plants	Elemental composition			References
		Increased/ accumulated	Decreased/ suppressed	Conditional variation	
Multiple stress factors	*Artemisia scoparia, A. absinthium, A. indica, A.santolinifolia, A. maritime, A. vulgaris, A. japonica, A. nilagirica, A. herba-alba, A. annua, A. brevifolia, A. moorcroftiana, A. dracunculus, A. roxburghiana*, and *A. dubia*	K, Na, Ca, Mg Cu, Cr, Cd, Ni, Pb, Mn, Fe		Zn, Co,	Ashraf et al. (2010)
	Achyranthes aspera, Alternanthera pungens, Brassica campestris, Cannabis sativa, Convolvulus arvensis, Hordeum vulgare, Justicia adhatoda, Parthenium hysterophorus, Ricinus communis, Withania somnifera	K, Na, Ca, Mg, Mn, Fe, Cu, Cr, Cd, Pb, Ni		Zn, Co,	Jabeen et al. (2010)
	Achillea wilhelmsii C. Koch, *Hertia intermedia* (Bioss) O. Ktze, *Nepeta praetervisa* Rech. F. *Peganum harmala* Linn, *Perovskia atriplicifolia* Benth, *Seriphidium quettense* (Podlech.) Ling., *Sophora mollis* (Royle) Baker	Cl, Cr, As, Al, Mn, Br, Lu		C, H, N, K, Ca, Cu S, Ce, Co, Cs, Eu, Fe, Hf, La, Na, Pb, Rb, Sb, Sc, Sm, Sn, Sr, Th, V, Yb, Zn	Anjum et al. (2019)
	Hawthorn, sea buckthorn, yarrow	Al, Fe			Abdykhalikova et al. (2018)

was disturbed because of intense mining, which developed ecological pressures on wild flora due to anthropogenic activities. Same plants were collected from another undisturbed natural habitat. Ash content was found comparatively higher in medicinal plant samples collected from industrial zone due to the presence of heavy metals such as aluminum (Al^{3+}) and iron (Fe^{3+}) ions. Furthermore, a significant reduction in tannin percentage of medicinal plants collected from industrial territory was also countered indicating detrimental effects of metal accumulation causing depletion of these phenolic compound derivatives in studied plant samples of disturbed area. For further reconfirmation, ash contents of respective samples were dissolved in water, and contrary to industrial area samples, medicinal plants collected from unpolluted natural habitat dissolved well in water since they contained no heavy metals. However, samples collected from industrial zone partially dissolved in acetic acid and completely dissolved in concentrated nitric acid, indicating the presence of heavy metals that can form compounds with organic acids. The overall findings clearly explained the accumulation of hazardous elements in herbal plants grown in polluted areas indicating the importance of collection site with contaminants and lacking the most important abiotic factors required for healthy plant growth.

Likewise, protected areas around the globe may serve as a key strategy for the protection and preservation of wild flora and could serve as a buffer zone against various environmental and anthropogenic stresses. In this context, a study was conducted by Anjum et al. (2019) to evaluate the effectiveness of protected area systems on elemental composition of medicinal plants (Table 8.1). Seven medicinal plants, viz. *Achillea wilhelmsii* C. Koch, *Hertia intermedia* (Bioss) O. Ktze, *Nepeta praetervisa* Rech.F. *Peganum harmala* Linn, *Perovskia atriplicifolia* Benth, *Seriphidium quettense* (Podlech.) Ling., and *Sophora mollis* (Royle) Baker, were collected from Hazarganji Chiltan National Park, accomplished as a well-known protected area for biodiversity conservation in Pakistan, for quantification of 33 elements, namely C, H, N, S, Al, As, Br, Ca, Ce, Cl, Co, Cr, Cu, Cs, Eu, Fe, Hf, K, La, Lu, Mn, Na, Pb, Rb, Sb, Sc, Sm, Sn, Sr, Th, V, Yb, and Zn. Same set of plant samples were also collected from another area exposed to several anthropogenic and environmental pressures. The results revealed that all the selected medicinal plant species were a good source of many essential elements such as K, Ca, Na, and Fe; however, significantly toxic levels of some trace elements were also documented, which may cause various other health hazards in human if consumed for a longer period of time. Furthermore, in plants from protected sites, relatively good quantities of essential elements were found compared to those samples collected from unprotected sites. The latter showed a comparatively higher affinity in accumulating some REEs and potentially dangerous elements such as Pb and As. These levels, although in relatively low concentrations, can highlight the positive impact that the national park has on plants. In fact, they can be considered medicinal plants with potential therapeutic effects. Furthermore, having protected sites contributes to the conservation of plant biodiversity. Wild medicinal plants growing in their natural ecosystem experience a variety of abiotic factors, and despite their importance and enormous curative and other economic uses, this natural flora is under a constant pressure of exploitation due to many man-made activities; thus, these valuable resources may face severe threats of extinction. In this regard, conservation practices must be considered for their lifelong preservation and sustainable utilization. Most of the studies focused on combine effects of environmental factors as under natural circumstances, medicinal plants are always exposed to multiples abiotic stresses simultaneously, whereas, rarely encounter a single stress factor with a greater magnitude due to some natural disaster or any unwanted anthropogenic activity.

Regardless of these combined effects of abiotic factors (natural and anthropogenic) on elemental composition of medicinal plants, a few studies, although in control, have been conducted to explore the effects of individual abiotic factors. Temperature as being one of the most crucial factors was taken into account by Ajagun et al. (2017) (Table 8.1), while manifesting the elemental composition of rhizome of *Zingiber officinale* at different temperatures. Inorganic elements such as sodium (Na), potassium (K), magnesium (Mg), calcium (Ca), phosphorous (P), and iron (Fe) were detected in different concentrations at room temperature 40, 50, and 60 °C. An increase in the levels of Na, K, Mg, Ca, and Fe was observed at 50 °C which was due to food processing techniques such as drying and soaking, and cooking would lead to an increase in the levels of mineral elements in food due to the deactivation of inhibitory anti-nutritional factors (Oluwalana et al. 2011). Further data explained a noteworthy increase in Ca concentration of medicinal plants at 50 °C when compared with other temperature ranges. Similarly a clear difference in Fe content of same plant across different temperatures was noted. However, the remaining elemental composition such as Mg and K contents further asserted the importance of temperature on processing of medicinal plants by exhibiting a significant different at varying temperatures (room temperature, 50 and 60 °C). Temperature fluctuations can seriously impair certain important biological processes that take place in living cells. High temperature can influence the production of reactive oxygen species (ROS), leading to oxidation damage of tissue or even death (Zinn et al. 2010). On the other hand, low temperature hinders the normal biochemical and physiological processes in plants, thereby causing severe symptoms such as chlorosis, necrosis, and wilting (Ruelland and Zachowski 2010). These alterations may result in loss of electrolytic content from the cytoplasm, which in consequence adapts for optional pathways to regulate electron flow in cell (Seo et al. 2010). According to Knight et al. (1998), temperature fluctuation caused changes in the concentration of calcium ions inside the cell content.

Human civilization in the name of industrial revolution imposed an imbalance in several abiotic factors. Elevated CO_2 concentrations are one such factor which may induce global warming (Allen et al. 2009; Meinshausen et al. 2009). When grown under elevated CO_2 concentrations, an increase in the amount of C stored in plants and soils due to higher photosynthetic rates was found (Tissue et al. 1997). Many studies have been documented to discuss the potential influence of elevated CO_2 on agricultural crops, while less effort has been made to explore medicinal plants with an emphasis on their mineral composition (Table 8.1). However, Al Jaouni et al. (2018) assessed the differences in mineral composition (K, Ca, Mg, P, Na Cu, Fe, Mn, Cd, Zn) and primary and secondary metabolite concentrations of two medicinal plants, viz., *Ocimum basilicum* L. and *Mentha piperita* L. grown under two CO_2 concentrations (ambient 360 ppm and elevated 620 ppm). A significant improvement in herbal biomass accompanied by an increase in inorganic elemental composition of K, Ca, Mg, Na Cu, Fe, Mn, Cd, Zn, and other phytochemicals under elevated CO_2 was observed during the study.

Cha et al. (2017) examined the effects of elevated CO_2 on the chemical composition of leaf waste of *Quercus acutissima* and *Fraxinus rhynchophylla* and investigated other parameters as well (Table 8.1). Their finding indicated complex pattern of elemental composition under elevated and ambient CO_2 conditions. C concentration in *Q. acutissima* leaf waste did not show any difference under elevated CO_2, while in *F. rhynchophylla*, a higher concentration of C was observed. In case of N content, a significant decline was observed in both plant species under elevated CO_2 conditions. The P content decreased in *Q. acutissima*, whereas it increased in *F. rhynchophylla*. The K and Ca content of leaf waste decreased in both plant species under elevated CO_2 conditions. However, more precise studies may evaluate the further effects of elevated CO_2 on these valuable natural resources.

Among all other stress factors, it has been established that drought stress is a very important stress factor and has a huge impact in limiting plant establishment at juvenile stage and also impairing plant growth and development processes (Anjum et al. 2003; Bhatt and Rao 2005; Kusaka et al. 2005; Shao et al. 2008). The metrological term drought commonly refers to the condition when there is a reduction in available water in soil, alongside a continuous loss of water via evaporation from soil or transpiration via plants. Tolerance to drought stress is exhibited at any metabolic stage in almost all plants but to which extent, it depends upon species as the level of tolerance varies from species to species. Prolonged drought stress may inhibit photosynthesis, arrest metabolic activities, and finally cause death of plants (Jaleel et al. 2008). It has been expected that more than 50% of fertile land may face plant growth problems by 2050 due to drought stress (Vinocur and Altman 2005). Furthermore, drought stress will severely affect nutrient availability, uptake, and translocation, as water is essential to conduct these processes (Vurukonda et al. 2016). Certain water-soluble nutrients will no longer be available due to lower rates of nutrient diffusion and mass flow, otherwise facilitated by water molecule (Barber 1995; Selvakumar et al. 2015). A decline in nitrate reductase activity due to low supply of nitrate from soil will also occur (Caravaca et al. 2005). Understanding plant responses to drought is of great importance and also a fundamental part for developing stress-tolerant plants (Reddy et al. 2004; Zhao et al. 2008). Plants with fresh and dry weights under water-limited conditions are desirable. A common adverse effect of water stress on crop plants is the reduction in fresh and dry biomass production (Farooq et al. 2009). Plant productivity under drought stress is strongly related to the processes of dry matter partitioning and temporal biomass distribution (Kage et al. 2004). According to Tahir and Mehid (2001), drought stress was the main limiting factor in biomass production of almost all genotypes of sunflower during their experiment. Similarly, a reduction in biomass productivity due to drought stress, was also observed in Poncirus trifoliatae seedlings by Wu et al. (2008) and in *Petroselinum crispum* by Petropoulos et al. (2008). Medicinal plants being sessile entities cannot escape from their natural habitat, yet they need to face every environmental extreme. Although commendable work has been carried out to understand the influence of drought as an abiotic factor on secondary metabolite composition, little is known about its effects on the mineral composition of herbal products (Table 8.1). To highlight the effects of drought stress on medicinal plants, Tadayyon et al. (2018) evaluated mineral (K, Mg, Ca, Na, Fe, Cu, and Zn) contents from leaves of different ecotypes of *Ricinus communis* L. (Esfahan, Ardestan, Arak, Naein, Yazd, and Ahwaz). Under drought stress, the concentration of K, Ca, and Na was induced, while the concentration of Fe, Cu, Mg, and Zn was inhibited. Further data revealed that ecotypes of same plant exhibited a variety of responses while combating a similar drought stress by showing alterations in their mineral composition. These findings are significant to understand medicinal plant responses, while interacting with different abiotic stresses. As for this study, Fe content varied significantly among different ecotypes of castor plant under stress conditions. It indicated complexity of plant response in terms of mineral accumulation to adapt environmental condition. Osuagwu and Edeoga (2012) examined the influence of drought stress on vitamin and nutritional aspects of the leaf samples of medicinal plant *Gongronema latifolium* (Benth). The study revealed a significant reduction of Ca and K in leaves under drought stress but an increase in N content. According to their findings, it was due to the mobilization of mineral contents such as Ca and K from leaf tissues to roots to serve as osmoprotectants in order to help stressed plant to withstand drought stress, while N moved from roots to leaves to help leaf tissues to produce more amino acids in order to synthesize proteins to resist stress. However, no significant variations in Na, Mg, and P content were observed. Six plants of family Lamiaceae, viz., *Lavandula latifolia* Med., *Mentha piperita* L., *Salvia lavandulifolia* Vahl., *Salvia sclarea* L., *Thymus capitatus* (L.) Hoff. et Link., and *Thymus mastichina* L., were analyzed for their mineral composition by García-Caparrós et al. (2019) in order to evaluate the effects of drought stress alongside different parameters. The results revealed a decrease in P concentration in all species under drought stress except *S. Sclarea*, whereas a reduction in N content was observed in *L. latifolia* and *T. Mastichina*. Likewise, Salehi et al. (2016) observed a decrease in P and K contents in *Matricaria chamomilla* L. under salt stress. Nevertheless, lack of the literature regarding the impacts of different abiotic stress on mineral composition of medicinal plants has created a gap to get an insight into plant responses.

The condition in which a plant grows in more than a normal amount of salt is called salt stress or salinity. Salinity is caused by saline soil or water. It is the main abiotic stress of agricultural lands around the world (Yan et al. 2005). According to Szabolcs (1994), salinity has a strong deleterious effect and this stress factor alone affected more than 954 million hectares of land worldwide, and it has a very damaging effect not only on soil but also on groundwater resources and wild flora. Other effects of salinity on plants include ion imbalance, oxidative harm, hyper-osmotic stress (Zhu 2001), reactive oxygen species production (Zhu et al. 1997), fall of leaf water potential, reduced stomata conductance, inhibition of photosynthesis, etc. (Baker and Rosenqvist 2004). In the last two decades, knowledge about the process of tolerance and adaptation to salinity in plants has increased, but still there is a need for further research (Shi et al. 2000; Kim et al. 2007). Medicinal plants like other plant resources are also severely affected by salt stress. Salt stress is recognized as one of the abiotic factors with profound effects on mineral composition of medicinal plants (Table 8.1). In this regard, Kamal Uddin et al. (2012) examined the influence of salt stress on the growth and mineral composition of *Portulaca oleraceae* L. Four salinity levels (0, 66, 132, and 264 mM) of NaCl were tested in an experimental setup. The accumulation of Na^+, Mg^{++}, and Cl^- increased with an increase in salt stress and the concentration of Ca^{++}, K^+, and Zn^+ declined. Furthermore, a difference in the accumulation capacity of different plant parts of *Portulaca oleraceae* L. under salt stress was also found as concentration of Ca^{++} and Zn^+ was higher in leaves and accumulation of Na^+ and K^+ was higher in stems . Relative ratios (Na^+/K^+, Na^+/Ca^{++}, Na^+/Mg^{++}, Mg^{++}/Ca^{++}, and Mg^{++}/K^+) were also found to be directly proportional to the increasing pattern of salt stress. Some elements including Na, Cl, and B have toxic effects on plants if accumulated in living tissues. According to Munns (2002), Na may cause osmotic stress and cell death if accumulated in excessive amounts in cell wall.

Salt stress further induced the accumulation of biochemical content and enhanced antioxidant activity in three selected accessions of *Amaranthus tricolor* (VA3, VA12, and VA14), when investigated in terms of nutrients, mineral, and phytochemical and antioxidant activities by Sarker et al. (2018) (Table 8.1). Mineral composition deciphered a great variation under varying salt stress parameter. Among the tested accessions of leave samples, VA14 was found to accumulate higher concentrations of Ca, K, Fe, Mn, Cu, and Zn contents, whereas VA3 and VA12, respectively, were found to accumulate high contents of Mg and Na. In contrast, a decrease in elemental content of Ca, K, Fe, Cu, and Zn was observed in VA3 samples. Similarly, lowest Mg and Na content was detected in VA14 and low Mn content in VA12. An increase in salt stress showed a dramatically significant reduction in K content and a sharp and significant increase in Ca, Mg, Fe, Mn, Cu, Zn, and Na contents in leaves (Table 8.1). Growth, mineral nutrition, yield, and chemical composition of essential oils of *Origanum majorana* L. were influenced in response to salt stress (Baatour et al. 2018). Plants were cultivated under different sodium chloride concentrations (0, 50, 100, 150 mM). The results of mineral content showed primary accumulation of Na^+ in roots at 50 mM NaCl concentration but a strong accumulation in leave tissues at 150 mM. A similar accumulation pattern was observed for Cl^-. However, Na^+ was accumulated more than Cl^-. Salt stress further caused a decrease in K^+ content in roots at 100 mM and a decrease in stems at 150 mM. A decrease in Ca^{2+} was also observed in plant roots. Essential oil yield and quality of *O. majorana* were severely affected. The overall plant growth was affected by salt stress due to an imbalance in the distribution, availability, partitioning, and transport of mineral nutrients in different plant parts. In arid and semi-arid regions where climate is characterized with prevailing conditions of high evapo-transpiration and low leaching water, it is a crucial step to understand abiotic threats, like salt stress, to achieve higher yield of medicinal plants with desirably good quality (Said-Al Ahl and Omer 2011). Nutritional imbalance in plant parts under salt stress is caused due to the competition of ions such as Na^+ and Cl^- with other ions such as

K^+, Ca^{2+}, and NO^{3-}. This competition results in ion toxicity in plant tissues and ultimately affects plant growth, yield, and metabolism (Grattan and Grieve 1998). Under salt stress, high availability of NaCl concentration induces higher accumulation of Na and Cl which ultimately caused unavailability of others sential elements such as N, P, Ca, K, and Mg in soil for several medicinal plants (Table 8.1) such as *Trachyspermum ammi* [L.] Sprague (Ashraf and Orooj 2006), Foeniculum vulgare Mill subsp. vulgare var. vulgare (Abd El-Wahab 2006), peppermint and lemon verbena (Tabatabaie and Nazari 2007), *Catharanthus roseus* (Linn.) G. (Jaleel et al. 2008), *Achillea fragrantissima* Forssk (Abd EL-Azim and Ahmed 2009), *Matricaria recutita* L. *Simmondsia chinensis* (Link) Schneider (Ali et al. 2013), and *Satureja hortensis* L. (Mohammadi et al. 2019). Salt stress imposes some major drawbacks on plants including ion toxicity, nutrient imbalance (N, Ca, K, P, Fe, and Zn), and oxidative stress that ultimately limits the water uptake of soil (Shrivastava and Kumar 2015). It was further reported that various tolerant and resistant varieties show different responses to different abiotic factors (Table 8.1). In this regard, Talei et al. (2012) investigated the effects of salt stress on 12 different accessions of medicinal herb *Andrographis paniculata* Nees. and found a clear reduction in the concentrations of N, P, K, Ca, Mg, and Mn and an increase in that of Na, Fe, Zn, and Cu under high salt stress. Furthermore, salt stress–tolerant accessions were able to accumulate higher concentrations of K^+, P, N, Mg^{2+}, and Ca^{2+} and relative lower concentrations of Na^+, Fe, Zn, and Cu than salt stress–sensitive accessions. Salinity is a type of abiotic stress that can ultimately alter mineral composition of plant tissues which may in turn alter metabolic activities including photosynthesis and respiration and thus may agitate several other vital activities such as protein synthesis.

8.3.2 Potential Influence of Biotic Factors on Elemental Composition of Medicinal Plants

Medicinal plants in their natural habitat are under vast array of biotic stress factors in nexus to abiotic stress factors. Biotic factors such as herbivore and pathogen attacks are equally stressful and detrimental (Dixon and Paiva 1995; Holopainen and Gershenzon 2010). Plant acclimation responses involve induction of stress signal to release biologically active secondary compound (Bernays and Chapman 2000; Hagerman and Butler 1991).

The literature is rich with plant-defense mechanisms against herbivores and all other pathogens. In addition, well-elaborated studies are conducted to explain anti-bacterial, anti-fungal, and anti-viral properties of medicinal plants. At present, relatively less is understood about the impacts of these biotic stress factors on mineral nutritional prospects of medicinal plants, despite countering plant defense. The need to carefully monitor changes in the quality and quantity of mineral composition of natural herbal products in response to biotic stress factors must be considered in the interest of public healthcare safety because accumulation of any hazardous element may cause serious life threats to human life.

8.3.3 Potential Influence of Edaphic Factors on Elemental Composition of Medicinal Plants

The earth is the reservoir of 92 elements, of which about 82 elements are utilized by many plant species (Reimann et al. 2001). Furthermore, it is an established fact that including oxygen, CO_2, and water, 14 mineral elements are considered essential for proper nutrition and development of plants (Mengel et al. 2001). Most of these elements are obtained from soil solution and any imbalance in soil properties may interfere with the absorption of these elements. Generally, six elements, namely, nitrogen, phosphorus, potassium, calcium, magnesium, and sulfur that are required in large amount are called macronutrients, while other elements including boron, iron, manganese, copper, zinc, nickel, and molybdenum that are required in trace or small amounts are called micronutrients. The importance of these nutrients on plant yield, biomass, productivity, and stress tolerance is often investigated through various experiments on plant responses under manipulated nutrient supply (Ncube et al. 2012). Kouki and Manetas (2002) also reported an increase in the level of proanthocyanin under limited phosphate availability. Furthermore, biosynthesis of certain phenolic compounds was stimulated under limited ion stress (Dixon and Paiva 1995). Soil being the primary source of many elements may serve as a potential factor that influences the elemental composition of medicinal plants. Certain characteristics of soil may change due to some hazardous practices which may result in availability of some elements, unavailability of certain other elements thus, disturbing concentration pool of elements for anchored vegetation. According to Obratov-Petković et al. (2006), the uptake of heavy metals is directly proportional to the concentration of heavy metals in soil solution and to the transfer rate from the solid phase into soil solution for heavy metals taken up by the plant roots. However, Kashem and Singh (2002) elaborated that Cd and Zn concentration increased with time in contrast to their concentration in soil solution.

In general, a source to sink relation exists in between soil solution and plant tissues but other physicochemical properties of soil may also influence elemental characterization of herbal products. Medicinal plants, when growing in contaminated soil, may accumulate toxic metals in their tissues and parts and transfer them from soil to the food, which may then be consumed by humans (Kabata-Pendias and Pendias 1992). The accumulation of these toxic elements may cause serious health hazards. Some of the elements like Zn, Fe, Cu, Cr, and Co may cause toxicity but only at higher concentrations; however, certain other elements like Pb, Hg, and Cd are toxic even in low concentrations (Radojevic and Bashkin 1999). Medicinal plants are considered as an excellent source of trace elements. Sulaiman et al. (2017) determined the concentration of nine elements, viz., Na, K, Zn, Fe, Co, Cu, Ni, Pb, and Cd from fourteen medicinal herbs, namely, *Matricaria chamomilla* L., Cinnamon, *Pimpinella anisum* L., *Zea maize*, *Anethum graveolens* L., Jeft, *Teucrium Polium* L., *Cassia italica*, *Echium talicum* L., *Ocimum basilcum* L., *Galeopsis sejetum*, *Nigella sativa* L., *Cyperus rotundus* L., and *Lupinus jaimehintoniana*, commonly used in Iraq, and every herb showed a variation in its elemental concentration which is attributed to mineral composition of associated soil. The results further concluded that although trace elements are beneficial, their concentration above the permissible limit may cause toxicity. This result was supported by Ashraf et al. (2010) who analyzed the elemental concentration of 4 major elements, namely, potassium, sodium, calcium, and magnesium and 9 trace elements, namely, zinc, copper, chromium, nickel, cobalt, lead, manganese, iron, and cadmium in 17 native medicinal plant species of genus *Artemisia*, viz., *A. scoparia, A. absinthium, A. indica, A. santolinifolia, A. maritime, A. vulgaris, A. japonica, A. nilagirica, A. herba-alba, A. annua, A. brevifolia, A. moorcroftiana, A. dracunculus, A. roxburghiana, A. dubia, A. kurramensis*, and *A. stanosephala*, which are commonly used by aboriginal communities of Pakistan. The Results showed the presence of toxic levels of certain elements which may cause metal poisoning in human beings, and accumulation of these hazardous elementals in *Artemisia* species is either because of the contaminated soil in which they are growing or may be these species are hyper-accumulators. Consequently, it was suggested that one should be cautious while utilizing these species in order to avoid any health hazards. Similar was the findings of Jabeen et al. (2010) who explored the therapeutic potentials of ten most commonly used medicinal plants (*Achyranthes aspera, Alternanthera pungens, Brassica campestris, Cannabis sativa, Convolvulus arvensis, Hordeum vulgare, Justicia adhatoda, Parthenium hysterophorus, Ricinus communis*, and *Withania somnifera*) by scanning their elemental concentration: K, Na, Ca, and Mg (major elements) and Zn, Cu, Cr, Ni, Co, Cd, Pb, Mn, and Fe (trace elements) (Table 8.1). Levels of elements such as iron, sodium, potassium, magnesium, and calcium were found to be excellent, but other trace elements were very high when compared to the international safety values for the utilization of human being. Soil nutritional status, therefore, may be considered as a determinant, but other properties must also be included while impacts on nutritional prospect of herbal products are investigated. Soil pH is one of those factors that have an enormous impact on soil biogeochemical process. Soil pH is, therefore, is called "master soil variable" that can control multitudes of different soil properties, which then may affect the overall plant growth and biomass yield (Brady and Weil 1999; Minasny et al. 2016). Mineral nutrition of plants has shown considerable fluctuation with soil pH. Wong and Wong (1990) observed a significant correlation of metal accumulation and reduction in plant tissues and pH of soil (Table 8.1). Their results deciphered a consistent decrease in Fe and Zn content in response to increasing soil pH, and a consistent accumulation of other elements such as Mo and Mn was evident in tested plant species such as *Brassica parachinensis* and *Brassica chinensis*. Trace elements present in soil depend upon pH for their solubility, mobility, and bioavailability, which, in turn, influencs their translocation to plant tissues (Förstner 1995). A lower pH level usually stimulates the solubility of trace elements due to low adsorption and high desorption properties. Inter mediate pH level facilitates the trend of trace element. Elemental adsorption increases from almost no adsorption to complete adsorption within a narrow pH range called the pH-adsorption edge. However, from this point onward, a complete adsorption of trace elements is evident (Bradl 2004). The literature further confirmed this reverse relation between the soil pH and solubility of trace elements. According to Kabata-Pendias (2011), increasing soil pH will decrease the solubility of most of the elements ultimately causing low concentrations in soil solutions. Further observed a significant hundred-fold decrease in the solubility of divalent metals and even a thousand-fold decrease in the solubility of trivalent metals. Availability of Zn, Cu, Mn, and Fe was greatly suppressed at high pH, while the availability of Mg, Ca and K was suppressed at low pH (Tinus 1980). In order to highlight the effects of soil pH on mineral composition, Mickelbart et al. (2012) performed an experiment by cultivating *Spiraea alba* Du Roi and *Spiraea tomentosa* L. at different soil pH levels. These plants are usually adapted to slightly acidic soils in their natural habitats. In this experiment the elemental sulfur was added in to soil with a pH of 7.2 to bring the pH 5.8 and 6.4. As a result of sulfur addition, a decrease in concentrations of exchangeable magnesium (Mg) and calcium (Ca) occurred in tested soil. The results of the study suggested the importance of soil pH by showing an alteration in concentrations of certain mineral nutrients. At low soil pH, concentrations of N, K, Zn, and Mn increased in leaf samples, whereas concentration of Mg decreased. Adamczyk-Szabela et al. (2015) estimated the influence of soil pH (5.1 and 10) on mineral composition (Cu, Zn, Mg) uptake by medicinal plant *Valeriana officinalis*. A clear reduction in zinc content was observed with increased pH of 10. Further results surprisingly exhibited a significantly higher concentration of Cu and Mn with an increase in soil pH. Wu et al. (2011) also reported that reduction in the concentration of heavy metals is not always coupled with higher soil pH levels. Kabata-Pendias and Pendias (1999) further explained that mobility of these metals at high pH often increased due to the formation of complexes with soil organic entities available to plants. The quality of medicinal plants may serve as an indicator of ecological parameters and cultivation techniques of their habitat. Above all those parameters, soil is one of the vital media that plays a major role in the promotion of plant growth, development of medicinal plants, and determination of final quality of herbal plant products (Liu et al. 2007). Among soil properties, nutritional elements are indispensable for the normal growth and production of medicinal plants. It plays keyrole in building up plants tissues and help in their normal metabolic activities (Al-Humaid 2005). Among other factors, soil type is one factor that influences mineral composition of medicinal plants. In this regard, the influences of four types of soil on the growth and physiological and biochemical characteristics of medicinal plant *Lycoris aurea* (L' Her.) were investigated by Quan and Liang (2017). In this study, four representative types of soil, including humus soil, sandy soil, garden soil, and yellow-brown soil were selected in order to identify suitable soil type for cultivation of this beneficial herb. Significant differences in elemental composition and activity of soil enzymes were observed. It was reported that humus soil was the most favorable as it contained higher contents of organic matter, alkali-hydrolysable nitrogen, Ca and Mg, and soil enzyme activity. Among the other soil types, yellow-brown soil was the least beneficial. Appropriate soil type is thus proved as another key factor to ensure medicinal plant quality. According to Antoniadis et al. (2017), soil type, soil pH, and trace elements may influence secondary metabolite composition of plants. Some of the recent research explored fascinating revelations about soil pH in many soil processes. Most importantly, soil pH was found to play a vital role in soil pollution control via distribution and removal of hazardous elements from soil ecosystems. For instance, the process of C and N mineralization and pesticide degradation facilitated at pH range 6.5–8, while degradation of petroleum and polycyclic aromatic hydrocarbons (PAHs) occurred at pH range 7–9. Eventually, soil pH can broadly be used to moderately monitor nutrient cycling, plant nutrition, and soil remediation (Neina 2019).

Soil ecosystem is flourished with biological components such as soil enzymes, which play a vital role in decomposition of organic matter and regulation of nutrient cycling (Wang et al. 2016). Soil enzymes accelerate nitrogen and carbon cycles in soil ecosystem by their biological activity; for example, hydroxylases (e.g., urease and sucrase) are used to hydrolyze proteins and polysaccharides to form simpler and smaller molecules that are easily absorbed by plants. It was reported that soil enzyme activity is highly dependent on other physicochemical properties of soil, soil type, cultivation, fertilizers, and agricultural practices (Han et al. 2014; Tamura et al. 2007). Soil microorganisms are considered as another biological component of soil system, which play a key role in the process of mineralization and breaking down of complex organic compounds in soil. In contrary, the functional diversity of microbial populations is greatly influenced by incorporation of organic amendments and quality and crop cover residues (Nair and Ngouajio 2012). One strategy to improve nutritional quality of soil in terms of soil enzyme activity would be to inoculate beneficial soil microorganism, while giving attention to mineral and other compounds. Several elements such as Fe, Na, and other heavy metals when present in soil may either influence or arrest enzymatic activity, thus; it is crucial to regulate their bioavailability (Turan et al. 2017). Medicinal plants involved in herbal therapies are usually prescribed for longer periods; therefore, even a small amount of heavy metal present in those plants may accumulate and cause hazardous effects (Kraft and Hobbs 2004). Medicinal plants cultivated in contaminated soil often experience stress due to undesirable soil properties and subsequently show alterations in plant constituents including accumulation of unwanted elements, which may obviously influence affect the therapeutic potential of valuable health care products. Nevertheless, availability and uptake of heavy metal from soils is a complex procedure that involves soil-plant interaction, where as significantly influenced by several factors including plant species, genotypes, mineral availability and allocation insoil, physicochemical properties of soil, biogeochemical cycling and microbial activities influencing mobility and availability of mineral to plant especially at rhizosphere level (Gebski 1998; Kabata-Pendias and Pendias 1999; Radanovic et al. 2001; Nadgórska-Socha et al. 2013). Nonetheless, knowledge regarding more precise impacts of edaphic factors on the regulation of elemental characterization of medicinal plants is still limited and needs more accurate research endeavors to explore their soil–mineral–plant interactions with emphasis on medicinal plants.

8.3.4 Potential Influence of Heavy Metal Stress on Elemental Composition of Medicinal Plants

Heavy metal stress is another widely reported abiotic factor related to medicinal plants. Toxic effect of heavy metal contamination of medicinal plants on human has caused numerous health hazards including kidney and liver failure leading to death (Street 2012). Plants may acclimate unfavorable environmental condition such as heavy metal stress by accumulating certain heavy metals in their tissues due to their adaption quality under undesirable environmental condition (Kabata-pendias 2011). However, it is important to explore these heavy metals that enter the food chain via such plant resources as potentially toxic elements and then become a part of food webs, thereby affecting living organisms (Boyd 2009). Accumulation of heavy metals in pharmaceutically active medicinal plants is a matter of great concern (Sarma et al. 2012). Denholm (2010) elaborated three major mechanisms that have been proposed to explain heavy metal contamination of medical plant-based products: contamination during cultivation, inadvertent cross-contamination during processing, and/or purposeful introduction of heavy metals for alleged medicinal purposes. Several medicinal plants are reported to show high accumulation affinity toward trace elements and REEs, such as *Nepeta praetervisa* reported by Anjum et al. (2019). *Senecio coronatus* (Thunb.) Harv., a medicinal plant native to Africa, was reported as a nickel (Ni)-hyper-accumulating plant (Przybyłowicz et al. 1995). Similarly, Ashraf et al. (2010) reported few species of *Artemisia* as hyper-accumulator of certain elements to toxic levels (Table 8.1). Two African medicinal plant species *Datura metal* L. and *Datura innoxia* Miller, Gard were also recommended as mellophytes due to their high accumulation tendency toward Co and Ni (Bhattacharjee et al. 2004; Kelly et al. 2002). Nkoane et al. (2005) indicated two medicinal plants namely *Helichrysum candolleanum* H. Buek and *Blepharis diversispina* (Nees) C.B. Clarke with high acclimation potential against heavy metal stress. Despite their importance in organic industries, use of these medicinal plants with regard to consumers' safety is not recommended (Street 2012). Thus, a thorough investigation of medicinal plants for their heavy metal content prior to consumption in order to prevent lethal heavy metal toxicity must be implemented (Sharma et al. 2009). It is also important to understand the influence of other abiotic factors such as pH, temperature, redox potential, cation exchange capacity, and organic matter content on the availability of heavy metal to plants (Gregor 2004). Furthermore, soil–plant and root–microbes interactions play vital roles in regulating heavy metal movement from the soil to edible plant parts (Islam et al. 2007). According to Gregor (2004), an increase in metal concentration in external medium reinforced metal accumulation in plant roots and leaves. A similar quandary was highlighted by Dghaim et al. (2015), while exclusively analyzing elements like Cu, Fe, Zn, Cd, and Pb in seven most popular herbal plants, viz, parsley (*Petroselinumcrispum*), basil (*Ocimum basilicum*), sage (*Salvia officinalis*), oregano (*Origanum vulgare*), mint (*Mentha spicata*), thyme (*Thymus vulgaris*), and chamomile (*Matricaria chamomilla*), used in United Arab Emirates on daily basis for either cooking or medicinal purposes (Table 8.1). Their results deciphered heavy metal contamination of these herbs as the levels were above the World Health Organization (WHO) permissible limit for herbal plants, and thus an urgent monitoring and testing plan for the quality control of these plants was suggested. Safety measures should be taken from the point of harvesting till they reach the consumers, and further studies must be conducted to find out elemental concentration especially in commonly used medicinal plants that are directly consumed in raw form by local indigenous communities.

Elemental composition of medicinal plants plays a vital role in structural composition of secondary metabolites, but it may have an undeniable effect on their regulation (Poutaraud and Girardin 2005). If accumulated in high concentration in medicinal plant parts, heavy metal may alter the ultra-morphological characteristic features and active organic constituents of therapeutic nature (Nasim and Dhir 2010). Heavy metal stress may also interfere with the uptake of other essential elements such as Cd stress causes interference with the uptake of Ca, Mg, K and P, which may cause an imbalance in nutritional composition (Benavides et al. 2005). Furthermore, Fe deficiency in roots of cucumber and sugarbeet was due to Cd-induced inhibition of the Fe (III) reductase (Alcántara et al. 1994). Chaffei et al. (2004) reported a decrease in the activity of nitrate and nitrite reductases in roots and leaves of Cd-treated tomato plants. Heavy metal contamination of soil and water in today's climate change scenario is a serious threat to human health and food safety. Deleterious anthropogenic activities further accelerated the occurrence and intensity of abiotic stresses such as heavy metal stress. As a consequence, plants are now exposed to toxicity of HMs more than any time in their history, since the beginning of their terrestrial life on planet earth. This necessitates making more efforts to deepen our appreciation of HMs and the way plants respond to their ever-growing presence (Emamverdian et al. 2015).

8.4 Conclusion

Human population is increasing and so are demands for better healthcare products. Medicinal plants and their traditional uses have secured enormous value globally due to their potential against serious diseases. Increasing commercialization and environmental contamination of medicinal plants pose several threats to their survival and conservation, especially in their natural ecosystems. Anthropogenic perturbation further accelerated and manifested a broad array of stresses on this nature's pharmacy. Natural and anthropogenic ongoing activities have not only wreaked the havoc on their distribution and availability but severely influence their quality. Alongside other quality parameters, mineral composition of medicinal plants must be monitored and assessed under varying ecological and environmental factors. Despite their undeniable importance in human health, little is known about the influence of environmental and edaphic factors on elemental composition of medicinal plants.

Any contamination/pollution by anthropogenic contributions or natural processes altering elemental composition of soil, air, or water ultimately influences the elemental profile of plants, which may alter the elemental profile of same medicinal plant species of different sites, indicating the importance of collection and cultivation sites. Different abiotic factors such as temperature, drought, salinity, and heavy metal stress imposed deleterious effects on elemental composition of medicinal plants, and though limited, data is still available to report the effects of these individual factors. Magnitude of multiple environmental factors on elemental composition of wild medicinal plants was also recognized. Edaphic factors such as soil pH are well elaborated for their influence on elemental profiles of plants. At present, relatively less is understood about the impacts of biotic stress factors on mineral nutritional prospects of medicinal plants. The need to carefully monitor changes in quality and quantity of mineral composition of natural herbal products in response to biotic stress factors must be considered in the interest of public healthcare safety because accumulation of any hazardous element may cause serious life threats to human life. Although not all active organic compounds of medicinal plants in relation to contaminated environment are discussed in this chapter, those discussed have substantial importance and may be considered as a sincere contribution to medicinal plants research, which may further be utilized in the selection of the right medicinal plants cultivation, as well as in their harvesting sites. The data discussed in this chapter can further be used in the discovery of new herbal drugs for the treatment of various diseases with minimal and non-redundant effects on human health.

References

Abd EL-Azim W, Ahmed ST (2009) Effect of salinity and cutting date on growth and chemical constituents of *Achillea fragrantissima* Forssk, under Ras Sudr conditions. Res J Agric Biol Sci 5(6):1121–1129

Abd El-Wahab M (2006) The efficiency of using saline and fresh water irrigation as alternating methods of irrigation on the productivity of *Foeniculum vulgare* Mill subsp. vulgare var. vulgare under North Sinai conditions. Res J Agric Biol Sci 2(6):571–577

Abdykhalikova K, Taurbaeva G, Chashkov V (2018) Influence of the industrial zone on the chemical composition of some medicinal plants. Norw J Develop Int Sci 17:3–6

Adamczyk-Szabela D, Markiewicz J, Wolf WM (2015) Heavy metal uptake by herbs. IV. Influence of soil pH on the content of heavy metals in *Valeriana officinalis* L. Water Air Soil Pollut 226(4):106

Adazabra A, Ntiforo A, Appiah-Kubi K (2012) Compositional elemental assessment of some core herbal plants used in northern Ghana by instrumental neutron activation analysis. Der Pharma Chemica 4:1962–1968

Ahmed F, Ishiga H (2006) Trace metal concentrations in street dusts of Dhaka city, Bangladesh. Atmos Environ 40(21):3835–3844

Ajagun EJ, Angalapele JA, Nwaiwu PN, Alabi MA, Oladimeji-Salami JA, Amba U (2017) Phytochemical screening and effect of temperature on proximate analysis and mineral composition of *Zingiber officinale* Rosc. Biotechnol J Int 18(3):1–7

Aksuner N, Henden E, Aker Z, Engin E, Satik S (2012) Determination of essential and non-essential elements in various tea leaves and tea infusions consumed in Turkey. Food Addit Contam B 5(2):126–132

Ali EF, Bazaid S, Hassan FAS (2013) Salt effects on growth and leaf chemical constituents of *Simmondsia chinensis* (Link) Schneider. J Med Plants Stud 1(3):22–34

Al Jaouni S, Saleh AM, Wadaan MA, Hozzein WN, Selim S, AbdElgawad H (2018) Elevated CO_2 induces a global metabolic change in basil (*Ocimum basilicum* L.) and peppermint (*Mentha piperita* L.) and improves their biological activity. J Plant Physiol 224:121–131

Al Moaruf Olukayode A, Bello MO, Ibrahim AO, Ogunwande IA, Olawore NO (2004) Heavy trace metals and macronutrients status in herbal plants of Nigeria. Food Chem 85(1):67–71

Alcántara E, Romera FJ, Cañete M, De la Guardia MD (1994) Effects of heavy metals on both induction and function of root Fe (lll) reductase in Fe-deficient cucumber (Cucumis sativus L.) plants. J Exp Bot 45(12):1893–1898

Al-Humaid A (2005) Effects of compound fertilization on growth and alkaloids of datura plants. J Plant Nutr 27(12):2203–2219

Allen MR, Frame DJ, Huntingford C, Jones CD, Lowe JA, Meinshausen M, Meinshausen N (2009) Warming caused by cumulative carbon emissions towards the trillionth tonne. Nature 458(7242):1163–1166

Anjum F, Yaseen M, Rasool E, Wahid A, Anjum S (2003) Water stress in barley (Hordeum vulgare L.) I. Effect on morpohological characters. Seeds 105:266–271

Anjum S, Bazai ZA, Rizwan S, Benincasa C, Mehmood K, Siddique N, Shaheen G, Mehmood Z, Azam M, Sajjad A (2019) Elemental characterization of medicinal plants and soils from Hazarganji Chiltan National Park and nearby unprotected areas of Balochistan, Pakistan. J Oleo Sci 68(5):443–461

Antoniadis V, Levizou E, Shaheen SM, Ok YS, Sebastian A, Baum C, Prasad MN, Wenzel WW, Rinklebe J (2017) Trace elements in the soil-plant interface: phytoavailability, translocation, and phytoremediation–a review. Earth Sci Rev 171:621–645

Arceusz A, Radecka I, Wesolowski M (2010) Identification of diversity in elements content in medicinal plants belonging to different plant families. Food Chem 120(1):52–58

Ashraf M, Orooj A (2006) Salt stress effects on growth, ion accumulation and seed oil concentration in an arid zone traditional medicinal plant ajwain (*Trachyspermum ammi* [L.] Sprague). J Arid Environ 64(2):209–220

Ashraf M, Hayat MQ, Mumtaz AS (2010) A study on elemental contents of medicinally important species of Artemisia L.(Asteraceae) found in Pakistan. J Med Plants Res 4(21):2256–2263

Baatour O, Maha Z, Nada B, Zeineb OA (2018) Effects of NaCl on plant growth and antioxidant activities in fenugreek (*Trigonella Foenum Graecum* L.). Biosci J 34(3):683–696

Bahadur A, Chaudhry Z, Jan G, Danish M, ur Rehman A, Ahmad R, Khalid S, Shah Z, Ali F, Mushtaq T (2011) Nutritional and elemental analyses of some selected fodder species used in traditional medicine. Afr J Pharm Pharmacol 5(8):1157–1161

Baker NR, Rosenqvist E (2004) Applications of chlorophyll fluorescence can improve crop production strategies: an examination of future possibilities. J Exp Bot 55:1607–1621

Barber SA (1995) Soil nutrient bioavailability: a mechanistic approach. Wiley, New York

Başgel S, Erdemoğlu S (2006) Determination of mineral and trace elements in some medicinal herbs and their infusions consumed in Turkey. Sci Total Environ 359(1–3):82–89

Benavides MP, Gallego SM, Tomaro ML (2005) Cadmium toxicity in plants. Braz J Plant Physiol 17(1):21–34

Bernays EA, Chapman RF (2000) Plant secondary compounds and grasshoppers: beyond plant defenses. J Chem Ecol 26(8):1773–1794

Bhatt R, Rao NS (2005) Influence of pod load on response of okra to water stress. Indian J Plant Physiol 10(1):54

Bhattacharjee S, Kar S, Chakravarty S (2004) Mineral compositions of Datura: a traditional tropical medicinal plant. Commun Soil Sci Plant Anal 35(7–8):937–946

Boyd RS (2009) High-nickel insects and nickel hyperaccumulator plants: a review. Insect Sci 16(1):19–31

Bradl HB (2004) Adsorption of heavy metal ions on soils and soils constituents. J Colloid Interface Sci 277(1):1–18

Brady N, Weil R (1999) Soil phosphorus, potassium, and micronutrients. In: Elements of the nature and properties of soils. Prentice-Hall, Upper Saddle River, pp 421–422

Brooks RR (1994) Plants that hyperaccumulate heavy metals. In: Plants and the chemical elements: biochemistry, uptake, tolerance and toxicity. Wiley, New York, pp 87–105

Caravaca F, Alguacil M, Diaz G, Marin P, Roldán A (2005) Nutrient acquisition and nitrate reductase activity of mycorrhizal Retama sphaerocarpa L. seedlings afforested in an amended semi-arid soil under two water regimes. Soil Use Manag 21(1):10–16

Cha S, Chae H-M, Lee S-H, Shim J-K (2017) Effect of elevated atmospheric CO_2 concentration on growth and leaf litter decomposition of Quercus acutissima and Fraxinus rhynchophylla. PLoS One 12(2):e0171197

Chaffei C, Pageau K, Suzuki A, Gouia H, Ghorbel MH, Masclaux-Daubresse C (2004) Cadmium toxicity induced changes in nitrogen management in Lycopersicon esculentum leading to a metabolic safeguard through an amino acid storage strategy. Plant Cell Physiol 45(11):1681–1693

Claeys H, Van Landeghem S, Dubois M, Maleux K, Inzé D (2014) What is stress? Dose-response effects in commonly used in vitro stress assays. Plant Physiol 165(2):519–527

Denholm J (2010) Complementary medicine and heavy metal toxicity in Australia. WebmedCentral 1:1–6

Dghaim R, Al Khatib S, Rasool H, Ali Khan M (2015) Determination of heavy metals concentration in traditional herbs commonly consumed in the United Arab Emirates. J Environ Public Health 2015:973878

Dixon RA, Paiva NL (1995) Stress-induced phenylpropanoid metabolism. Plant Cell 7(7):1085

Emamverdian A, Ding Y, Mokhberdoran F, Xie Y (2015) Heavy metal stress and some mechanisms of plant defense response. Sci World J 2015:756120

Ernst W, Verkleij J, Schat H (1992) Metal tolerance in plants. Acta Bot Neerl 41(3):229–248

Farooq M, Wahid A, Kobayashi N, Fujita D, Basra S (2009) Plant drought stress: effects, mechanisms and management. In: Sustainable agriculture. Springer, Cham, pp 153–188

Fernandes MDFL, Moraes SMD, Sousa PHMD, Magalhães CEDC, Almeida MMB, Silva MGDV (2019) Characterization of leaves used in infusion preparation grown in Northeastern Brazil by chemometric methods based on their multi-elemental composition. Food Sci Technol 39:309–315

Fifield FW, Haines PJ (2000) Environmental analytical chemistry. Wiley-Blackwell, London

Förstner U (1995) Land contamination by metals: global scope and magnitude of problem. In: Metal speciation and contamination of soil. Lewis Publishers (CRC Press), Boca Raton, pp 1–33

García-Caparrós P, Romero MJ, Llanderal A, Cermeño P, Lao MT, Segura ML (2019) Effects of drought stress on biomass, essential oil content, nutritional parameters, and costs of production in six Lamiaceae species. Water 11(3):573

Gaspar T, Franck T, Bisbis B, Kevers C, Jouve L, Hausman J-F, Dommes J (2002) Concepts in plant stress physiology. Application to plant tissue cultures. Plant Growth Regul 37(3):263–285

Gebski M (1998) Soil and fertilizer factors affecting the uptake of heavy metals by plants. Postepy Nauk Rolniczych 5:3–16

Ghani A, Mustafa I, Ali Z, Ishtiaq M, Ahmed I, Hasa N (2014) Ethno-medicinal survey of plants of Soon valley, Khushab district, Punjab, Pakistan. J Med Plant Res 8:1031–1034

Gilani AH (2005) Trends in ethnopharmacology. J Ethnopharmacol 100(1–2):43–49

Grattan S, Grieve C (1998) Salinity–mineral nutrient relations in horticultural crops. Sci Hortic 78(1–4):127–157

Gregor M (2004) Metal availability, uptake, transport and accumulation in plants. In: Heavy metal stress in plants from biomolecules to ecosystems. Springer, Cham, pp 1–27

Hagerman A, Butler LG (1991) Tannins and lignins. In: Herbivores: their interactions with secondary plant metabolites, vol 1. Elsevier, Amsterdam, pp 355–388

Haidu D, Párkányi D, Moldovan RI, Savii C, Pinzaru I, Dehelean C, Kurunczi L (2017) Elemental characterization of Romanian crop medicinal plants by neutron activation analysis. J Anal Methods Chem 2017

Hamilton AC (2004) Medicinal plants, conservation and livelihoods. Biodivers Conserv 13(8):1477–1517

Han F, Wang L, Wang F (2014) Analysis of relationship between soil enzymes activities and fertilities in the field of grain for green in downstream of the Shiyang River. Chinese J Soil Sci 45:1396–1401

Holopainen JK, Gershenzon J (2010) Multiple stress factors and the emission of plant VOCs. Trends Plant Sci 15(3):176–184
Islam E, Yang X-e, He Z-l, Mahmood Q (2007) Assessing potential dietary toxicity of heavy metals in selected vegetables and food crops. J Zhejiang Univ Sci B 8(1):1–13
Jabeen S, Shah MT, Khan S, Hayat MQ (2010) Determination of major and trace elements in ten important folk therapeutic plants of Haripur basin. Pak J Med Plants Res 4(7):559–566
Jaleel CA, Sankar B, Murali P, Gomathinayagam M, Lakshmanan G, Panneerselvam R (2008) Water deficit stress effects on reactive oxygen metabolism in Catharanthus roseus; impacts on ajmalicine accumulation. Colloids Surf B: Biointerfaces 62(1):105–111
Kabata-Pendias A (2011) Trace elements in soils and plants, 4th edn. CRC Press, Boca Raton
Kabata-Pendias A, Pendias H (1984) Trace elements in plants and soils. CRC Press, Boca Raton, pp 233–237
Kabata-Pendias A, Pendias H (1992) Soil and plants. Lewis Publication Inc, Boca Raton
Kabata-Pendias A, Pendias H (1999) Biogeochemistry of trace elements. Polish Scientific Publishing Company, Varsov, p 400. (Kabata-Pendias, 2011 #542)
Kage H, Kochler M, Stützel H (2004) Root growth and dry matter partitioning of cauliflower under drought stress conditions: measurement and simulation. Eur J Agron 20(4):379–394
Kanias GD, Loukis A (1987) Determination and correlation of active constituents and trace elements in the medicinal plant Thymus capitatus Hoffm. and Link. Fresenius' Z Anal Chem 327(3–4):355–357
Kashem M, Singh B (2002) Distribution and mobility of Cd, Ni, Zn in contaminated tropical soil profiles. 17th WCSS, 14–21 August 2002. Thailand(625–1)-(625–9)
Kelly RA, Andrews JC, DeWitt JG (2002) An X-ray absorption spectroscopic investigation of the nature of the zinc complex accumulated in Datura innoxia plant tissue culture. Microchem J 71(2–3):231–245
Khan KY, Khan MA, Niamat R, Munir M, Fazal H, Mazari P, Seema N, Bashir T, Kanwal A, Ahmed SN (2011) Element content analysis of plants of genus Ficus using atomic absorption spectrometer. Afr J Pharm Pharmacol 5(3):317–332
Kim BG, Waadt R, Cheong YH, Pandey GK, Dominguez-Solis JR, Schültke S, Lee SC (2007) The calcium sensor CBL10 mediates salt tolerance by regulating ion homeostasis in Arabidopsis. 52(3):473–484
Kloke A, Sauerbeck D, Vetter H (1984) The contamination of plants and soils with heavy metals and the transport of metals in terrestrial food chains. In: Changing metal cycles and human health. Springer, Cham, pp 113–141
Knight H, Brandt S, Knight MR (1998) A history of stress alters drought calcium signalling pathways in Arabidopsis. Plant J 16(6):681–687
Koc H, Sari H (2009) Trace metal contents of some medicinal, aromatic plants and soil samples in the Mediterranean region, Turkey. J Appl Chem Res 8(4):52–57
Kohlmünzer S (2003) Farmakognozja (in Polish – Pharmacognosy). PZWL, Warsaw
Kolasani A, Xu H, Millikan M (2011) Evaluation of mineral content of Chinese medicinal herbs used to improve kidney function with chemometrics. Food Chem 127(4):1465–1471
Kouki M, Manetas Y (2002) Resource availability affects differentially the levels of gallotannins and condensed tannins in Ceratonia siliqua. Biochem Syst Ecol 30(7):631–639
Kraft K, Hobbs C (2004) Pocket guide to herbal medicine. Georg Thieme Verlag, Stuttgart
Kusaka M, Ohta M, Fujimura T (2005) Contribution of inorganic components to osmotic adjustment and leaf folding for drought tolerance in pearl millet. Physiol Plant 125(4):474–489
Liang C, Tao L (2004) Element selenium and human body health [J]. Stud Trace Elem Health 3
Liu Y, Zhang Z, He Y, Zhang B, Li X (2007) Quality of crude traditional Chinese drugs and ecological environment. Modern Tradit Chin Med Materia Materia-World Sci Technol 9:65–69
Mathew S, Abraham TE (2006) Studies on the antioxidant activities of cinnamon (Cinnamomum verum) bark extracts, through various in vitro models. Food Chem 94(4):520–528
Meinshausen M, Meinshausen N, Hare W, Raper SC, Frieler K, Knutti R, Frame DJ, Allen MR (2009) Greenhouse-gas emission targets for limiting global warming to 2 C. Nature 458(7242):1158–1162

Mengel K, Kirkby EA, Kosegarten H, Appel T (2001) Nitrogen. In: Principles of plant nutrition. Springer, Cham, pp 397–434

Mickelbart MV, Gosney MJ, Camberato J, Stanton KM (2012) Soil pH effects on growth and foliar nutrient concentrations of Spiraea alba Du Roi and Spiraea tomentosa L. HortScience 47(7):902–906

Minasny B, Hong SY, Hartemink AE, Kim YH, Kang SS (2016) Soil pH increase under paddy in South Korea between 2000 and 2012. Agric Ecosyst Environ 221:205–213

Mohammadi H, Hazrati S, Parviz L (2019) Morphophysiological and biochemical response of savory medicinal plant using silicon under salt stress. Ann Univ Mariae Curie-Sklodowska Sectio C–Biol 2:29–40

Munns R (2002) Comparative physiology of salt and water stress. Plant Cell Environ 25(2):239–250

Nadgórska-Socha A, Kafel A, Kandziora-Ciupa M, Gospodarek J, Zawisza-Raszka A (2013) Accumulation of heavy metals and antioxidant responses in Vicia faba plants grown on monometallic contaminated soil. Environ Sci Pollut Res 20(2):1124–1134

Nair A, Ngouajio M (2012) Soil microbial biomass, functional microbial diversity, and nematode community structure as affected by cover crops and compost in an organic vegetable production system. Appl Soil Ecol 58:45–55

Nasim SA, Dhir B (2010) Heavy metals alter the potency of medicinal plants. In: Reviews of environmental contamination and toxicology. Springer, Cham, pp 139–149

Ncube B, Finnie J, Van Staden J (2012) Quality from the field: the impact of environmental factors as quality determinants in medicinal plants. S Afr J Bot 82:11–20

Neina D (2019) The role of soil pH in plant nutrition and soil remediation. Appl Environ Soil Sci 2019

Nkoane BB, Sawula GM, Wibetoe G, Lund W (2005) Identification of Cu and Ni indicator plants from mineralised locations in Botswana. J Geochem Explor 86(3):130–142

Obratov-Petković D, Popović I, Belanović S, Kadović R (2006) Ecobiological study of medicinal plants in some regions of Serbia. Plant Soil Environ 52(10):459–467

Oluwalana I, Oluwamukomi M, Fagbemi T, Oluwafemi G (2011) Effects of temperature and period of blanching on the pasting and functional properties of plantain (Musa parasidiaca) flour. J Stored Prod Postharv Res 2(8):164–169

Osuagwu G, Edeoga H (2012) The influence of water stress (drought) on the mineral and vitamin content of the leaves of Gongronema latifolium (Benth). Int J Med Aromat Plants 2(2):301–309

Petropoulos SA, Daferera D, Polissiou M, Passam H (2008) The effect of water deficit stress on the growth, yield and composition of essential oils of parsley. Sci Hortic 115(4):393–397

Poutaraud A, Girardin P (2005) Improvement of medicinal plant quality: a Hypericum perforatum literature review as an example. Plant Genet Resour 3(2):178–189

Przybyłowicz W, Pineda C, Prozesky V, Mesjasz-Przybyłowicz J (1995) Investigation of Ni hyperaccumulation by true elemental imaging. Nucl Instrum Methods Phys Res, Sect B 104(1–4):176–181

Quan M, Liang J (2017) The influences of four types of soil on the growth, physiological and biochemical characteristics of Lycoris aurea (L'Her.) Herb. Sci Rep 7:43284

Radanovic D, Antic-Mladenovic S, Jakovljevic M (2001) Influence of some soil characteristics on heavy metal content in Hypericum perforatum L. and Achillea millefolium L. In: International conference on medicinal and aromatic plants. Possibilities and limitations of medicinal and aromatic plant, vol 576, pp 295–301

Radojevic M, Bashkin V (1999) Organic matter. In: Practical environmental analysis. Royal Society of Chemistry, London, pp 325–329

Raju MK, Madhusudhana Rao PV, Brahmaji Rao JS, Venkatasubramani CR (2016) Elemental analysis of medicinal plants from different sites by instrumental neutron activation analysis. Int J Bioassays 5(3):4892–4896

Reddy AR, Chaitanya KV, Vivekanandan M (2004) Drought-induced responses of photosynthesis and antioxidant metabolism in higher plants. J Plant Physiol 161(11):1189–1202

Reimann C, Koller F, Frengstad B, Kashulina G, Niskavaara H, Englmaier P (2001) Comparison of the element composition in several plant species and their substrate from a 1 500 000-km^2 area in Northern Europe. Sci Total Environ 278(1–3):87–112

Robinson MM, Zhang X (2011) The world medicines situation 2011, traditional medicines: global situation, issues and challenges. World Health Organization, Geneva, pp 1–2
Ruelland E, Zachowski A (2010) How plants sense temperature. Environ Exp Bot 69(3):225–232
Said-Al Ahl H, Omer E (2011) Medicinal and aromatic plants production under salt stress. A review. Herba Pol 57(2):72–87
Salehi A, Tasdighi H, Gholamhoseini M (2016) Evaluation of proline, chlorophyll, soluble sugar content and uptake of nutrients in the German chamomile (Matricaria chamomilla L.) under drought stress and organic fertilizer treatments. Asian Pac J Trop Biomed 6(10):886–891
Sarker U, Islam MT, Oba S (2018) Salinity stress accelerates nutrients, dietary fiber, minerals, phytochemicals and antioxidant activity in Amaranthus tricolor leaves. PLoS One 13(11):e0206388
Sarma H, Deka S, Deka H, Saikia RR (2012) Accumulation of heavy metals in selected medicinal plants. In: Reviews of environmental contamination and toxicology. Springer, Cham, pp 63–86
Schroeder HA (1973) The trace elements and man. Book Sales Inc, New York
Selvakumar G, Bhatt RM, Upreti KK, Bindu GH, Shweta K (2015) Citricoccus zhacaiensis B-4 (MTCC 12119) a novel osmotolerant plant growth promoting actinobacterium enhances onion (Allium cepa L.) seed germination under osmotic stress conditions. World J Microbiol Biotechnol 31(5):833–839
Seo AY, Joseph A-M, Dutta D, Hwang JC, Aris JP, Leeuwenburgh C (2010) New insights into the role of mitochondria in aging: mitochondrial dynamics and more. J Cell Sci 123(15):2533–2542
Serfor-Armah Y, Akaho E, Nyarko B, Kyere A, Oppon-Boachie K (2003) Application of instrumental neutron activation analysis to plant medicines in Ghana: a review. J Radioanal Nucl Chem 257(1):125–128
Shao H-b, Chu L-y, Shao M-a, Jaleel CA, Hong-mei M (2008) Higher plant antioxidants and redox signaling under environmental stresses. C R Biol 331(6):433–441
Sharma RK, Agrawal M, Marshall FM (2009) Heavy metals in vegetables collected from production and market sites of a tropical urban area of India. Food Chem Toxicol 47(3):583–591
Sheded M, Pulford I, Hamed A (2006) Presence of major and trace elements in seven medicinal plants growing in the South-Eastern Desert, Egypt. J Arid Environ 66(2):210–217
Shi H, Ishitani M, Kim C, Zhu J-K (2000) The Arabidopsis thaliana salt tolerance gene SOS1 encodes a putative Na+/H+ antiporter. Proc Natl Acad Sci 97(12):6896–6901
Shrivastava P, Kumar R (2015) Soil salinity: a serious environmental issue and plant growth promoting bacteria as one of the tools for its alleviation. Saudi J Biol Sci 22(2):123–131
Somers E (1974) The toxic potential of trace metals in foods. A review. J Food Sci 39(2):215–217
Street R (2012) Heavy metals in medicinal plant products—an African perspective. S Afr J Bot 82:67–74
Sulaiman ID, AlKubaisy RK, Salman JD (2017) Determination of some essential elements in Iraqi medicinal herbs. Ibn Al-Haitham J Pure Appl Sci 26(1):234–240
Szabolcs I (1994) Soils and salinisation. In: Pessarakali M (ed) Handbook of plant and crop stress. Marcel Dekker, New York, pp 3–11
Tabatabaie SJ, Nazari J (2007) Influence of nutrient concentrations and NaCl salinity on the growth, photosynthesis, and essential oil content of peppermint and lemon verbena. Turk J Agric For 31(4):245–253
Tadayyon A, Nikneshan P, Pessarakli M (2018) Effects of drought stress on concentration of macro-and micro-nutrients in Castor (Ricinus communis L.) plant. J Plant Nutr 41(3):304–310
Tahir MHN, Mehid S (2001) Evaluation of open pollinated sunflower (Helianthus annuus L.) populations under water stress and normal conditions. Int J Agric Biol 3:236–238
Talei D, Kadir MA, Yusop MK, Valdiani A, Abdullah MP (2012) Salinity effects on macro and micronutrients uptake in medicinal plant King of Bitters (Andrographis paniculata Nees.). Plant Omics J 5(3):271–278
Tamura K, Dudley J, Nei M, Kumar S (2007) MEGA4: molecular evolutionary genetics analysis (MEGA) software version 4.0. Mol Biol Evol 24(8):1596–1599
Tinus RW (1980) Nature and management of soil pH and salinity. In: Proceeding North American forest tree nursery soils workshop; July 1980. pp 72–86

Tissue D, Thomas R, Strain B (1997) Atmospheric CO_2 enrichment increases growth and photosynthesis of Pinus taeda: a 4 year experiment in the field. Plant Cell Environ 20(9):1123–1134

Turan M, Nikerel E, Kaya K, Kitir N, Gunes A, Mokhtari NEP, Tüfenkçi Ş, Karaman MR, Çimrin KM (2017) Enzyme dynamic in plant nutrition uptake and plant nutrition. In: Enzyme inhibitors and activators. IntechOpen Limited, London, p 225

Uddin K, Juraimi AS, Anwar F, Hossain MA, Alam MA (2012) Effect of salinity on proximate mineral composition of purslane ('Portulca oleracea'L.). Aust J Crop Sci 6(12):1732

Valls J, Richard T, Trotin F, Monti J-P, Mérillon J-M, Vitrac X (2007) Carbon-14 biolabeling of flavanols and chlorogenic acids in Crataegus monogyna cell suspension cultures. Food Chem 105(2):879–882

Vinocur B, Altman A (2005) Recent advances in engineering plant tolerance to abiotic stress: achievements and limitations. Curr Opin Biotechnol 16(2):123–132

Vtorova V (1987) Substantiation of methods and objects of observations over chemical composition of plants during monitoring of forest ecosystem. Inf Bull Probl III Counc Mutual Econ Help 1:1–2

Vurukonda SSKP, Vardharajula S, Shrivastava M, SkZ A (2016) Enhancement of drought stress tolerance in crops by plant growth promoting rhizobacteria. Microbiol Res 184:13–24

Wang L, Wang F, Guo C, Han F, Wei L, Li F (2016) Review: progress of soil enzymology. Soil 48(1):12–21

Wilkinson K Use of traditional healers vastly exaggerated. Mail and Guardian. 2013 Aug 1. [Internet] [Cited 2016 Feb 21] Available from https://mg.co.za/article/2013-08-01-usage-of-traditional-healers-vastly-exaggerated/:1–9

Wong JW, Wong MH (1990) Effects of fly ash on yields and elemental composition of two vegetables, Brassica parachinensis and B. chinensis. Agric Ecosyst Environ 30(3–4):251–264

Wong M, Tan P, Wee Y (1993) Heavy metals in some Chinese herbal plants. Biol Trace Elem Res 36(2):135–142

Wu Q-S, Xia R-X, Zou Y-N (2008) Improved soil structure and citrus growth after inoculation with three arbuscular mycorrhizal fungi under drought stress. Eur J Soil Biol 44(1):122–128

Wu D, Liu Y, Liu Y, Dong Y, Chen J, Feng Y, Du J (2011) Study on available copper content in cultivated soils and its affecting factors in Zhuhai. In: 2011 international conference on computer distributed control and intelligent environmental monitoring, IEEE, pp 1350–1353

Yan S, Tang Z, Su W, Sun W (2005) Proteomic analysis of salt stress-responsive proteins in rice root. Proteomics 5(1):235–244

Zhao C, Guo L, Jaleel CA, Shao H, Yang H (2008) Prospects for dissecting plant-adaptive molecular mechanisms to improve wheat cultivars in drought environments. Comp Rend Biol 331(5):579–586

Zhu J-K (2001) Cell signaling under salt, water and cold stresses. Curr Opin Plant Biol 4(5):401–406

Zhu J-K, Hasegawa PM, Bressan RA, Bohnert HJ (1997) Molecular aspects of osmotic stress in plants. Crit Rev Plant Sci 16(3):253–277

Zinn KE, Tunc-Ozdemir M, Harper JF (2010) Temperature stress and plant sexual reproduction: uncovering the weakest links. J Exp Bot 61(7):1959–1968

Chapter 9
Climate Change and Climate-Smart Agriculture

Aneeba Rashid and Safdar Ali Mirza

9.1 Introduction

Climate-smart agriculture is basically a technique used for the transformation and reorientation of different agricultural systems to support food security, keeping in mind the harsh facts of changing climate. The extensive and prevalent variations in patterns of precipitation and temperature jeopardize agricultural production. These changes also increase the susceptibility of individuals who depend on farming for their livelihood, belonging mostly to the poor populations of the world. The changing climate disrupts the food marketplace, posing risks to the supply of food community wide (Arslan 2015). Such intimidations can be diminished by enhancing the adaptive capability of local farmers along with the growth of resilience and use of resources in agrarian production systems.

The overall significance of the respective objectives of CSA fluctuates at different locations and circumstances. Acknowledgment of achieving trade-offs is predominantly significant in unindustrialized countries (FAO 2018b): these countries have set developments in agriculture as a top-priority through adaptation for food security and fiscal growth (FAO 2018). Farmers are the most important contributor to their economy, yet they are the most affected by the changing climate. Climate-smart agriculture highlights the significance of structuring evidence in identifying workable choices and essential empowering actions. CSA offers tools for evaluating the tendency of diverse approaches, methodologies, technologies, and practices regarding their impacts on nationwide progress and food security, keeping in mind the location-specific impacts of changing climate. It shares prevailing learning experience and information about sustainable development in the agriculture sector.

A. Rashid · S. A. Mirza (✉)
Botany Department GC University, Lahore, Pakistan
e-mail: safdaralimirza@gcu.edu.pk

Q. Mahmood (ed.), *Sustainable Plant Nutrition under Contaminated Environments*, Sustainable Plant Nutrition in a Changing World,
https://doi.org/10.1007/978-3-030-91499-8_9

9.2 The Perplexities of Global Climate

The expansion of urban areas and human activities is effecting the climate of biosphere. Researchers provide much assurance that the rise in global temperature and production of greenhouse gases is a consequence of anthropogenic activities. The rise in atmospheric temperature illustrates an important, wide-reaching issue of changing climate. Data are available regarding changing temperatures at the local level, which prove to be important to augment data available at the global level. A study conducted in Japan showed that the average temperatures in two of its cities (Gifu and Ogaki) were alike with a slight variance of 1 °C (Ali and Mutolib 2016). The average daily temperature range was substantial in Gifu as compared to Ogaki city. In the past 30 years, the average temperature in these two cities was moderately increased; the reason for this increase is the significant nature of the terrestrial area of these two cities. It has been estimated that rise in temperature up to 0.31 °C in the twenty-second century (IPCC 2013) poses a severe threat to available freshwater resources on Earth. Melting glaciers increase sea level, which affects the habitats of many animals and plants. According to IPCC (2007), taken as a whole, the range of published evidence indicates that the net damage costs of climate change are likely to be significant and to increase over time. The increase in temperature is not constant across the globe, so every continent is facing a different magnitude of variations in climate. The timespan can only be estimated if the release of greenhouse gases is controlled. In the past century, average precipitation has shown drastic changes. In some regions of the world where the level of precipitation has increased, the trend of increased precipitation is likely even in those areas where a decrease in precipitation level was expected (USGCRP 2017). Since the beginning of this century, the intensity, rate of recurrence, and time period of tornadoes, cyclones, and hurricanes all over the planet have been drastically increased through anthropogenic activities. Rain and the intensity of rainfall storms are likely to increase as the climate becomes warmer. Further, intense droughts and heat waves and less intense cold waves, increased summer temperatures, wildfires in certain areas, and decreases in soil moisture are predictable for many regions around the globe (NASA 2020).

The oceans and other water bodies, being a dynamic source for carbon sinks, assimilate massive volumes of carbon dioxide (CO_2) gas, safeguarding the normal limits of CO_2 in the higher atmosphere layers. These water bodies are also now facing fluctuations of temperatures for 1 °C, with the expectation that the temperatures can rise by 1.5 °C or above (WWF 2020). The increased temperatures of water bodies and increased volumes of CO_2 gas in oceans, which is making them more acidic, is one of the worse impacts that climate change is inflicting on the oceans. The barrier reefs deep in the oceans are predicted to deteriorate by 70%–90% with such increased temperatures. A temperature increase as much as 2 °C may cause all barrier reefs to vanish, and these reefs are an important source of protein for many living organisms (WWF 2020). Recalling the previous record, it is probable the rise in sea level will increase by another one to four feet by the start of the next century. Consequently, the seawaters will continue to expand and the surface of the Earth will be more prone to flooding (NASA 2020). After the oceans, the forestlands are greatly significant as they also absorb huge amounts of CO_2 gas. CO_2 is foremost among the greenhouse gases that account for this increasing global warming. The forestlands with their carbon sinks are habitats of many flora and fauna, including wildlife. Such plant materials after being burned release assimilated CO_2 back into the atmosphere.

The change in climate is also augmented in the Arctic and Antarctic continents, as these continents are predominantly exposed to the effects of continuously increasing global warming. Up to about 2050, the Arctic Ocean is likely to lose all its surface ice during summer seasons (NASA 2020). Continued global warming is also expected to be the leading reason for the endangerment and loss of wildlife. The expected increase in temperature (1.5 °C) may put 30% of wildlife at danger of becoming extinct, and warming of 2 °C may be destructive of the ecosystems. Under such rapid changes in climate, most species are unable to adapt to such drastic climatic conditions (IPCC 2019).

There is a dire need to understand the perplexity of this global climate change as we may be the last possible generation that has the time to do something productive about this danger. The effects of global climate change on diverse ecological and social structures may be mitigated or adapted in future, such as the increase in average global temperature, which will benefit some countries of the world and at the same time harm other countries. Overall yearly expenses may tend to increase with time because of the increases in temperatures worldwide.

9.3 Climate Change and the Agricultural System

The increasing burden of crop production, of living beings, and the use of freshwater sources are some of the factors of agriculture that are constantly affecting the Earth. Burning of fossil fuels and changing water cycles are leading causes affecting the agriculture sector. The compounded climatic aspects may reduce the productivity of food and cash crops, which may result in price increases of those crops in some countries. The rising temperatures have mainly exaggerated the duration of the growing seasons. The early blossoming and harvest of crops is being widely observed round the globe. Climate change has positive effects on agriculture in certain regions, such as northern Europe, where agricultural yield is observed to be increased owing to the extended growing season as well as the lengthier frost-free time periods (EEA 2015). However, the situation is quite different in southern Europe. The extreme heat waves with decreased accessibility of freshwater and low rainfall are projected to reduce the agricultural yield in years to come because of the harsh weather conditions as well as the presence of pests and diseases (EEA 2015).

Considering the example of Mediterranean climates, harsh heat waves and water shortages are quite common in the summer season; some summer crops might be grown in the winter season with this climate change (EEA 2015). Southeastern Europe is also facing decreased production owing to the warmer and drier summer season. However, there is no possibility of transferring crop production from summer to the winter season. The extension in the growing season caused by the increase in warm temperatures might affect the proliferation and spread of diseases, pests, or invasive weeds, which in return reduces agricultural production. In 2010, it appeared that the higher night temperatures drastically affected the corn yields all over the USA (USGCRP 2017); in 2012, the corn yield was also badly affected owing to the warmer winter, causing a loss of about 220 million dollars (USGCRP 2014). Dealing with drought in areas where rising summer temperatures cause soils to become drier is a challenge. Less water is available for irrigation than is needed in some places. Other food sources affected by climate change are in the northeast Atlantic region, where the stocks of fish have also altered. This alteration is distressing the local people who depend upon these stocks all over the logistics network (EEA 2015). Owing to the warmer water temperatures, oceanic transport and shipment are also increased in some regions. These warmer temperatures also support the formation of invasive marine species that harm local fish markets.

The impacts of climatic change on the lives of farmers fluctuate by region depending on access to agriculture-related technologies and how the farmers use the available technologies. This high-tech complexity regulates agricultural productivity way beyond than the agronomic and climatic legacies. Consequently, the food deficiency is not merely an invention from climatic determinism; it can be managed through developments in fiscal, radical, and agrarian plans at local and international levels. Presently, the zones with food shortage have labor-intensive approach of agribusiness with weeds largely controlled by hoeing. The disparities are fairly inconsistent between the production of these labor-intensive farming lands and the agricultural soils where synthetic fertilizers, insecticides, and genetically modified crop varieties are used. The change in climate has a diverse influence on tropic ecosystems from the warmer temperatures; the farmers working in tropic areas are mostly incapable of handling the alterations in climate, because they have fewer fiscal and technical options in their agronomic structures. The C3 plants (wheat, rice, etc.) include more than 90% of all higher plants in the world, so their response to this CO_2 shift will be quite problematic. Until now, the wild C3 plants have responded well to this CO_2 level increase, which indicates the likelihood that greater weed pressure on neighboring crop plants will reduce crop production. The climate change and its consequences are intimately associated with the agriculture sector.

The overall impact of climate change on agriculture is predicted to be negative worldwide. Even though a few colder regions and their summer crops will see an advantage, most of the warmer regions and their winter crops will not. Although these increases in CO_2 concentrations in fewer crops are likely to increase their growth and efficacy of water usage, climatic influences, for instance, extreme heat waves, recurrent droughts, and flooding, are expected to reduce or even diminish the agricultural yield potential in most crops. There are some indirect climatic impacts as well; for instance, the amplified resistance of weeds to grow, the increasing spread of pathogens and pests, and the altered seasons as well as additional variations in agroecosystems for crops. The frequency and spreading of wild plants and insects are expected to rise with climatic change. This rise can cause more difficulties for farmers as those crops have not been exposed to these wild plants formerly. The increasing concentrations of CO_2 lessen the amounts of protein and other vital nutrients in most crop plants (mainly C3 plants such as wheat). This unswerving consequence of rising CO_2 concentration on the dietary value of crop plants characterizes a possible risk to human well-being. Likewise, this well-being is vulnerable to increased usage of pesticides on crop plants.

9.4 Agricultural Instability and Economic Relevance

Agricultural instability may eventuate several economic consequences. First, there may be an increase in price for the most important cultivated crops, such as rice, wheat, maize, and soybeans, leading to higher feed and meat prices. This increase may decrease the trend of meat consumption, resulting in an increase in cereals consumption, and may lead to food insecurity (Fig. 9.1).

Climate change may disrupt food security in terms of food accessibility, availability, quality and spoilage (USDA 2015) combined with increasing population pressure. Universally, the effects of changing climate on farming, livestock, agriculture, and food supply are expected to be similar to the trend observed by the USA (USGCRP 2017). Climate change may also cause several disturbances that affect distribution and transport of food locally or globally, rendering substantial influences on the safety, quality, and availability of food. In

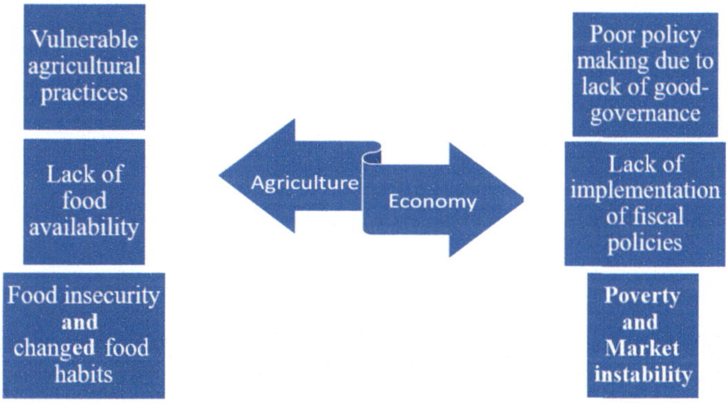

Fig. 9.1 Relationship between agriculture and economy

the USA, for instance, the system for food transportation often transfers large quantities of grain by water (USGCRP 2017). During the summer of 2012, higher temperatures with less rainfall led to the worst droughts the USA has seen. This drought caused serious effects on the basin of the Mississippi River. This river is considered as a most important coast-to-coast shipping route for the Midwestern agricultural trade. The drought affected the availability of good-quality food, and financial losses were incurred because of the reduction in regular barge traffic flow, the capacity of goods being transported, and the number of citizens working in the US tugboat industry. In spring 2013, that summer drought of 2012 was followed by massive floods all over the Mississippi River basin (Ashley and Ashley 2008). This flooding resulted in disturbances of barge traffic and the transportation of food. These changes in regular transportation reduce the capability of local farmers to export their cultivated food products to the global market, which in return affects food prices globally. Food prices are also increased locally (USGCRP 2017).

Over the next few years, the most direct influence of changing climate will be on the agricultural and food systems. Lobell et al. (2008) showed that increase in temperatures and reduction in rainfall are expected to decrease yields for corn, rice, wheat, and other primary crops grown in semiarid regions in the upcoming 20 years. Such fluctuations could have a considerable effect on global food security. Areas that lack proper food security depend on locally cultivated products to meet their nutritional demands. The tropical and subtropical areas are considerably overrun by global climate changes and variations in food prices. In regions with unstable food security, a few farmers eat their own cultivated products and also trade in the local markets, exposing farmers to climate change in various ways. These changes can develop hunger on a massive scale even when there is enough food available in the markets that has been traded in from other countries.

The national income of a country can be affected by massive droughts that limit the capacity of low-budget countries to buy food from the international market. Consequently, huge upsurges in food prices decrease food availability for the poor. For instance, Tanzania competes with the ethanol manufacturers and hog farmers of the USA for corn. Overall, half of the malnourishment in Tanzania is driven by factors not related to food: one factor is diseases such as AIDS and malaria (FAO 2018), which have become more acute and prevalent with increased temperatures.

9.5 Water Management and Food Security

Water is considered a significant factor for food security, the systematic access to sufficient good-quality nutrition by people to achieve energetic and healthy lives (FAOWATER 2017). The lack of water can be a foremost reason for drought and malnutrition, particularly in those regions where individuals rely on local agriculture and farming for their basic necessities of life, that is, food and revenue (Singh et al. 2013). Furthermore, in terms of water accessibility, irregular precipitation and alterations in seasons cause an impermanent lack of food. Overflows and famines cause most of the severe food crises. The accessibility of water fluctuates intensely from region-to-region, and even regions with an inadequate or irregular supply of water maximize their utilization, which immensely increases agricultural production. Maximization of water use is crucial for the improvement of food security (Singh et al. 2013). It is also significant in reducing poverty in those countryside regions where most of the world's hungry people reside.

Human nutritional needs rely 80% on plant products alone (FAO 2017). Intrinsically, plants are important for food security. Plant foods are also crucial for the enduring availability of enough, reasonably priced, safe and healthy food for the active and fit survival of human beings. The production of additional food from lands already cultivated for agriculture frequently requires nitrogen-based fertilizers in bulk. This usage releases nitrous oxide emissions in the atmosphere and becomes a cause for climate change. Similarly, the excessive usage of fertilizer in agriculture releases nitrates to the soil and groundwater (FAOWATER 2017). Although this phenomenon is not directly related to climatic change, the higher amounts of nutrients such as phosphates and nitrates in the water lead to the process of eutrophication. This process, in turn, encourages algal growth and diminishes the amount of oxygen in the water. This whole phenomenon has severe effects on life below water's surface and life on land in terms of water quality. At the moment, agriculture is under strong pressure to halt the destruction of our environment. This destruction occurs in various ways: by reducing water sources, by contaminating the water channels, and by causing infertility and erosion of soil. There is a need to carefully manage irrigation to avoid or reverse the already-extensive damage to the environment and the spread of water-borne diseases. Similarly, the overuse of water in one region leads to the deprivation of water in another region in the world.

It is expected that the global food production will increase 50% by 2050 to meet the predictable needs of the increasing global population. This challenge will be difficult to meet as climatic change is melting a huge part of the Himalayan glaciers, which, in turn, is affecting about 25% of global cereal production through affecting water accessibility in the Asian continent only (FAOWATER 2017). Additionally, despite a significant effort by the pest and plant disease administration in the previous 40 years to double food production, the pests and diseases are still affect about 10–16% of global food production. Water management, plant growth, and food security are complex phenomena that are interlinked owing to the changes in climate. There is a dire need to understand the significance of these phenomena and applications to ensure healthy living in the world (Chakraborty and Newton 2011).

9.6 Case Studies: South Asia, Africa, China

Bangladesh Floating Gardens (South Asia): Case Study:
The climate change has adversely affected Bangladesh in terms of accumulation of heavy rainfalls. It is also seen in the form of more recurrent storms and severe flooding resulting from rainfall and the sea-level rise. In Bangladesh, incessant water-logged conditions were observed causing harvests to be lost, leading agricultural land to be scarce. These excessive rainfalls, in areas with lower altitude (southern coastal and southcentral) in Bangladesh, cause persistent immersion for more than half the year, mainly in the rainy season (FAO 2018). Therefore, harvesting of crops is not likely to be possible on such lands. In these situations, adaptation based on specific locations has become a top-priority to improve food security in the vulnerable populations of the country along with resilience actions to combat the change in climate.

The Food and Agriculture Organization (FAO) led research (2015) on an efficacious climate-smart production structure that was adapted in areas of lower altitude in Bangladesh. These farmers transformed the huge flooding times of that year into practical floating gardens. From local organic materials, the farmers made floating plots that could be used for growing and raising various vegetables or seedlings for sale. The study identified the key factors that contributed to the resilience and livelihoods of the local people. This research was carried out in three districts, namely, Barisal, Gopalganj, and Pirojpur. Experts, traders, and government representatives were interviewed, and the experts confirmed that the farmers had made the rectangular beds before the start of the rainy season (FAO 2018). Then, they sowed seeds of about 30 different kinds of vegetables and crops into these beds, seeds that were able to grow in such a hydroponic production system. The most frequently grown vegetables, crops, and spices were brinjal, cabbage, cauliflower, chili, cucumber, Indian spinach, okra, papaya, red amaranth, ribbed gourd, stem amaranth, tomato, turmeric, turnip, and wax gourd. Growing this kind of mixed crops was considered as the most widespread system.

This work confirmed that growing these floating gardens have numerous benefits such as using waterlogged areas for cultivation; almost no synthetic fertilizers or compost were required for this kind of cultivation. The biomass generated after cultivation was used as an organic fertilizer source, and throughout the floods, the floating gardens were used as shelters for livestock. Additionally, local fishermen were able to cultivate these food crops and to do fishing at the same time, because the floating gardens were constructed on beds made up of bamboos and other plant materials. This change also permitted these plots to change with the surface level of the river water. Overall, the FAO concluded that floating gardens have proven to be ecofriendly while supporting food security and nourishment. The production of vegetables organically is important for local, urban, and export markets (FAO 2018).

India's Climate-Smart Villages: South Asia Case Study:
India has been seen to build climate-smart villages (CSVs) for the first time. Local agriculturalists were assisted to sustainably harvest more crops. In such a system, agriculturalists and scientists did assessments and implementations of climate-smart agronomic techniques, skills, and facilities. Considering the example of Haryana (state), the local farmers executed climate-smart actions (FAO 2014) such as leveling of the land, substitutive moistening and aeration of rice, lessening the usage of water, refining the fertility of soil, and carrying out financial benefits. After experiencing this accomplishment in climate-smart villages, the local government started an initiative to promote the building of a further 500 such villages for rice and wheat structures in the state up to 2015. The initiative was instigated by the department of agriculture in Haryana through unified funding. Five other states of India have also approved initiatives to build these climate-smart villages on 237,000 hectares. The five states are predominantly those most susceptible to the changing climate. To combat the challenge, the International Food Policy Research Institute (IFPRI) has established and acquiesced about 140 million USD for the climate-smart villages initiatives (FAO 2014).

Burundi's Climate-Smart Landscape-Level Planning: Africa Case Study:
Referring to the estimates provided by climate-modeling, the African country Burundi is inclined to become more dry and hot owing to the change in climate (FAO 2018). The country is projected to have prolonged hotter seasons with increased precipitation in the monsoon. The model has also predicted that the agricultural sector of the country should essentially acclimatize to the climate change. FAO approved a project (2010–2014) for the Kagera river basin, the "Kagera Transboundry Agro-Ecosystem Management Project," which was funded by the Global Environment Facility (GEF). This river basin is shared by three other African countries: Rwanda, Tanzania, and Uganda. The initiative utilized the "Land degradation assessment in drylands: world overview of conservation approaches and technologies (LADA-WOCAT)" method (FAO 2018). The method helped in devising agro-environmental approaches and combating the issues of governance in this territory.

The multidisciplinary team developed a proper map for the sites. Information on the type and nature of vegetation cover, biodiversity, and water was collected. The effects were evaluated referring to the local inhabitants livings and crucial ecosystem services (FAO 2018). The vital activities for the climate-smart production were expansion of the intercession toolset, recognition of micro-watersheds, biophysical analyses of the ecosystems, expansion of community planning activities, execution of plan activities, and endogenous assessment and monitoring.

The outcome of this initiative was that the local inhabitants successfully managed 50 project locations. The training for this management and handling was given to a total of 1,200 people, 60% women. A few of the techniques that they learned were utilization of biological pest control, endorsing agrobiodiversity, and capably reaping water resources, plus the utilization of better-quality seeds. Owing to these trainings, more than 4600 ha of ruined and watershed area was restored (FAO 2018). Another significant outcome was that the local people became capable of improving their management of the vegetation coverage; they also self-created an ultimate transformation in their attitude toward prolonged development actions. The LADA-WOCAT method was intended to be replicated for climate-smart landscape-level integration (FAO 2018), and the attributes of land-use system mapping consisted of several significant constraints unconditionally associated with the degradation of land along with the management of soil and water. Additionally, to become accustomed and adapt resilience toward changing climate, improved quality seeds were made available to assist the farmers to manage their lands in better ways. This step improved and stabilized agronomic production even during global climate change (FAO 2018).

CSA for Local Farmers in Kenya and Tanzania, Africa: Case Study:
The Food and Agriculture Organization (FAO) of the United Nations launched an initiative in 2010, the Mitigation of Climate Change in Agriculture (MICCA) Program. The program was intended to make agriculture climate smart. This project took into consideration assorted climate-smart agricultural methodologies for small-scale farmers. In Tanzania and Kenya, two pilot projects were based on the evaluation and review procedures at several stages. Assuming the location specificity of CSA, the methodologies are acknowledged to be founded on the specified agro-ecological and socioeconomic conditions for both projects.

The farmers who worked under these projects partook in some sessions in the fields to categorize prevailing agricultural methodologies and their imaginable outcomes. This practice permitted the local farmers to constitute a set menu of possibly appropriate climate-smart methodologies that were willingly combined in their up-to-date farming structures. The capacity building of such climate-smart methodologies was considered to be strongly linked to extended previously used methodologies and inducement mechanisms to endorse the approval of these newly-implemented methodologies, e.g., dairy farmers in Kenya and farmer-field schools in Tanzania (FAO 2014).

In both pilot projects, about 2,500 farmers received the proper training on agricultural climate-smart methodologies and learned how to implement them. Almost half of these farmers were women. Three hundred cooking stoves were used to reduce deforestation; and nurseries (44), stocked plantlets (134,381), planted seedlings of trees (<33,500), and terraces (235) were established to preserve soil and water using cow dung, two digesters of biogas being constructed. This program was, undoubtedly, an interdisciplinary task that was accomplished based on the FAO's long-held statement of addressing climate change. The task was carried out by its diverse technical divisions, working in partnership with international and national administrations (FAO 2014).

China's IMAD Project: Case Study:

In recent years, even though the overall volume of water utilized for irrigation has not increased, China still remains the main water user. China has merged a few climate-smart considerations into its agricultural investment projects. For example, the Integrated Modern Agriculture Development Project (IMAD), started in April 2014, has incorporated climate change considerations into the identification, design, and implementation cycle of their project. This project was supported by the World Bank with a total investment of 313 million USD for a period of 5 years (FAO 2014).

The major tasks for project identification were extensive strategies, institutional and practical consultations that emphasized the element that change in climate remained a foremost challenge toward achieving sustainable agriculture development in the region. These tasks also highlighted the need to respond to these climatic challenges in the best possible way. Consequently, the development of this project set a few objectives: developing sustainable and climate-resilient agriculture systems in designated zones were the most significant.

Changing climate vulnerability and water resource management assessments took place during the preparation of the project. The effects of climate change on the availability and demand of water were conducted. The effects of changing climate were set as target goals for the monitoring indicators during the project. The productivities obtained from these evaluations were joined with the information obtained through other climate change-linked initiatives. This combination directed the formulation of three parts (FAO 2014). The first was the upgrading of the agricultural structure of irrigation to form resiliency against the frequent harsh droughts and mighty floods that are mainly brought through changing climate. Second was the advancement of climate-smart practices in the field of agriculture that contained enhanced management of the crops and lands along with improved varieties and latest technologies, and third was the firming-up of the main institutions and to ensuring capacity building among staff members.

During the implementation of the project, the effects of changing climate on hydrological environments were evaluated at particular locations. In addition, the modifications for designing the water infrastructure were introduced. After 2 years, training workshops were organized on the topic of sustainable and CSA. The aim of this workshop was to revive and apprise the knowledge of staff regarding the accustomed idea and design for the project (FAO 2014).

9.7 Strategies Relevant to Climate-Smart Agriculture

In 2010, the concept of CSA was introduced for the first time by FAO. In the Hague Conference focusing on agriculture, food security, and climate change, a paper was presented regarding this concept since then, the concept has expanded and been driven in the development plan both county-wide as in the Africa CSA-Alliance and universally as the Global CSA-Alliance (Arslan 2015). The Economic and Policy Analysis of Climate Change (EPIC) program has been working on a project under FAO in partnership with three countries, Malawi, Vietnam, and Zambia. The project was entitled as "CSA: Capturing the Synergies between Mitigation, Adaptation and Food Security," initiated in 2012 (FAO 2018b), and mainly funded by the European Commission. At the national level, the CSA strategy applied by FAO-EPIC consists of three key phases of actions: constructing the base of evidence, forming and permitting the technical, institutional, and policy agendas, and (the third phase) improving funding choices and building in response circles amid investigation and policy. Assuming the fact that CSA is vastly location-specific, it is significant to form locally corresponding drive to provide sustenance to the nationwide policies. This initiative has led to implementing on the three most important parts of activities. The local initiatives aid rationalized developments to improve the incorporation of CSA into nationwide progress strategies and agrarian sectoral plans. They also ease the input of systematic evidence to assist the nationwide and local plans and policies for improving food security. FAO is backing up the Climate Change, Agriculture and Food Security (CCAFS) for building an international CSA database (FAO 2018).

Climate-smart agriculture endorses synchronized activities via agriculturalists, scientists, the private division, civil society, and policy makers on the way to climate-resilient ways. There are four key action zones that include compilation of evidence, boosting the efficiency of local institutions, encouragement of unity between climatic and agriculture policies, and connecting climatic and agriculture-based funding. Climate-smart agriculture varies from normal business-based strategies through highlighting the capability of executing flexible, background-specific explanations, reinforced by ground-breaking plan, policy, and funding activities. Mitigation is frequently an important mutual advantage of activities accomplished to improve national food security and adaptation. Nevertheless, this must be realized that the advantage may take in extra budgets. If, somehow, the correct amounts of budgets of low-emission advancement plans can be estimated in comparison with the old high-emission advancement plans, this will benefit the nonindustrialized countries to associate the efforts being done for agricultural development to create the mitigation that mutually benefits the foundations of climatic economics. Intensification of sustainability is considered as a keystone. Well-organized resource utilization pays toward the adaptation and mitigation through impacts on farm yield and revenues in addition to the lower releases of greenhouse gases as part of the production. This intensification taking place for prevailing agricultural lands has substantial potential for mitigation through decreasing the rate of converting forestlands and marshlands, even though extra safety procedures are mandatory.

Climate-smart agriculture employs ecological services for the sustenance of productivity, adaptation and mitigation (FAO 2018). One example consists of aquaculture, agroforestry, cohesive crops and livestock. Another example is high-quality management of pests, water and nutrients. Landscape approaches are also among examples, in addition to upgraded grassland and forestry management (FAO 2012). Assimilating the trees into agricultural systems, reinstating damaged land, enhancing the efficacy of usage of water and nitrogen (N) fertilizer, management of compost, and the utilization of anaerobic bio-digesters are all good examples of ecological services included in agricultural systems. Improving the quality of soil provides benefits of production, adaptation, and mitigation through maintaining levels of carbon (C), oxygen (O), and plant-nutrient cycles, which leads to improved resilience toward famine and overproduction as well as carbon sequestration.

The ambiguity that climate change is imposing demands more flexibility and quick-response capability, structuring resilience highlights about reduction in the threat of being food insecure and improving the quick-response capability to handle with the hazards. This management encompasses progressive and transformative activities such as improved facility of information, on-time accessibility of inputs for production, alterations and modifications in production methods for improving ecological services and sustaining productivity during climatic tremors, enhanced governance of marketplaces to reduce cost instability, and long-range insurance with protection programs. These transformative activities take in the most important shifts in production of agriculture and sources of living. The production of agriculture includes activities from crops to livestock, and mainly among crops. The sources of living include amplified reliance taking place from non-farm revenues.

Climate-smart agriculture strategies help and educate the farmers to have access to definite inputs, for instance, tree seedlings, seeds, and fertilizers, and to adopt farming approaches such as changing the time of sowing seeds in agreement with warm temperature and rainfall, rotating crops for water accessibility, and consuming those varieties or cultivars that are well suited for the new environments. Advanced methods that depend on information and communication technologies (ICTs) can cope with these problems. There are four major areas where immediate action is required from public, private, and civil society stakeholders at global to regional levels (FAO 2018). The first is constructing the evidence and evaluating the tools; the second is firming up state and regional institutions in the best possible way; the area is about development of synchronized and evidenced policies; and the fourth area is enhancing finances and their efficacy.

Spatial and temporal scales for assessment of climate change impacts on agriculture are not suitable for planning at national and local levels. For this reason, downscaling of these models to scales more suitable for supporting decisions is necessary. Adequate statistics on forthcoming changes in climatic unpredictability at these scales, and their effects on agriculture, are more significant than prolonged trends in climatic variables. The advancement and practical implementation of problem-based methodologies for planning adaptations have a substantial tendency in recognizing vigorous actions in the aspect of uncertainty. Cooperation amid worldwide, regional, and local information is also subjugated.

Tools are desirable for assessing the potential for adaptation and mitigation of diverse strategies and skills from local to worldwide scales. This potential covers the effects of extreme events regarding agriculture and food security, evaluating ways for enhancing flexibility in agriculture and food systems, and recognizing choices for and budgets of reducing emissions of greenhouse gases. Numerous landscape methodologies and investigation of the feasible options in present far-sightedness and set-up initiatives significantly increase the effectiveness of research exertions from local to worldwide levels.

One more significant gap in the evidence base is recognizing the barriers toward the implementation of agricultural approaches and ways of incapacitating these barricades, concentrating on the susceptible, as well as marginalized, groups, small-scale farmers and women. Even though the local farmers are becoming accustomed to the changing climatic environments, yet the implementation of potentially valuable approaches is frequently very low. There is a specific requirement for full-bodied learning and research for progress in the consideration of working in diverse agroecosystems and agricultural structures, enabling recognition of constituents of climatic smartness in diverse biophysical and socioeconomic frameworks.

The next top-priority area for action regarding CSA is the consolidation of state and local institutes to sustain the adaptive capacity by improving access of people to resources together with statistics (FAO 2018). Official advancement has long been a most important drive of policies pertinent to agricultural development, even though insufficient strategy or funding has provided some varied success. Experimental evidence proposes that four extents are requiring public backing to balance private exertions. The first area is the augmentation and propagation of information, predominantly by means of evidence to become accustomed to approaches according to local circumstances. The second area is regarding synchronized energies where approaches produce constructive spillover profits, such as lessening risks of overflow and pest epidemics and conserving biodiversity. The third area constitutes the wide-ranging approaches of risk management aimed at handling extreme weather conditions that are distressing many farmers simultaneously. The fourth area is the consistent, appropriate, and unbiased access for contributions to sustain resource-use efficacy.

Another top-priority area for CSA is forming empowering policy in addition to the governing structure by improved synchronization of policies related to agriculture, the changing climate, the ecosystem, and the food system. A qualifying planning situation necessitates configuration across plan areas, which is aided by the exchange of ideas across related offices to discourse exchanges, gaps, and connections. Coordination is predominantly significant between nationwide agronomic rules, investment plans, strategies and changing climatic tools, together with National Adaptation Programs (NAPs), Nationally Appropriate Mitigation Actions (NAMAs), and Climate Change Investment Plans (CCIP). About 44 states were involved in the planning of NAMAs, of which 18 have recognized the agricultural actions as a top priority (UNCC 2010). The last, but not least, top-priority area is the increase and improvement in targeting funding sources for the provision of transition to CSA. Involving the climatic funding to the old sources of agronomic finances is a vital constituent of this exertion. Becoming accustomed to the agriculture systems necessitate increasing the open investment. In addition, recognizing and crediting the mutual benefits of mitigation that are obtained by the process of adaptation is a significant way of enhancing fiscal capital.

Changing climate is leading toward a new and huge demand for such joint collaboration. For attaining the scale needed to meaningfully decrease hazards related with frequent extreme weather conditions, synchronized exertions are mandatory through the numbers of those farmers who are involved in handling public resources and lands. The discussions of multi-shareholders for backing up the amended supremacy of tenure systems that considers the benefits of women, underprivileged, and sidelined farmers as a top-priority are a promising way along with more traditional exertions to raise the safety of tenure over privately managed land. Wide-ranging strategies for risk-management necessitate an improved consideration of the sturdiness of diverse risk-management tools under climatic variability and synchronization of activities by people belonging to public, private, and civil society from global to regional to local levels.

A country's government may perhaps deliver such approaches and policies for practical and unified risk-management by, for instance, the formation of a national panel that synchronizes strategies for management and institutes for monitoring, prevention, and response related to the risks. The private sector too can have a main part in risk-management. Nonetheless, the operative involvement should essentially enable through clear, effective, and implementable protocols and advanced partnerships between public and private representatives. Societal protection programs that provide the least income or food availability similarly disturb the exposure to risk by means of possible effects on production choices. There has been a substantial growth regarding the development of similar protection programs in past years internationally.

9.8 Financial Assistance to CSA: A Mandatory Step

The overall allocated finance budget is inadequate to fulfill the demand for agriculture worldwide. It is poorly targeted as compared to other sectors in the world. Even though climatic finance has increased prominently in recent years, it is still expected to encounter only a fairly small portion of the overall agriculture-based investments.

An estimate of acquiring 209 billion USD annually to raise production will fulfill the increased demand by the world population only until 2050. Financial assistance to CSA is a mandatory and crucial step to achieve sustainability in food security and agricultural production. The utmost capable financing sources for CSA consist of the Adaptation Fund, Global Environment Fund (GEF), and Green Climate Fund (GCF) (IISD 2017). Some other funds available to support the local agriculturalists and fishermen under the European Union to acclimatize with the climate change are EAFRD (European Agricultural Fund for Rural Development), CAP (Common Agricultural Policy), and EIB (European Investment Bank) are some of these funds' initiatives under EU (EEA 2015).

The adaption fund is a ground-breaking supporting scheme that emphasizes the requirements of the utmost susceptible people and the option of straight accessibility. During their sixth period (2014–2018), the GEF top-up was a total amount of 4.43 billion USD. Of this amount, about 1.26 billion USD was meant to be used for the climatic change program. Similarly, about 431 million USD was assigned for the area of land degradation. The Green Climate Fund expected to distribute 100 billion USD per annum until 2020 with the aim of covering adaptation and mitigation among all areas, that is, public and private capitals (FAO 2018): economic amenities comprising loaning, leasing, equity capital, interim capital-loans, and long-standing investment loans. In addition, the utilization of savings and deposits and fiscal intermediation, for instance, online banking services, transfers, transactions, and insurance, are also included in financial services.

Ethiopia's Farmers Accessing Finance Through Credit Guarantee Services: Case Study:

The manufacture of coffee is a backbone for agriculture in Ethiopia. The companies that handle the production and selling of coffee are weaker and not well managed in the country. There are several debtors in the coffee sector who are frequently restricted via price instability and other financial limitations. These schemes purpose to cope with these challenges through helping companies to get access to the money in a more supportive way. The finance scheme provided a surety scheme for credit with the amount of 2.25 million USD to permit the banks to consider the surety (guarantee) as a substitute for old collateral (WBG 2016). A risk-distribution pact was signed between the Rabobank Foundation and CFC that covered half the amount of any losses experienced as a result of loaning to local farmers. About 42 companies in Ethiopia were recognized at the beginning. Of these, only 22 were chosen to take part in the credit scheme (WBG 2016). The model and strategies for formulating the business plan were designed so that these companies might apply for the finances in the form of loans for coffee production time. The surety scheme assisted in providing workable money loans to 11 companies with worth more than 700,000 USD (WBG 2016). Altogether these companies received their credits directly from the bank, which was the first time in history that they accepted money straight from the bank. During the closing phase of the loan, the bank reported possessing a recovery rate of 98%. The companies had not had such experience with obtaining loans in the past. So, they were feeling authorized to get the finances autonomously. Since the beginning of this scheme in Ethiopia, the important features were capacity development and initiation of educational events. The surety schemes also supported loaning to the client companies and farmers.

International support is essential for nationwide efforts to build synchronized methodologies for policy making regarding changing climate, agriculture, and food security. Another important point to highlight is that capacity building, expertise development, and funding empower the national climate-smart agricultural activities, which involves more consistency across multidimensional processes for policy making. Some significant initiatives of this kind include the United Nations Framework Convention on Climate Change (UNFCCC), Sustainable Development Goals (SDGs), and the Committee on World Food Security (CFS). The inferences were approved and recommended through the Subsidiary Body for Scientific and Technological Advice (SBSTA) at the UNFCCC Climate Talks (UNFCCC 2014). Previous dialogues about food security and climate change (by the CFS) and debates over unified methodologies for land (at the UNFCCC) altogether assist to form impactful international policies.

9.9 Climate-Smart Agriculture: A Solution to the Climate Change Perplexities

Changing climate brings more uncertainty and ambiguity for farmers and local administrators. A cohesive, indication-based, and life-changing innovative approach to address local and global climate and food security involves synchronized activities at all stages. These stages range from research methodologies to the strategies and finances across public, private, and civil society. By using accurate methodologies, strategies, and finances, the agriculture sector moves onto pathways leading toward CSA; this results in reducing food insecurity and poverty in the short run whereas reducing the changing climate as a risk to food security in the longer run. CSA is the most appropriate approach to renovate the existing agriculture systems for the provision of food security and poverty reduction globally. A recent overview of climate-smart agriculture is presented in Table 9.1.

Table 9.1 Year-wise growth in climate-smart agriculture (CSA)

Year	Places	Happenings	References
2009	UN-FAO	The term CSA development was first discussed.	FAO (2009)
2010	The Hague, Netherlands	First Global Conference on Food Security, Agriculture and Climate was organized. The term CSA was first defined here by the FAO.	FAO (2010)
2011	Wageningen, Netherlands	A conference on Global Science–CSA was organized. CSA–Science for Action was first published.	FAO (2011)
2012	Hanoi, Vietnam	Second Global Conference on Food Security, Agriculture and Climate was organized.	FAO (2012)
2013	Johannesburg, South Africa Davis, California Rome	Third Global Conference on Food Security, Agriculture and Climate was organized. A conference on CSA-Global Science was organized. First module of Climate-Smart Agriculture Sourcebook was launched.	FAO (2013) FAO (2013) FAO (2013)
2014	New York, USA	Climate Summit took place. The Global Alliance for the CSA plan was proposed in the summit.	FAO. (2014)
2015	Montpellier, France	A Conference on CSA-Global Science was organized.	FAO (2020)
2016	Marrakech, Morocco	UN Climate Change Conference COP22 was organized and its 22nd session was based on climate change, agriculture and food security.	FAO (2016)
2017	Johannesburg, South Africa Rome	Fourth Global Conference on Food Security, Agriculture and Climate was organized. Second module of Climate-Smart Agriculture Source book was launched.	FAO (2017)
2018	Washington, D.C., USA	The World Bank Group invested $4.5 billion in agribusiness.	WBG (2019)
2019	Bali, Indonesia Washington, D.C., USA	Fifth Global Conference on Food Security, Agriculture and Climate was organized. World Bank Group launched "CSA Investment Plans: Bringing CSA to Life."	FAO (2020)

9.10 Productivity, Adaptation, and Mitigation: Pillars of CAA

Climate-smart agriculture is grounded on three pillars (Fig. 9.2). The first pillar is sustainable maintenance or increase in agriculture production. The second pillar is the adaptation of agricultural approaches according to changing climatic conditions. The third pillar is the mitigation of emissions of greenhouse gases from agriculture (Saj et al. 2017).

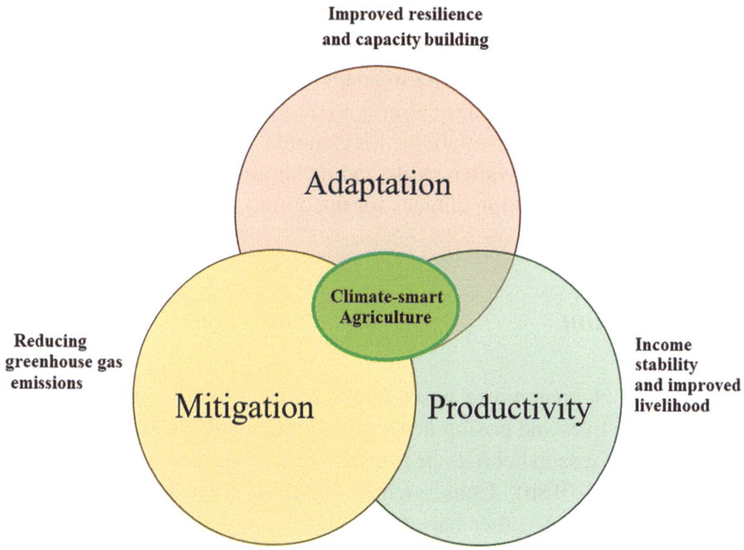

Fig. 9.2 Pillars of CSA

9.10.1 Productivity

It is expected that the world population will be around 9 billion by 2050, so there is a crucial need to make sufficient healthy food available for future populations. The available foods need enrichment in nutrients, for instance, vitamins, mineral elements, and proteins, along with consideration of various factors such as energy requirement, age, gender, and condition of pregnancy. A supplementary task for the forthcoming is regarding the increasing need of animal products mainly in non-industrialized countries, for example, milk, butter, cheese, and meat. Food security is not only an issue of growing production. The issue is to avoid the decay or wastage of food with the value chain, specifically the variety of actions that are essential to provide the product to the buyers (Camagni and Kherallah 2014). The quantity of diet produced is only considerable when this reaches the people for consumption. Maintaining food security depends on increasing production sustainably for upcoming generations.

9.10.2 Adaptation

Agriculture is urgent through several problems during the climate change with extreme weather actions. Internationally, significant variations are observed in the pattern of changing seasons and regular temperatures. Few countries with longer growth seasons and warmer temperature conditions have ideal agriculture from the climate change; most countries show negative effects of changing climate on agriculture. There is an ultimate necessity to adapt farming with the intention of ensuring the resiliency of current agricultural systems affected by climate change. Adaptation includes practical variations, for example, altering the crops or livestock being farmed, usage of innovative skills and technologies, and using climatic and weather-related data for making choices for upcoming generations.

9.10.3 *Mitigation*

Increasing levels of greenhouse gases and the changing climate and its effects, remain, impacting lives and posing new issues for humankind. At present, the agriculture sector is the reason behind one fourth of human-induced emissions of greenhouse gases (FAO 2018b). Consequently, there is great potential to decrease emissions and to mitigate other harmful impacts of the agricultural sector on the atmosphere. Mitigation of resource input and growing efficacy frequently go side by side with mitigating greenhouse gas emissions. If the release of greenhouse gas emissions to the atmosphere is reduced, then there will be fewer climate-related issues for future generations. Adaptation of agricultural methodologies and maintenance of food production caused by changing climate support is solving recent environmental and nutritional issues. Moreover, by mitigation it will be easier for farmers to adapt to the changing climate. One more significant feature of mitigation is the upward movement of carbon in plants and soils. This uptake helps in reduction of the concentration of CO_2 in the surrounding environment.

9.11 The Way Forward

In achieving the aim of agricultural production and food security, improved resilience of farmers and the lowest emissions of greenhouse gases, CSA need to implement several systems, which consist of landscapes and bio-networks in addition to value chains (Vermeulen 2015). It is important to follow interactions between the diverse components of the system, to evaluate and handle trade-offs and to do cost-and-benefit analyses from the systems perspective (FAO 2018b). Following this path, we can identify those actions that lead to the anticipated consequences most resourcefully on the basis of previous cooperation of shareholders' aims (Dinesh 2019). CSA can outdo the advanced methodologies, skills, and practices, for example, with famine-resistant crop varieties and precision farming. The core recommendations for moving forward in execution of CSA and accelerating transformed food systems are the following.

1. *Unusual partnerships*

Several CSA stakeholder groups are mostly relaxed in their own particular groups. They mostly feel uncomfortable engaging with other shareholder groups; language is one of the barriers. This division amid various shareholders should be linked. So, there is a need to build more unusual partnerships, as a new norm.

2. *Beyond agriculture*

Several sectors such as supply chains, food trade, selling, purchasing, and procurement are identified as the main factors under agricultural production, but there is a need to include cross-food system associations, full-cost accounting, and all-inclusive evaluation of opportunities and dangers of reshaping supply chains in this field. This practice may help to transform the agriculture and food systems on a broader spectrum.

3. *Effective communication*

Several innovative approaches that help in the advancement of agriculture are left to be associated with the controversial and developing matters. There should be an effective communication between researchers, agriculturists, farmers, stakeholders, and local traders to ensure transformation in food systems.

4. *Leadership and levers*

All changes associated with transformation occur only if a strong leadership is there to drive the change along with the levers. Following these main levers helps the leader to carry out the transforming actions: empowering farmer and consumer organizations, women, and youth; digitally enabled climate-informed services; climate-resilient and low-emission practices and technologies; innovative finance to leverage public and private sector investments; reshaping supply chains, food retail, marketing, and procurement; and fostering enabling policies and institutions (Dinesh 2019).

5. *Beyond business*

The success of CSA needs to go beyond usual business practices with the intention of realizing the transformative consequences in general. The identified contributing activities consist of focusing on the amendment of Nationally Determined Contributions to update national and universal commitments, employing results-based payments, and innovative research models, addressing the policy execution to handle the implementation gap and digitalizing the field of agriculture.

References

Ali R, Mutolib A (2016) Comparison of air temperature under global climate change issue in gifu city and ogaki city, Japan. Indonesian Journal of Science & Technology 1(1):37–46. https://www.researchgate.net/publication/305402063_Comparison_of_Air_Temperature_under_Global_Climate_Change_Issue_in_Gifu_city_and_Ogaki_city_Japan

Arslan A (2015) Climate-smart agricultural initiatives in Africa. The Sahel & West Africa Week, Milan, Italy. Economics and Policy Innovations for Climate-Smart Agriculture (EPIC), Agricultural Development Economics Division (ESA) of the FAO

Ashley ST, Ashley WS (2008) Flood fatalities in the United States. J Appl Meteorol Climatol 47:805–818. https://doi.org/10.1175/2007JAMX1611.1

Camagni M, Kherallah M (2014) Commodity value chain development projects. International Fund for Agricultural Development, Rome

Chakraborty S, Newton AC (2011) Climate change, plant diseases and food security: an overview. Plant Pathol 60(1):2–14. https://doi.org/10.1111/j.1365-3059.2010.02411.x

Dinesh D (2019) The way forward for climate-smart agriculture. 5th Global science conference on climate-smart agriculture. Climate Change, Agriculture and Food Security (CCAFS). https://ccafs.cgiar.org/blog/way-forward-climate-smart-agriculture#.XrQDA6hKjIU

EEA (2015) Agriculture and climate change. European Union. https://www.eea.europa.eu/tr/isaretler/isaretler-2015/makaleler/tarim-ve-iklim-degisikligi

FAO (2009) Food security and agricultural mitigation in developing countries: options for capturing synergies. FAO, Rome. http://www.fao.org/docrep/012/i1318e/i1318e00.pdf

FAO (2010) Climate-smart agriculture: policies, practices and financing for food security, adaptation and mitigation. FAO, Rome

FAO (2011) Climate change mitigation finance for smallholder agriculture: a guide book to harvesting soil carbon sequestration benefits. FAO, Rome. http://www.fao.org/docrep/015/i2485e/i2485e00.pdf

FAO (2012) Mainstreaming climate-smart agriculture into a broader landscape approach. Food and Agricultural Organization of the United Nations, Rome, Italy

FAO (2013) Climate smart agriculture sourcebook. FAO Rome. http://www.fao.org/3/i3325e/i3325e.pdf

FAO (2014) FAO success stories on climate-smart agriculture. Food and Agricultural Organization of the United Nations, Rome, Italy

FAO (2016) State of food and agriculture report. Climate change Agriculture and Food Security. FAO Rome. http://www.fao.org/3/a-i6030e.pdf

FAO (2017) Plant health and food security. International Plant Protection Convention (IPPC). Rome, Italy

FAO (2018) Climate-Smart Agriculture Case studies 2018. Successful approaches from different regions. Food and Agriculture Organization of the United Nations, Rome, Italy, pp 1–44

FAO (2018b) The 5 implementation steps of climate-smart agriculture and the sustainable development goals: synergies & trade-offs

FAO (2020) Three big moves for GACSA's Action Groups. http://www.fao.org/climate-smart-agriculture/news/detail/en/c/1205425/

FAOWATER (2017) Water at a glance: the relationship between water, agriculture, food security and poverty. Rome. http://www.fao.org/3/ap505e/ap505e.pdf

IISD (2017) Climate finance institutional update: new financing from climate funds, green bonds. SDG Knowledge Hub. http://sdg.iisd.org/news/climate-finance-institutional-update-new-financing-from-climate-funds-green-bonds/

IPCC (2007) Summary for policymakers, in climate change 2007: impacts, adaptation and vulnerability. Contribution of Working Group II to the Fourth Assessment Report of the Intergovernmental Panel on Climate Change. Cambridge University Press, Cambridge, UK, p 17

IPCC (2013) Summary for policymakers. In: Stocker TF, Qin D, Plattner G-K, Tignor M, Allen SK, Boschung J, Nauels A, Xia Y, Bex V, Midgley PM (eds) Climate change 2013: the physical science basis. Contribution of Working Group I to the Fifth Assessment Report of the Intergovernmental Panel on Climate Change. Cambridge University Press, Cambridge/New York

IPCC (2019) Summary for policymakers. In: Shukla PR, Skea J, Buendia EC, Masson-Delmotte V, Pörtner HO, Roberts DC, Zhai P, Slade R, Connors S, van Diemen R, Ferrat M, Haughey E, Luz S, Neogi S, Pathak M, Petzold J, Pereira JP, Vyas P, Huntley E, Kissick K, Belkacemi M, Malley J (eds) Climate change and land: an IPCC special report on climate change, desertification, land degradation, sustainable land management, food security, and greenhouse gas fluxes in terrestrial ecosystems. In press

Lobell DB, Burke MB, Tebaldi C, Mastrandrea MD, Falcon WP, Naylor RL (2008) Prioritizing climate change adaptation needs for food security in 2030. Science 319:607

NASA (2020) The effects of climate change. NASA's jet propulsion laboratory. California Institute of Technology, Palo Alto

Saj S, Torquebiau E, Hainzelin E, Pages J, Maraux F (2017) The way forward: an agroecological perspective for climate-smart agriculture. Agric Ecosyst Environ 250:20–24

Singh SP, Singh B, Kumar U (2013) Water management strategies for achieving food security. APCBEE Procedia 5:423–428. https://doi.org/10.1016/j.apcbee.2013.05.073

UNCC (2010) Report of the Conference of the parties on its fifteenth session, Copenhagen 2009. Appendix II: Nationally appropriate mitigation actions of developing country parties. http://unfccc.int/meetings/cop_15/copenhagen_accord/items/5265.php

UNFCCC (2014) Bonn climate change conference. United Nations framework convention on climate change. https://unfccc.int/process/conferences/pastconferences/bonn-climate-change-conference-june-2014/statements-and-resources/4

USDA, Brown ME, Antle JM, Backlund P, Carr ER, Easterling WE, Walsh MK, Ammann C, Attavanich W, Barrett CB, Bellemare MF, Dancheck V, Funk C, Grace K, Ingram JSI, Jiang H, Maletta H, Mata T, Murray A, Ngugi M, Ojima D, O'Neill B, Tebaldi C (2015) Climate change, Global Food Security, and the U.S. Food System

USGCRP (2014) Climate science special report. Third National Climate Assessment (NCA3), USA

USGCRP (2017) Climate Science Special Report. Fourth National Climate Assessment (NCA4), volume I

Vermeulen SJ (2015) How to access climate change risks in value chain projects. International Fund for Agricultural Development, Rome, Italy

WBG (World Bank Group) (2016) Making climate finance work in agriculture. Discussion paper. FAO, USA

WBG (World Bank Group) (2019) Climate smart agriculture investment plans: bringing CSA to Life. Discussion paper, FAO, USA. https://www.worldbank.org/en/topic/agriculture/publication/climate-smart-agriculture-investment-plans-bringing-climate-smart-agriculture-to-life

WWF-UK (2020) Effects of climate change. https://www.wwf.org.uk/learn/effects-of/climate-change

Chapter 10
Effects of Polychlorinated Biphenyls on Plant Growth

Nadeem Iqbal, Nida Nazir, Muhammad Numan, Malik Tahir Hayat, Qaisar Mahmood, Bibi Saima Zeb, Bin Ma, and Zaigham Abbas

10.1 Introduction

Polychlorinated biphenyls (PCBs) are a set of chlorinated organic contaminants listed in the Stockholm Convention owing to their omnipresence in environmental and biological specimens (Quintanilla-López et al. 2020). It is because of their chemical stability, cost-effective manufacturing, and non-flammability that PCBs have been universally utilized as lubricants, fluids, and electrical equipment coolant (Reddy et al. 2020). For years, the encounter of humans to environmental organic halogen pollutants has been a global issue, ultimately leading to the publishment of

N. Iqbal
Department of Environmental Sciences, COMSATS University Islamabad, Abbottabad Campus, Pakistan

Doctoral School of Environmental Sciences, University of Szeged, Szeged, Hungary

N. Nazir · M. Numan · M. T. Hayat (✉) · B. S. Zeb
Department of Environmental Sciences, COMSATS University Islamabad, Abbottabad Campus, Pakistan
e-mail: mtahir@cuiatd.edu.pk

Q. Mahmood
Department of Environmental Sciences, COMSATS University Islamabad, Abbottabad, Pakistan

Department of Biology, College of Science, University of Bahrain, Sakhir, Bahrain

B. Ma
College of Environmental and Natural Resource Sciences, Zhejiang University, Hangzhou, China

Z. Abbas
Ministry of Climate Change, Islamabad, Pakistan

Q. Mahmood (ed.), *Sustainable Plant Nutrition under Contaminated Environments*, Sustainable Plant Nutrition in a Changing World,
https://doi.org/10.1007/978-3-030-91499-8_10

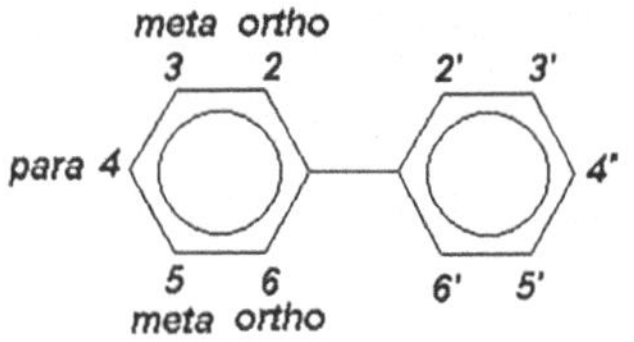

Fig. 10.1 Structure of PCBs

multiple reports on certain potential drawbacks and risk analysis (Van den Berg et al. 2006; Wittsiepe et al. 2007). Biphenyls are chlorinated to manufacture PCBs, which encompass a set of 209 distinct chemicals, each having 12 carbon atoms. The PCB anatomy is shown in Fig. 10.1. Hydrogen atoms are replaced by chlorine at any of the 10 numbered sites, in the given structure (Beyer and Biziuk 2009).

Owing to the fat-soluble characteristics and steady biotransformation of PCBs, they are being discovered in human as well as animal tissues, despite the fact that both their making and utilization was prohibited in the 1970s (Zhao et al. 2017; Grimm et al. 2015). The release of immense volumes of PCBs into natural habitats and their tenacity along with their perilous influence on human health collectively accounted them as contaminants that are troublesome to manage (Reddy et al. 2020). To prevent the effect of PCBs on living creatures and their surroundings, it is imperative to have profound information related to the sources from where they are being discharged into the environment. Consequently, in the past 20 years, much research was reported regarding the outcome and dissemination along with plausible proposals for mitigation (Reddy et al. 2019; Jing et al. 2018). Lately, to analyze PCBs from all kinds of specimens, encompassing biota and environment, a plethora of inquisitive strategies have been stated. However, most of the advanced methodologies utilized contemporary chromatographic techniques having mass spectrometry (MS) recognition, which are not only comparatively exorbitant but also difficult to use in emergent nations because of the expense. Thus, some other approaches with relatively simpler techniques and equipment should be developed, so that the contribution of underdeveloped nations in the scrutiny of PCBs can be improved (Reddy et al. 2020).

10.2 Sources of PCBs

Even though the universal existence of PCBs can be observed in every aspect of the natural habitat involving flora, fauna, soil, air, and water, its definite origin in the ecosystem is yet unrevealed because of the abrupt conversion (Markowitz and Rosner 2018). Usually, the sediments are large reservoirs of PCBs, and the PCBs will remain for many years until reallocated into a water body. Then, PCBs in soil and water organisms can be detected through the food chain (Ehsani and Hayes 2018). On the other hand, PCBs absorbed by crops through soil and also from irrigation water can also be the part of the food chain and thus can accumulate in organisms (de Souza et al. 2018). In addition, PCBs tend to bind to fats, and as they are very stable through the food chain, they can accumulate in livestock. In general, food is the most likely way for humans to be exposed to PCBs (Cetin et al. 2018). The toxicity of polychlorinated biphenyls to all organisms must address their presence in almost all kinds of environments and ecosystems (Reddy et al. 2020). As well as commercial production, PCBs are also produced as a thermodynamically stable by-product of specific combustion processes (incineration of wastes containing PCB). EPA regulations on the storage and disposal of PCBs includes waste incineration, the only method acceptable for PCBs unless the EPA has obtained permission to dispose of it in another way (EPA 1978).

The PCBs are spread through two main sources.

1. Commercial and industrial production.
2. By-products produced by the combustion of thermodynamically stable compounds.

The commercial production of PCBs was started about 60 years ago by the chlorination of biphenyls. In 1992, the estimated amount of PCBs was 1.5 million tons throughout the world (Van Metre et al. 2000). By weight, the percentage composition of chlorine present in PCBs was the key factor based on which they were marketed. Different trade names were used under which PCBs were available: for example, aroclor was available in the USA, kanegafuchi in Japan, and kanechlor and clophen in Bayer and Germany. PCBs were produced commercially in plants established in Austria, Germany, France, Great Britain, Italy, Spain, Japan, USSR, and the USA (Teng et al. 2008). Statistical data on losses of PCBs could not be generated accurately. To meet this problem, a special disposal system having the capacity of 10 million compounds per year was set up by Monsanto (Sharma et al. 2015).

In the USA, the market leader company was the Monsanto Industrial Chemical Company. Its production included 98% of PCBs and based on eddy-current testing (ECT) technique of its whole production (Hu et al. 2008). In Japan, Kanegafuchi Chemical Co. Ltd. was the first company that started production of PCBs. In 1954 Kaneka also marketed PCBs under the name of Kaneclor. In 1972, production as well as the import and export of PCBs were banned in Japan. The degree of chlorination in PCBs was used to determine physiochemical properties such as electric conductivity and inflammability (Delgado et al. 2002). PCBs were also used as cutting oils, electric fluids in capacitors and transformers, adhesives, pesticides extenders, hydraulic lubricants, paints, and carbonless copy papers. These applications mainly resulted in release of PCBs into the environment directly or indirectly (Samara et al. 2006). Main sources in the environment included accidents, leakage of chemicals from industrial facilities, and inappropriate disposal practices. In most of the industrial chemical leakages, PCBs were practically unmanageable (Gasic et al. 2010). Possible sources through which PCBs enter the environment include the following:

(i) Leakage from heat exchangers and transformers
(ii) Partially sealed hydraulic fluid systems
(iii) Spills or losses during manufacturing processes of PCBs or fluids containing PCBs
(iv) Vaporization or leaching of chemicals containing PCBs

(v) Inappropriate disposal practices of waste containing PCBs or fluids
(vi) Use of waste electrical insulator oil/solvent in herbicide treatments

10.3 Exposure and Effects of PCBs

PCBs are produced worldwide. Because of their excellent stability, viscosity, and electrical insulation and nonflammable properties, they are often used in enclosed applications such as transfers and capacitors. Being lipophilic in nature, PCBs can easily enter the food chain. In addition to their lipophilic nature, PCBs are environmentally stable and can resist very high temperatures, so they can survive for a long time in the environment (Yao et al. 2017). Hence, the danger of PCBs, particularly PCBs such as dioxin, has attracted the attention of environmental toxicologists and biomedical scientists. The PCB structure has a biphenyl part composed of two benzene rings (Bruner-Tran and Osteen 2010). Environments with PCBs in its surroundings pose serious health threats for humans (Fu et al. 2011). Because of PCB presence in the environment, in the USA commercial production of PCBs was banned in the early 1970s, with growing concerns about potential adverse health effects, and production eventually ended in 1977.

PCBs are related to health effects in laboratory animals, but compared to actual human environmental risks, such PCB doses are usually much higher. Under multiple-dose and lifetime exposure conditions, PCBs are observed to cause tumors in laboratory animals. For regulatory purposes, some governments or consulting agencies, including the US EPA, claimed that there is multiple evidence for considering PCSs carcinogenic for living organisms. However, studies show that high-level exposure in occupational settings to long-term PCBs through skin contact and inhalation exposure has not increased cancer risk. In fact, sensory (eye-skin) irritation is still the only persistent health effect among workers that can be attributed to PCBs (Ross 2004).

10.3.1 Effect of PCBs on Plants

It is an established fact that PCBs cause growth inhibition of plants (Anyasi and Atagana 2011). It is also documented that PCBs at concentrations of 0.3–10 ppm resulted in fewer algal cell numbers as compared to controls. Some data are available for the phytotoxic effects of PCBs on higher plants. Another study focused on the internal chloroplastic disorganization in *Spirodela oligorrhiza* when this plant was exposed to a 5 ppm Aroclor 1242 concentration (Furukawa et al. 2004). Similarly, a phytotoxic effect of PCBs was reported in beets, pigweed, and soybeans, resulting in reduced plant height and fresh weight loss (Anyasi and Atagana

2011). Similarly, Aroclor 1254 (at 1000 ppm) caused growth inhibition of soybeans by about 47%. Nevertheless, cumulative water usage is recorded more sensitive to PCB as compared to plant growth. There are two general routes for PCBs uptake in the plants: one is via the root system, and the other is adsorption on stems and leaves. Symplastic and apoplastic pathways translocate the movement of PCBs through epidermal layers. The symplastic pathway is the most crucial way for the uptake of provided PCBs. In contrast, the apoplastic pathway is used by terrestrial plants for the uptake of airborne PCBs, whereas dissolved PCBs are taken up by hydrophytes and microbes (Mackova et al. 2007).

It is well documented that PCB concentrations in plants depend on PCB concentrations in the soil. In addition, temperature, clay, and organic content can affect mobility and PCB availability in the soil. Similarly, some plant species can uptake PCBs differently from others (Borja et al. 2005). Previously, biphenyl metabolites were not accorded so much importance. Various plant species have been found to remove mono- and the dehydroxylated biphenyls (Aken et al. 2010). Several plant species were used to remove PCBs from the soil such as eastern gamagrass, prairie cord, lake sedge, and river bulrush, and different PCBs concentrations were found in the plants after the experiment (Anyasi and Atagana 2011). Similarly, another report investigated the effective remediation of a chlorinated PCB (Aroclor 1260) by reductive dechlorination. In this process, high soil water content and low transpiration rate are provided, which would cause the deposition of less chlorinated PCBs that are usually lost under the aerobic stage of the work (Aken et al. 2010). Using the reductive dechlorination process, the *meta*- and *para*-positions in the 2,3′,4,4′,5′-pentachloro biphenyls are removed. Under maintained water saturation in the sediments, cell storage of chlorinated PCBs results. Therefore, for the enhanced removal of low chlorinated PCBs, only those plant species are preferentially used which can eliminate water from the sediments and oxygenate the rhizospheric soil through aerenchyma cells (Smith et al. 2009).

Microorganisms choose the rhizospheric zone of plants as a preferred environment. The abundance of microbes around a plant rhizosphere is 1.2 × 1011 cells/cm^3 within 1 mm distance from the roots (Pal et al. 2007). In addition to that, up to 10% of the root system is covered with microbes. Bacteria form an association with fungal species and live in a symbiotic relationship, that is, the mycorrhiza (association between fungi and plant roots). Plants also provide nutrients to the microbes in addition to shelter. Fungi provide N and P to the plants and in return obtain sugars from them (Anyasi and Atagana 2011). It is reported that mulberries grown at PCBS-contaminated sites secreted a significant number of phenolic compounds, which are essential for the growth of bacteria responsible for the degradation of PCBs. Similarly, the root system can also excrete organic compounds that can solubilize the soil contaminants (Hatamian-Zarmi et al. 2009).

Both plants and microbes can increase the removal of PCBs (Aroclor 1248) significantly in the combined form as compared to individually conducted experiments (Dzantor et al. 2000). In this report, a substratum was used in the rhizosphere of canary grass and the flat pea, and for plant growth, a moisture replacement system was deployed. These plants had also been studied well in terms of inherited dissipation of PCBs under controlled condition experiments (Dzantor and Woolston 2001). Another plant, namely, burr medic, was also used to explore rhizospheric degradation as compared to other selected plants. After 100 days, about 69% of previously added PCB was found in the unplanted soil. Flat pea, burr medic, and reed canary grass removed 65%, 54%, and 59% of PCBs, respectively. The difference was not encountered in the plants inherited with the dissipation of PCBs and unstudied plants. There could be other factors responsible for this variability in the results (Dzantor et al. 2000).

Another study revealed that the use of organic residues could increase PCB degradation in the rhizospheric soil because organic residues might contain stimulators for PCB degradation. Plants and rhizospheric microbial interactions resulted in considerable removal or change in PCBs concentrations in the presence of alfalfa and inoculants in the rhizosphere. Direct or indirect biotransformation, adsorption of the pollutant on the surface of alfalfa, and biotranslocation can bring a significant reduction in PCB levels (Anyasi and Atagana 2011). However, the biotransformation of PCBs in the presence of alfalfa plants increased over the use of bioaugmentation alone. On the other hand, use of the bioaugmentation process reduced alfalfa growth by causing hardness in the soil. The reduced biomass growth led to lower PCB transformations. Hence, it is concluded that bacteria and PCBs can have toxic effects on the plants because of enhanced solubility, bioavailability, and biotransformation. Further research is required to understand these mechanisms fully and to optimize the conditions for phytoremediation of PCBs in the soil (Mehmannavaz et al. 2002; Olson et al. 2007).

10.4 Uses of PCBs

PCBs had been used in a wide range of electrical appliances, especially in transformers and capacitors. Some other common uses include lubricants, plasticizers, hydraulic fluids, and heat transfer fluids. PCBs were commercially produced by the companies as different products. These products consist of a different range of PCBs and their congeners. In the United States, the most used form of PCBs was aroclor (Sofuoglu et al. 2013). Aroclor is the mixture of PCB that was mainly produced in the United States mainly from 1930 to 1979. Different types of aroclors were used. For distinction, different suffix numbers were used for their naming. These numbers indicated the level of concentration of chlorine present in each compound (García et al. 2012). There is a possibility that PCBs may be present in the environment and in products that were produced commercially before the ban of PCBs in 1979. PCB maybe present in the following products. Different chemical mixtures of PCBs and its congeners were used in the aforementioned products. PCBs were produced and used commercially before 1979 (Davies and Delistraty, 2016).

- Capacitors
- Transformer
- Voltage regulator
- Oils used in in hydraulic systems and motors
- Other electrical equipment including switches, re-closers, electromagnets, and bushings
- Cable insulation
- Thermal insulation materials that mostly consist of felt, fiberglass, cork, and foam
- Tape and adhesives
- Oil-based paints
- Plastics
- Floor finish
- Carbonless copy paper
- Fluorescent light ballast

10.4.1 Usage in Agricultural Land

Today's society is basically dependent on agriculture, and it is the essential component of the economy of a country. Most of the food being provided to the world's population is from agricultural products (Halfadji et al. 2019). Also, these products are responsible for the pressure on natural resources throughout the globe:12% of the total land surface is occupied by food and fiber crops, which are also responsible for the consumption of 82% of freshwater supplies throughout the world. Intense agricultural practices rely on the use of a wide range of pesticides and fertilizers, causing different levels of degradation in the environment (Omwoma et al. 2019).

Chlorinated organic compounds, especially PCBs, were present in different pesticides. Organochlorine pesticides such as dichlorodiphenyltrichloroethane (DDT), Aldrin, and some others have been used in agricultural practices as well as in the control of insect-borne diseases from the 1930s until their ban in the 1970s (Sun et al. 2016). Species at the top of the food chain were adversely impacted using these pesticides and insecticides. Because of their environmental persistence and hazards, these pesticides were phased out from most of the developed countries during the 1970s. But after these pesticides were banned in developed countries, they were still in use in developing countries (Li et al. 2018). Large amounts of these pesticides including PCBs have been deposited in the agricultural lands causing damage to the plants as well as to soil quality. PCBs are considered one of the most extensive damage-causing agents to the environment. As PCBs are highly resistant to degradation and very slightly soluble in water, they combine with organic matter and soil particles and thus remain in the soil for many years. These polychlorinated biphenyls also are lipophilic in nature, which property helps them pass into the food chain and are committed into the organism's body, causing damage to their health. Consequently, these chemicals have a very wide range of adverse impacts on the soil, plant health, soil microorganisms, and the whole ecosystem, as well as finally threatening human life (Antolín et al. 2016).

Significant quantities of PCBs deposited during the 1970s are also present in sediments as well as in surface waters. A considerable number of PCBs also settle on or become attached to the surface of clays, microparticles, or organic matter or become suspended in the soil. These chemicals stay in the soil for a longer period and are slowly released into the water, soil, and environment Needham & Ghosh (2019). Also, PCBs that are present in the atmosphere deposit on the land through precipitation and snow and become attached to particulate matter such as soot and dust particles, then settle down to the ground in the form of dry precipitation.

In general, the levels of PCBs present in the environment are much higher than the concentrations of other persistent organic pollutants, mainly because of the lipophilic nature of PCBs. These chemicals enter the soil by spills, disposal of sewage sludge to agricultural land, and also by dry and wet deposition. Soil organic matter is a natural sink for the deposition of lipophilic and polar compounds. Thus, PCBs are held still in soil for long time and diffuse to the environment slowly, which largely impacts plant growth. The other main reason is the use of sludge as fertilizer (Hu et al. 2009).

10.5 Plant-Mediated PCB Metabolism

In plants, PCB metabolism takes place in three different stages. In the first process, the PCBs are oxidized to hydroxylated compounds. These products are reactive and soluble. In the second process, the activated compounds are conjugated with plant molecules, forming more soluble and lesser toxic compounds. In the last process, sequestration occurs in which conjugate molecules are adsorbed onto the plant organelles (Rezek et al. 2007). Several studies have documented the transformation of PCBs in plant cells (Mackova et al. 2007). The most studied metabolites of PCBs (mono- and dihydroxylated) are metabolized in significant amounts. The transformation ability of PCBs differs under influencing factors such as microbial strains. The capability of plants to oxidize mono- or dechlorinated PCBs and conversion in mono- and dehydroxylated biphenyls is also well illustrated. Other research has also supported PCB metabolism by plant cell culture (Aken et al. 2010). In addition to microbial strains, the degree of chlorination is also a determining factor for PCB metabolism. Moreover, various enzymes have been reported to facilitate PCB metabolism such as peroxidase, transferase, oxidase, and oxygenase. The initial process of PCB metabolism is rendered by peroxidases and cytochrome P-450 (Aken 2008). Further, other enzymes that are commercially available have shown the capability to transform dichloro- and tetrachlorobiphenyls by horseradish peroxidase (HRP) and in vitro cell cultures of plants by Remazol Brilliant Blue R (RBBR) oxidases. Extracts of *Medicago sativa* and *Zea mays*, containing reductase enzyme, also showed dechlorination of biphenyl (Magee et al. 2008).

10.6 PCBs Uptake By Plants

Uptake of PCBs by plants is dependent on the octane rating of the contaminant, ranging from 0.5 to 4.5 log K_{ow}. In this study, the hydrophobic compounds are mostly taken up by plants and transferred to plant cells. The efficiency of PCBs uptake in plants with 4.5 to 8.2 K_{ow} values could result in a reduced degree of chlorination. In another relevant study, higher concentrations of Aroclor 1260 were found in the roots as compared to the shoots (Zeeb et al. 2006). PCBs containing more than four chlorines, such as tetra and hexas, hepta, and nonas, were also found in small concentrations. Therefore, the translocated PCBs taken up by plants through the root system in higher concentrations are also documented. PCBs with a low number of chlorines have more opportunity for the translocation from roots to the aerial parts of the plants (Liu and Schnoor 2008). Another investigation inferred the enhanced PCBs concentration in the leaves and stems of pumpkin plants whereas in the roots PCBs concentrations were unchanged (Aslund et al. 2008). However, the uptake of PCBs by plants is mostly carried out by the roots while other transport mechanisms result in negligible effects (Aken et al. 2010).

10.7 Phytoremediation of PCBs

Remediation of contaminated sites using plants is well established (Awan et al. 2020). Contaminated sites such as sludge, polluted groundwater, sediments, contaminated air, and soil can be treated by in situ and ex situ phytoremediation through pollutant degradation, stabilization, and removal (Chekol et al. 2004; Macková et al. 2007). In addition to that, various contaminants, such as polyaromatic hydrocarbons, pesticides, petroleum hydrocarbon, metals, landfill leaches, and solvents, can be treated via phytoremediation (Burken 2003; Chroma et al. 2003). Phytoremediation is mediated through different mechanisms: phytoextraction, phytodegradation, phytostabilization, and rhizospheric degradation. Rhizospheric degradation occurs in the rhizospheric soil immediately around the plant roots. Plants release different types of compounds into the rhizosphere, which are consumed by microorganisms to increase their activities and growth (Macková et al. 2007). These stimulated microbes speed up the degradation process of the PCBs by their enhanced enzymatic system. Therefore, the combined plants and soil microbial approach can overcome many individual limitations, degrade many PCBs, and provide a beneficial basis for increasing remediation of polluted sites (Leigh et al. 2006). For instance, relevant research was conducted in which PCBs were transformed by combining different interacting processes among bacteria and plants. It was observed that the enzymatic biphenyl operon pathway transformed two plant PCB intermediate products such as 2-chloro-4-hydroxy biphenyl and 3-chloro-4-hydroxybiphenyl, respectively. It was reported that horseradish and black nightshade transformed more than 90% of 2-chlorobenzoate and 20–40% of dichlorobenzoates. These results open the path for utilization of the biological system for the degradation of PCBs through plant–microbe interactions (Macková et al. 2007).

PCBs are persistent compounds that remain in the environment over a long period of time. PCBs do not have the ability to leach down but adsorb on the soil particles as the number of chlorines increases. Therefore, lower chlorinated PCBs can mobilize in the soil to some extent. In plants, PCBs can also be adsorbed in smaller amounts because these contaminants pose harmful effects on plants and hinder the biochemical processes of photosynthesis and respiration (Toro et al. 2006). However, the contradiction also exists between the active transport of PCBs by plants. According to some studies, plants absorb PCBs via active transport whereas other plants do not. PCBs in the soil can be treated using consortia of various anaerobic or aerobic bacteria, but the active binding site of the PCBs molecules in the soil are usually blocked by indigenous contaminant-degrading microbes. Therefore, for significant success in the bioremediation, the selected and specific microbial inoculation is preferred. So, the microbial inoculation is done by direct supply of microbial consortia especially selected for bioremediation or by using plant–microbial interactions (Toro et al. 2006; Anyasi and Atagana 2011).

10.8 Microbial Degradation of PCBs

The presence of high chlorine components in PCBs makes them chemically stable and less water-soluble. PCBs with higher numbers of chlorine constituents are not biodegradable because of this strong chemical stability (Furukawa and Fujihara 2008). Later, several scientific investigations found the way to remediate some PCBs using microorganisms (Pieper and Seeger 2008). There are basically two metabolic routes of microorganisms for the degradation of PCBs, anaerobic and aerobic. These metabolic pathways mainly depend upon microbial diversity, the degree of chlorination of the congeners, and the redox conditions (Aken et al. 2010). Rhizoremediation is the most promising mechanism as the interacting process between microbes and plants to enhance remediation kinetics. The synergistic interaction between plants and microbes can overcome the limitations of bioremediation. Therefore, plants can withstand the presence of toxic contaminants because of the microbes present in the vicinity of the roots (Chaudhry et al. 2005). In this case, microbial degradation of PCBs results from a lack of reduced energy sources in the rhizosphere of plants, which is compensated by the excretion of plant exudates. Various processes for the removal of contaminants are used intensively, such as bacterial inocula responsible for pollutant degradation, plant growth-promoting rhizobacteria (PGPR), and land forming (Anyasi and Atagana 2011). However, the integrated approach can control the demerits of individual techniques. In contrast, some plant species are very sensitive to pollutants. Therefore, these plants grow slowly in the polluted soils and the remediation potential will also be reduced by low biomass production (Idris and Ahmed 2003).

10.8.1 Degradation of PCBs By Plant–Microbial Interactions

Plant–microbial interactions have been proved useful to degrade several organic and inorganic contaminants. Microbes and plants interact among themselves to obtain certain essential nutrients, defend against pathogenic attacks, and cause sustainable plant growth. In most cases, the transformation of PCBs is mediated abiotically, and in others, microbes and other organisms also intervene in the transformation process (Montone et al. 2001; Macková et al. 2007). The biotransformation converts potentially harmful compounds into less toxic or completely nontoxic compounds. The biotransformed compounds are no longer harmful to living organisms and persistent in the environment. Different approaches must be studied for the determination of intermediate and other products for the comparison with the biotransformation of PCBs (Golden et al. 2003).

Therefore, some studies have finally determined found that microorganisms have the potential to degrade some congeners of PCBs and have recommended this environmental approach to implement the bioremediation process of PCBs-contaminated sites (Sierra et al. 2003). Bacteria can degrade PCBs through two pathways: (1) aerobic degradation through the biphenyl pathway and (2) anaerobic dechlorination. The former technique is widely used and reported in the literature. In 1973, Ahmad and Focht reported two species of *Achromobacter* that had the ability for the degradation of PCBs (Trapp and Kulhanek 2006). After that, several bacterial strains were isolated from other PCBs-polluted sites, sequenced, and their biphenyl pathways studied for degrading the PCBs. Later, gene expression was studied for encoding the specific enzymes responsible for the degradation of PCBs, and these enzymes were produced in large quantities and used in in vitro experiments for particular PCBs or their mixture (Schnoor 2002). Plants handle organic compounds differently as compared to microbes. The contaminants are directly taken up the plants by their roots, transformed, transported, and stored in forms nontoxic to the plants (Macek et al. 2002, 2004). However, these nonphytotoxic forms could be harmful to humans. Plants prefer to store the polar products in the vacuoles or convert them into insoluble cellular structures by oxidative polymerization instead of excretion. It is also crucial that the stored metabolites must be in nontoxic or less toxic forms as compared to the parent chemical (Macek et al. 2005).

10.9 Aerobic and Anaerobic Transformation of PCBs

Numerous studies have reported that PCBs congeners containing fewer than five chlorine atoms are degraded significantly. PCBs congeners containing more than four chlorine atoms are biorefractory to aerobic microbes. However, there are certain reports that penta- and hexa-chlorobiphenyls can be degraded (Borja et al. 2005). Anaerobic transformation of PCBs has also been documented by several studies. Therefore, the preferential removal of *meta*- and *para*-chlorine atoms from potentially toxic PCBs showed biotransformation under aerobic environments. After that, the resultant compounds are easily degraded aerobically. This kind of biotransformation is mostly exhibited by halogenated aromatic compounds (Aken et al. 2010). PCBs with hexachlorobiphenyl have been degraded into penta- and tetra-chlorobiphenyl by soil microorganisms under sequenced aerobic and anaerobic treatments. In both aerobic and aerobic treatments, PCBs are degraded preferentially by different microbes (Mackova et al. 2007). The other determining factor in degradation is the degree of chlorination in PCBs congeners. In addition, there are several other influencing environmental factors, such as microbial growth, the presence of other substrates, pH, and temperature. However, these factors can be optimized to obtain effective degradation potential (Borja et al. 2005).

10.10 The Effect of Chemical and Organic Amendments on PCBs Degradation

Because of their toxic nature, PCBs have harmful effects on both plants and animals. Soils, waters, food webs, or sediments could be sources of PCB contaminations through biochemical cycles (Hayat et al. 2019). In this report, an older contaminated soil was investigated for the rhizoremediation of PCBs. *Lolium multiflorum* planted in contaminated soil in the presence of green garden waste and beech woodchip biochars was analyzed for remediation ability. Sodium dodecyl sulfate as a surfactant was used in this study. The efficiency of the bioremediation was determined based on PCBs degradation, growth of the degrading microbes, and plant growth. The results showed that the planted soils had more removal efficiency as compared to unplanted soils. Moreover, green garden waste biochar removed about 80% of the total PCBs concentration; lesser removal was observed in the treatments in which the surfactant was applied. PCBs-degrading microbes and plant growth also had a significant and consistent role in PCBs degradation. Hence, it is concluded from this study that the use of different vegetation, PCBs-degrading bacteria, and amended soils can considerably increase the efficiency of PCBs degradation (Wiszniewska et al. 2016; Hayat et al. 2019).

Sewage sludge and sediments containing PCBs were treated using cucumber plants. Such contaminated soils must be treated before their disposal. The Cucurbitaceae family is well known for its huge potential to store a high number of PCBs as compared to other plants. However, this accumulation can result in oxidative stress that might limit their values. In this report, the effect of these PCBs was investigated on an antioxidative system and soil toxicity levels after 5 days of the experiment. The highest percentage was recorded for sewage sludge at around 38.3%; in sediment-amended soil it was 27.38%. Moreover, guaiacol peroxidase activity was reduced to 49% in sewage sludge. On the other hand, glutathione *S*-transferase was enhanced to 172% as compared to control (Wyrwicka et al. 2014).

10.10.1 The Use of Plant Terpenes for the Degradation of PCBs

The effect of plant terpenes was explored to observe the degradation of PCBs via the biphenyl pathway. Four different plant materials containing terpenes were used: tangerine peels, pine needles, ivy leaves, orange peels, and two synthetic terpenes, limonene and carvone, to degrade the mixture of PCBs (DELOR 103) via a bacterial strain (*Pseudomonas stutzeri*) isolated from PCBs-polluted soil after long-term exposure. The results showed the highest degradation potential of PCBs mediated in the presence of pine needles (50.5%) and ivy leaves (52.8%) when the mineral medium was used. The same outcome to biphenyl was obtained in the case of synthetic terpenes. It is concluded from this research that the terpenes play a significant role in the degradation of PCBs because these terpenes act as biphenyl dioxygenase, an essential enzyme capable of PCB degradation (Dudášová et al. 2012).

10.10.2 Effect of PCBs-Contaminated Soil Amendments on Plants

The phytotoxic effect of PCBs-contaminated soil amendments with different fractions (0%, 10%, 25%, 50%, 75%, and 100%) collected from the upper Hudson River (USA) was investigated on the soil characteristics and on physiological and biometric levels in cucumber as well as zucchini. The amended soil with PCBs enhanced the PCB concentration from 13.7 μg/kg to 255 μg/kg in zucchini (Urbaniak et al. 2019). The plants were used as phytoremediators in the experiment, conducted for 4 weeks. Initially, the plants did not show any toxic effects but latterly growth inhibition was observed especially in zucchini-grown soil and soil without any plants. The results showed an increase in the cucurbit fresh biomass as compared to control. The leaf surface area was increased in zucchini in 50%–75% mixtures whereas no changes were observed in cucumber under lower doses of admixture. The total chlorophyll content was decreased more in cucumber as compared to zucchini (Urbaniak et al. 2019). In additiont, the soil amended with PCBs had reduced nitrogen (N) content whereas phosphorus (P) was significantly enhanced in 50–100% sediment samples. Similarly, the potassium (K) quantity was enhanced in a 10% sediment mixture while in more than 50% mixtures, it was reduced significantly. The results revealed the growth increase in cucumber in 10–25% sediment mixtures whereas a higher PCBs-contaminated sediment inhibited the growth. Anthocyanin, phenylpropanoid, and flavonol contents were significantly decreased in higher soils amended with PCBs containing sediments (Urbaniak et al. 2020).

10.10.3 Biochar Application Reduced Soil PCBs Availability

Biochar is the by-product of organic matter under low oxygen availability. In this process, biochar is heated to decompose the organic matter and this biochar is used to improve health quality and productivity. Recently, many researchers investigated the capability of biochar to increase soil health and quality in agriculture soils but in a sustainable way for carbon sequestration. This scientific report documented the availability of PCBs in the soil for plants and focused on the reduced availability to *Cucurbita pepo* (*C. pepo*). The results revealed that the 2.8% biochar-amended soil reduced PCB availability to roots in previously PCBs-added soils. In 11.1% biochar-amended soil, more PCBs reductions of 89% and 83% were observed, while in the shoot system, PCBs reductions were 22% and 54%. The application of biochar also increased the overall plant biomass and higher survival rate from worm attacks. Conclusively, biochar application has a significant role to reduce PCBs bioavailability in the soils, decrease potential risks for humans, and improve soil health by the reduction of CO_2 emissions (Denyes et al. 2012).

Another study explored the potential of activated carbon and biochar to understand the availability of organic compounds in the vicinity of the rhizosphere of *C. pepo*. Granular activated carbon and biochar were found equal reducers of PCBs uptake by plants. The results showed that there were 64%, 72%, and 74% PCB concentrations in the *C. pepo* root tissue with the addition of Blueleaf biochar, Burt's biochar, and 2.8% granular activated carbon, respectively. In this research, the earthworm (*Eisenia fetida*) was also used to mechanically mix the carbon amendments, which significantly reduced the PCBs concentrations to around 39%, 59%, and 66% lower as compared to the manually mixed experiment. Therefore, the addition of earthworms or other organisms, which increase soil fertility of the mixture well, could overcome the deficiency of short-term ability to reduce PCBs concentrations (Denyes et al. 2013).

10.10.4 Mixed Plant Cultivation for the Remediation of PCBs-Contaminated Soils

The phytoremediation of PCBs-contaminated soils has been investigated using two different plant species (tall fescue and alfalfa) both separately and in combination. The effect of this phytoremediation approach was evaluated on soil enzymatic activities as well as catabolic genes. The soil was collected from transformers and a capacitor storage locality. The experiment was conducted in the greenhouse with fertilizers provided to improve soil fertility in the pots. Soil catalase, dehydrogenase activities, and bacterial number were determined, and real-time polymerase chain reaction (PCR) was performed. Total bacterial number and dehydrogenase activity were significantly enhanced in planting treatments with mixed cropping soil. The mixed cropping strategy revealed its significance for increased bacterial biomass production. Nevertheless, the tall fescue single plant treatment showed the highest PCB removal because of its high biomass. The mixed plant cultivation followed the tall fescue single plant treatment and removed the second-highest amounts of PCBs from the contaminated soil. Conclusively, the mixed cropping approach can be used to remediate contaminated soils because it provides a greater number of bacteria, PCBs degraders, and enhanced enzymatic activities (Li et al. 2013).

10.11 Relationship Between Plant Age and PCBs Accumulation

Phytoremediation is an environmentally safe and cost-effective approach for the treatment of potentially harmful PCBs as compared to other traditional techniques such as landfilling and incineration. These traditional approaches are expensive, time consuming, and impose very large expenses for the transportation and excavation of soil (Low et al. 2010). Further, in the phytoextraction process, the pollutant is moved from the soil through roots and translocated to the shoot system, where it can be treated as a polluted waste, whereas the disposal of contaminated plants is easier and more cost-effective than that of polluted soil (Arthur et al. 2005). To date, it is feasible to treat PCBs through phytoremediation processes using advances in technology and scientific knowledge. Contaminants such as PCBs, polyaromatic hydrocarbons, dichlorodiphenyltrichloroethane (DDT) and its metabolites, and chlordane can be treated by phytoremediation (Whitfield Åslund et al. 2007; Kobayashi et al. 2008).

In this respect, relevant research work was published to understand PCB uptake by pumpkin plants over time. A soil contaminated with PCB at a 6.5 mg/kg concentration (Aroclor 1248) from a polluted industrial site was collected, and plants were grown in that soil and harvested five times over the experiment total of 42 days. After the harvesting, the root and shoot systems of plants were used for the determination of total PCBs. No significant difference was observed between the harvests. Moreover, the plants were also able to stabilize PCBs concentration over time. However, the biomass of the shoot was increased significantly from 4.3 g (12 days) to 59 g (42 days), suggesting the accumulation of PCBs throughout the plant growth period. Conclusively, the pumpkins can be harvested for the accumulation of the PCBs when the plant shoot has achieved maximum biomass (Low et al. 2010). In another relevant study in which pumpkin plants were analyzed for the accumulation of PCBs (Aroclors 1254 and 1260), in some plants, the lower stems contained about 43 μg/g PCBs concentration. This bioaccumulation reveals the use of pumpkin plants as phytoremediators. Also, plant density was significantly reduced in terms of plant biomass. Hence, this study provided evidence that pumpkin primarily translocated the PCBs contaminant from the roots to the shoot. Other pollutant transfer routes such as atmospheric deposition, direct soil, redeposition on shoots, and volatilization from soil posed negligible influence on the overall PCB concentration in pumpkin plants (Åslund et al. 2008).

10.12 Combined Application of Phytoremediation and Bioaugmentation

A PCB-polluted site was treated using an integrated approach of phytoremediation and bioaugmentation combined. Four plant species, namely, *Medicago sativa, Avena sativa*, *Brassica juncea*, and *Brachiaria decumbens*, were used as phytoremediators. Biochar and alignate were utilized as carrying materials. Several congeners of PCBS such as 44, 66, 118, 138, 153, 170, and 180 were selected, and the concentration of each congener was determined after 60 days of the experiment. The highest PCB removal was recorded in *B. juncea* with biochar amendment. Congener PCB66 had the highest removal percentage in both types of carrier materials and immobilized and free microbes; PCB170 had the lowest removal percentage in all treatments. Biochar as a carrier material showed a positive effect of PCBs removal. There was no significant effect on microbial inoculation using both carrier materials as confirmed by real-time PCR (Pino et al. 2016). Another similar study reported the use of rhizospheric and non-rhizospheric soil to remove different congeners of PCBs. Removal efficiencies vary according to plant type, congener type, and soil type. Different plants such as alfalfa, mustard, and oats have an essential role in the removal of PCBs through rhizodegradation. *Brachiaria decumbens* also showed the capability to absorb some PCBs congeners. The use of alfalfa and oats enhanced microbial growth with *bphA1* gene expression. Alfalfa influenced the *bphA1* gene and showed potential to remediate PCBs in the rhizospheric soil (Pino et al. 2019).

Hence, PCBs pose serious threats to all living organisms because of their persistent nature in the environment. PCBs contamination is a global issue, and full attention is required to address this problem. Traditional methods can hardly be deployed to remediate PCBs-contaminated sites because of their low efficiency and high costs. Comparably, rhizoremediation has an advantage over other treatment approaches because of its environment-friendly approach and cost-effectiveness. Indigenous microbes are stimulated in the rhizospheric zone and their potential is enhanced to degrade PCBs in aged, contaminated soils. These rhizospheric microorganisms also promote root development and plant growth. In this report, aerobic bacteria were isolated from potentially toxic soil and bioaugmented in the rhizosphere of *Phalaris arundinacea* (canary grass). Bacteria isolated were selected based on biphenyl activity, emulsification property, 2,3-catechol dioxygenase activities, and bphA1 gene expression required for oxidation of PCBs. The results revealed that *Rhodococcus* strains showed great potential to degrade many PCB congeners and reduced plant stress, promoting the emergence of lateral roots in *Arabidopsis thaliana*. Moreover, these bacterial strains also increase plant biomass to survive in soils highly contaminated with PCBs. Conclusively, the characteristics of reduced plant stress, PCBs degradation, plant growth, and root development constitute a suitable and sustainable approach of microbially assisted rhizoremediation using the bioaugmentation technique (Vergani et al. 2019).

10.13 Comparative Studies on Mycoaugmentation, Biostimulation, and Natural Attenuation

To explore the most suitable and efficient approach for PCB remediation, three different remediation approaches were studied and compared. In this study, the bioremediation capability of PCBs was determined in the soil with different amendments in microcosms. In the mycoaugmentation approach, two fungal species, *Trametes sanguinea* and *Pleurotus sajor-caju*, were inoculated, and in the biostimulation, sugarcane bagasse was supplemented. Natural attenuation treatment was taken without any kind of amendments. The remediation potential of these approaches was analyzed based on soil properties, toxicological tests, and contaminant content. After 90 days of the experiment, mycoaugmentation and biostimulation approaches revealed approximately 90% PCBs remediation efficiency. Different enzymatic activities such as endoxylanase, laccases, and manganese-peroxidase were enhanced in amended soils after 60 days of the experiment. Sugarcane bagasse emerged as a suitable substrate for fungal colonization in PCBs-contaminated soils. Phosphorus and oxidable organic matter were enhanced in mycoaugmentation and biostimulation approaches. Conclusively, both biostimulation and mycoaugmentation approaches could be used together for effective bioremediation of PCBs-contaminated soils (Sadañoski et al. 2020).

10.14 Conclusion

It is concluded from the foregoing discussion that PCBs were used extensively in agriculture and industry for a prolonged period because of their unique characteristics, which eventually resulted in the devastation of environmental health and quality. This environmental deterioration affected all forms of living organisms including humans. These PCBs have been reported as carcinogenic in nature and can have harmful effects on plants. PCBs were present in different pesticides. Organochlorine pesticides such as dichlorodiphenyltrichloroethane (DDT) and Aldrin have been used in agricultural practices as well as in the control of insect-borne diseases since the 1930s until the ban in the 1970s. It is an established fact that PCBs cause growth inhibition of plants. Moreover, symplastic and apoplastic pathways translocate the movement of PCBs through epidermal layers. The symplastic pathway is the most crucial path for the uptake of provided PCBs. In contrast, the apoplastic pathway is used by terrestrial plants for the uptake of airborne PCBs whereas dissolved PCBs are taken up by hydrophytes and microbes. It is well known that PCB concentrations in plants depend upon PCB concentrations in the soil. Temperature, and clay, and organic content, also can affect mobility and PCBs availability in the soil. To minimize these harmful effects, several treatment technologies have been utilized. Hence, phytoremediation is an environmentally safe and cost-effective approach for the treatment of potentially harmful PCBs as compared to other traditional techniques such as landfilling and incineration.

References

Aken BV (2008) Transgenic plants for phytoremediation: helping nature to clean up environmental pollution. Trends Biotechnol 26:225–227

Aken BV, Correa PA, Schnoor JL (2010) Phytoremediation of polychlorinated biphenyls: new trends and promises. Environ Sci Technol 44:2767–2776

Antolín-Rodríguez JM, Sánchez-Báscones M, Martín-Ramos P, Bravo-Sánchez CT, Martín-Gil J (2016) Estimation of PCB content in agricultural soils associated with long-term fertilization with organic waste. Environ Sci Pollut Res 23(12):12372–12383

Anyasi RO, Atagana HL (2011) Biological remediation of polychlorinated biphenyls (PCB) in the environment by microorganisms and plants. Afr J Biotechnol 10(82):18916–18938

Arthur EL, Rice PJ, Rice PJ, Anderson TA, Baladi SM, Henderson KL, Coats JR (2005) Phytoremediation—an overview. Crit Rev Plant Sci 24(2):109–122

Åslund MLW, Zeeb BA, Rutter A, Reimer KJ (2007) In situ phytoextraction of polychlorinated biphenyl (PCB)-contaminated soil. Sci Total Environ 374(1):1–12

Åslund MLW, Rutter A, Reimer KJ, Zeeb BA (2008) The effects of repeated planting, planting density, and specific transfer pathways on PCB uptake by *Cucurbita pepo* grown in field conditions. Sci Total Environ 405(1–3):14–25

Awan B, Sabeen M, Shaheen S, Mahmood Q, Ebadi AG, Toughani. M (2020) Phytoremediation of zinc contaminated water by marigold (*Tagetes minuta* L). Cent Asian J Environ Sci Technol Innov 1(3):150–158. https://doi.org/10.22034/CAJESTI.2020.03.04

Beyer A, Biziuk M (2009) Environmental fate and global distribution of polychlorinated biphenyls. In: Reviews of environmental contamination and toxicology, vol 201. Springer, Boston, pp 137–158

Borja J, Marie-Teleon D, Auresenia J, Gallardo S (2005) Polychlorinated biphenyls and their biodegradation. Process Biochem 40:1999–2013

Bruner-Tran KL, Osteen KG (2010) Dioxin-like PCBs and endometriosis. Syst Biol Reprod Med 56(2):132–146

Burken JG (2003) Uptake and metabolism of organic compounds: green liver model. In: McCutcheon SC, Schnoor JL (eds) Phytoremediation. Transformation and control of contaminants Wiley interscience. John Wiley Sons, Inc, Hoboken, pp 59–84

Cetin B, Yurdakul S, Gungormus E, Ozturk F, Sofuoglu SC (2018) Source apportionment and carcinogenic risk assessment of passive air sampler-derived PAHs and PCBs in a heavily industrialized region. Sci Total Environ 633:30–41

Chaudhry Q, Blom-Zandstra M, Cupta S, Joner EJ (2005) Utilizing the synergy between plants and rhizosphere microorganisms to enhance breakdown of organic pollutants in the environment. Environ Sci Pollut Res 12(1):34–48

Chekol T, Vough LR, Chaney RL (2004) Phytoremediation of polychlorinated biphenyl-contaminated soils: the rhizosphere effect. Environ Int 30(6):799–804

Chroma L, Moeder M, Kucerova P, Macek T, Mackova M (2003) Plant enzymes in metabolism of polychlorinated biphenyls. Fresen Environ Bull 12(3):291–295

Davies H, Delistraty D (2016) Evaluation of PCB sources and releases for identifying priorities to reduce PCBs in Washington state (USA). Environ Sci Pollut Res 23(3):2033–2041

de Souza AC, Taniguchi S, Figueira RCL, Montone RC, Bícego MC, Martins CC (2018) Historical records and spatial distribution of high hazard PCBs levels in sediments around a large South American industrial coastal area (Santos Estuary, Brazil). J Hazard Mater 360:428–435

Delgado IF, Barretto HH, Kussumi TA, Alleluia IB, Baggio CDA, Paumgartten FJR (2002) Serum levels of organochlorine pesticides and polychlorinated biphenyls among inhabitants of Greater Metropolitan Rio de Janeiro, Brazil. Cad Saude Publica 18(2):519–524.5

Denyes MJ, Langlois VS, Rutter A, Zeeb BA (2012) The use of biochar to reduce soil PCB bioavailability to *Cucurbita pepo* and *Eisenia fetida*. Sci Total Environ 437:76–82

Denyes MJ, Rutter A, Zeeb BA (2013) In situ application of activated carbon and biochar to PCB-contaminated soil and the effects of mixing regime. Environ Pollut 182:201–208

Dudášová H, Lukáčová L, Murínová S, Dercová K (2012) Effects of plant terpenes on biodegradation of polychlorinated biphenyls (PCBs). Int Biodeter Biodegr 69:23–27

Dzantor EK, Woolston J (2001) Enhancing dissipation of aroclor 1248 using substrate amendment in rhizosphere soil. J Environ Sci Health A36(10):1861–1871

Dzantor EK, Chekol T, Vough., L. R. (2000) Feasibility of using forage grasses and legumes for phytoremediation of organic pollutants. J Environ Sci Health A35(9):1645–1661

Ehsani S, Hayes D (2018) Efficacy of dredging for remediating polychlorinated biphenyl-contaminated sediments of the lower Fox River. Environ Eng Sci 35(12):1387–1393

EPA (1978) U.S. Environmental Protection Agency. Support document: Draft voluntary environmental impact statement for polychlorinated biphenyls (PCBs) manufacturing, processing, distribution in commerce and use ban regulation

Fu JJ, Wang YW, Zhang AQ et al (2011) Spatial distribution of polychlorinated biphenyls (PCBs) and polybrominated biphenyl ethers (PBDEs) in an e-waste dismantling region in Southeast China: use of apple snail (Ampullariidae) as a bioindicator. Chemosphere 82:648–655

Furukawa K, Fujihara H (2008) Microbial degradation of polychlorinated biphenyls: biochemical and molecular features. J Biosci Bioeng 105:433–449

Furukawa K, Hikaru S, Masatoshi G (2004) Biphenyl dioxygenases. Functional versatilities and directed evolution. J Bacteriol 186:5189–5196

García-de la Parra LM, Cervantes-Mojica LJ, González-Valdivia C, Martínez-Cordero FJ, Aguilar-Zárate G, Bastidas-Bastidas P, Betancourt-Lozano M (2012) Distribution of pesticides and PCBs in sediments of agricultural drains in the Culiacan Valley, Sinaloa, Mexico. Arch Environ Contam Toxicol 63(3):323–336

Gasic B, MacLeod M, Klanova J, Scheringer M, Ilic P, Lammel G, Hungerbühler K (2010) Quantification of sources of PCBs to the atmosphere in urban areas: a comparison of cities in North America, Western Europe and former Yugoslavia. Environ Pollut 158(10):3230–3235

Golden R, Doull J, Waddell W, Mandel J (2003) Potential human cancer risks from exposure to PCBs: a tale of two evaluations. Crit Rev Toxicol 33(5):543–580

Grimm FA, Hu D, Kania-Korwel I, Lehmler HJ, Ludewig G, Hornbuckle KC, Duffel MW, Bergman A, Robertso LW (2015) Metabolism and metabolites of polychlorinated biphenyls (PCBs). Crit Rev Toxicol 45:245–272

Halfadji A, Touabet A, Portet-Koltalo F, Le Derf F, Merlet-Machour N (2019) Concentrations and source identification of polycyclic aromatic hydrocarbons (PAHs) and polychlorinated biphenyls (PCBs) in agricultural, urban/residential, and industrial soils, east of Oran (Northwest Algeria). Polycycl Aromat Compd 39(4):299–310

Hatamian-Zarmi A, Shojaosadati SA, Vasheghani-Farahani E, Hosseinkhani S, Emamzadeh A (2009) Extensive biodegradation of highly chlorinated biphenyl and Aroclor 1242 by *Pseudomonas aeruginosa* TMU56 isolated from contaminated soil. Int Biodeter Biodegr 63:788–794

Hayat A, Hussain I, Soja G, Iqbal M, Shahid N, Syed JH, Yousaf S (2019) Organic and chemical amendments positively modulate the bacterial proliferation for effective rhizoremediation of PCBs-contaminated soil. Ecol Eng 138:412–419

Hu GC, Luo XJ, Dai JY, Zhang XL, Wu H, Zhang CL, Wei FW (2008) Brominated flame retardants, polychlorinated biphenyls, and organochlorine pesticides in captive giant panda (*Ailuropoda melanoleuca*) and red panda (*Ailurus fulgens*) from China. Environ Sci Technol 42(13):4704–4709

Hu GJ, Chen SL, Zhao YG, Sun C, Li J, Wang H (2009) Persistent toxic substances in agricultural soils of Lishui County, Jiangsu Province, China. Bull Environ Contam Toxicol 82(1):48–54

Idris A, Ahmed M (2003) Treatment of polluted soil using bioremediation–a review. Faculty of Chemical and Environmental Engineering, University of Putra, Malaysia, pp 1–18

Jing R, Fusi S, Kjellerup BV (2018) Remediation of polychlorinated biphenyls (PCBs) in contaminated soils and sediment: state of knowledge and perspectives. Front Environ Sci 6:79

Kobayashi T, Navarro RR, Tatsumi K, Iimura Y (2008) Influence of compost amendment on pyrene availability from artificially spiked soil to two subspecies of *Cucurbita pepo*. Sci Total Environ 404(1):1–9

Leigh MB, Prouzová P, Macková M, Macek T, Nagle DP, Fletcher JS (2006) Polychlorinated biphenyl (PCB)-degrading bacteria associated with trees in a PCB-contaminated site. Appl Environ Microbiol 72(4):2331–2342

Li Y, Liang F, Zhu Y, Wang F (2013) Phytoremediation of a PCB-contaminated soil by alfalfa and tall fescue single and mixed plants cultivation. J Soil Sediment 13(5):925–931

Li Q, Lu Y, Wang P, Wang T, Suriyanarayanan S, Liang R, Khan K (2018) Distribution, source, and risk of organochlorine pesticides (OCPs) and polychlorinated biphenyls (PCBs) in urban and rural soils around the Yellow and Bohai Seas, China. Environ Pollut 239:233–241

Liu JY, Schnoor JL (2008) Uptake and translocation of lesser chlorinated polychlorinated biphenyls (PCBs) in whole hybrid poplar plants after hydroponic exposure. Chemosphere 73:1608–1616

Low JE, Whitfield Åslund ML, Rutter A, Zeeb BA (2010) Effect of plant age on PCB accumulation by *Cucurbita pepo* ssp. *pepo*. J Environ Qual 39(1):245–250

Macek T, Macková M, Kucerová P, Chromá L, Burkhard J, Demnerova K (2002) Phytoremediation. In: Agathos SN, Reineke W (eds) Biotechnology for the environment: soil remediation. Kluwer Academic Publishers, Brussels, pp 115–137

Macek T, Francova K, Kochánková L, Lovecká P, Ryslava E, Rezek J, Mackova M (2004) Phytoremediation—biological cleaning of a polluted environment. Rev Environ Health 19(1):63–82

Macek T, Surá M, Pavliková D, Francová K, Scouten WH, Szekeres M, Macková M (2005) Can tobacco have a potentially beneficial effect to our health? Zeitschrift fur Naturforschung C (J Biosci) 60(3-4):292–299

Macková M, Vrchotová B, Francová K, Sylvestre M, Tomaniová M, Lovecká P, Macek T (2007) Biotransformation of PCBs by plants and bacteria–consequences of plant–microbe interactions. Eur J Soil Biol 43(4):233–241

Magee KD, Michael A, Ullah H, Dutta SK (2008) Dechlorination of PCB in the presence of plant nitrate reductase. Environ Toxicol Pharmacol 25(2):144–147

Markowitz G, Rosner D (2018) Monsanto, PCBs, and the creation of a "world-wide ecological problem". J Public Health Policy 39(4):463–540

Mehmannavaz R, Prasher SO, Ahmad D (2002) Rhizospheric effects of alfalfa on biotransformation of polychlorinated biphenyls in a contaminated soil augmented with *Sinorhizobium meliloti*. Process Biochem 37(9):955–963

Montone RC, Taniguchi S, Weber RR (2001) Polychlorinated biphenyls in marine sediments of Admiralty Bbay, King George Island, Antarctica. Mar Pollut Bull 42(7):611–614

Needham TP, Ghosh U (2019) Four decades since the ban, old urban wastewater treatment plant remains a dominant source of PCBs to the environment. Environ Pollut 246:390–397

Olson PE, Castro A, Joern M, DuTeau NM, Pilon-Smits EA, Reardon KF (2007) Comparison of plant families in a greenhouse phytoremediation study on an aged polycyclic aromatic hydrocarbon-contaminated soil. J Environ Qual 36(5):1461–1469

Omwoma S, Mbithi BM, Pandelova M, Ssebugere P, Lalah JO, Wang Y, Schramm KW (2019) Comparative exposomics of persistent organic pollutants (PCBs, OCPs, MCCPs and SCCPs) and polycyclic aromatic hydrocarbons (PAHs) in Lake Victoria (Africa) and Three Gorges Reservoir (China). Sci Total Environ 695:133789

Pal A, Wauters G, Paul AK (2007) Nickel tolerance and accumulation by bacteria from rhizosphere of nickel hyperaccumulators in serpentine soil ecosystem of Andaman, India. Plant and Soil, 293(1):37–48

Pieper DH, Seeger M (2008) Bacterial metabolism of polychlorinated biphenyls. J Mol Microbiol Biotechnol 15(2–3):121–138

Pino NJ, Muñera LM, Peñuela GA (2016) Bioaugmentation with immobilized microorganisms to enhance phytoremediation of PCB-contaminated soil. Soil Sediment Contam Int J 25(4):419–430

Pino NJ, Múnera LM, Peñuela GA (2019) Phytoremediation of soil contaminated with PCBs using different plants and their associated microbial communities. Int J Phytoremediation 21(4):316–324

Quintanilla-López JE, Galindo-Iranzo P, Lebrón-Aguilar R (2020) Congener-specific determination of hydroxylated polychlorinated biphenyls by polar-embedded reversed-phase liquid chromatography-tandem mass spectrometry. J Chromatogr A 1626:461353

Reddy AVB, Moniruzzaman M, Aminabhavi TM (2019) Polychlorinated biphenyls (PCBs) in the environment: recent updates on sampling, pretreatment, cleanup technologies and their analysis. Chem Eng J 358:1186–1207

Reddy AVB, Moniruzzaman M, Madhavi G, Aminabhavi TM (2020) Modern approaches in separation, identification and quantification of polychlorinated biphenyls. Curr Opin Environ Sci Health 18:26–39

Rezek J, Macek T, Mackova M, Triska J (2007) Plants metabolites of polychlorinated biphenyls in hairy root culture of black nightshade *Solanum nigrum* SNC-90. Chemosphere 69:1221–1227

Ross G (2004) The public health implications of polychlorinated biphenyls (PCBs) in the environment. Ecotoxicol Environ Saf 59(3):275–291

Sadañoski MA, Tatarin AS, Barchuk ML, Gonzalez M, Pegoraro CN, Fonseca MI, Villalba LL (2020) Evaluation of bioremediation strategies for treating recalcitrant halo-organic pollutants in soil environments. Ecotoxicol Environ Saf 202:110929

Samara F, Tsai CW, Aga DS (2006) Determination of potential sources of PCBs and PBDEs in sediments of the Niagara River. Environ Pollut 139(3):489–497

Schnoor JL (2002) Phytoremediation of soil and groundwater. GWRT Series E-Series TE-02-01:1–45

Sharma BM, Nizzetto L, Bharat GK, Tayal S, Melymuk L, Sáňka O, Larssen T (2015) Melting Himalayan glaciers contaminated by legacy atmospheric depositions are important sources of PCBs and high-molecular-weight PAHs for the Ganges floodplain during dry periods. Environ Pollut 206:588–596

Sierra I, Valera JL, Marina ML, Laborda F (2003) Study of the biodegradation process of polychlorinated biphenyls in liquid medium and soil by a new isolated aerobic bacterium (*Janibacter* sp.). Chemosphere 53(6):609–618

Smith KE, Banks MK, Schwab AP (2009) Dewatering of contaminated sediments: greenhouse and field studies. Ecol Eng 35(10):1523–1528

Sofuoglu SC, Sofuoglu A, Holsen TM, Alexander CM, Pagano JJ (2013) Atmospheric concentrations and potential sources of PCBs, PBDEs, and pesticides to Acadia National Park. Environ Pollut 177:116–124

Sun J, Pan L, Tsang DC, Zhan Y, Liu W, Wang X, Li X (2016) Polychlorinated biphenyls in agricultural soils from the Yangtze River Delta of China: regional contamination characteristics, combined ecological effects and human health risks. Chemosphere 163:422–428

Teng Y, Zheng MK, Luo YM, Gao J, Li ZG, Wu LH (2008) Spatial distribution of soil PCBs congeners in typical area of Yangtze River delta region. Huan jing ke xue (Huanjing kexue) 29(12):3477–3482

Toro SD, Zanaroli G, Fava F (2006) Aerobic bioremediation of an actual site soil historically contaminated by polychlorinated biphenyls (PCBs) through bioaugmentation with a non-acclimated, complex source of microorganisms. Microb Cell Fact 5:11

Trapp S, Kulhanek A (2006) Human exposure for food: one equation for all crops is not enough. In: Macková M, Macek T, Dowling D (eds) Phytoremediation and rhizoremediation: focus on biotechnology, vol 9A. Springer, Berlin Heidelberg

Urbaniak M, Lee S, Takazawa M, Mierzejewska E, Baran A, Kannan K (2019) Effects of soil amendment with PCB-contaminated sediment on the growth of two cucurbit species. Environ Sci Pollut Res Int:1–13

Urbaniak M, Baran A, Lee S, Kannan K (2020) Effects of amendments of PCB-containing Hudson River sediment on soil quality and biochemical and growth response of cucumber (*Cucumis sativus L.* cv 'Wisconsin SMR 58') Int J Phytoremediation:1–9

Van den Berg M, Birnbaum LS, Denison M, De Vito M, Farland W, Feeley M, Rose M (2006) The 2005 World Health Organization reevaluation of human and mammalian toxic equivalency factors for dioxins and dioxin-like compounds. Toxicol Sci 93(2):223–241

Van Metre PC, Mahler BJ, Furlong ET (2000) Urban sprawl leaves its PAH signature. Environ Sci Technol 34(19):4064–4070

Vergani L, Mapelli F, Suman J, Cajthaml T, Uhlik O, Borin S (2019) Novel PCB-degrading *Rhodococcus* strains able to promote plant growth for assisted rhizoremediation of historically polluted soils. PLoS One 14(8):0221253

Wiszniewska A, Hanus-Fajerska E, Muszyńska E, Ciarkowska K (2016) Natural organic amendments for improved phytoremediation of polluted soils: a review of recent progress. Pedosphere 26(1):1–12

Wittsiepe J, Fürst P, Wilhelm M (2007) The 2005 World Health Organization re-evaluation of TEFs for dioxins and dioxin-like compounds—what are the consequences for German human background levels? Int J Hyg Environ Health 210(3–4):335–339

Wyrwicka A, Steffani S, Urbaniak M (2014) The effect of PCB-contaminated sewage sludge and sediment on metabolism of cucumber plants (*Cucumis sativus* L.)´. Ecohydrol Hydrobiol 14(1):75–82

Yao M, Hu T, Wang Y, Du Y, Hu C, Wu R (2017) Polychlorinated biphenyls and its potential role in endometriosis. Environ Pollut 229:837–845

Zeeb BA, Amphlett JS, Rutter A, Reimer KJ (2006) Potential for phytoremediation of polychlorinated biphenyl (PCB) contaminated soil. Int J Phytoremediation 8(3):199–221

Zhao Y, Zhan J, Liu G, Ren Z, Zheng M, Jin R, Zhang X (2017) Field study and theoretical evidence for the profiles and underlying mechanisms of PCDD/F formation in cement kilns co-incinerating municipal solid waste and sewage sludge. Waste Manag 61:337–344

Chapter 11
Uptake of Organic Pollutants and the Effects on Plants

Bibi Saima Zeb, Malik Tahir Hayat, Tahseen Zeb, Faisal Younas Khan, Haleema Zeb Abbasi, Iffat Nawaz, and AbdolGhaffar Ebadi

Abbreviations

ABC	ATP-binding cassette
BCF	Bioconcentration factor
BPA	Bisphenol A
CTC	Chlor-tetracycline
DOC	Dissolved organic carbon
GF	Griseofulvin
GTN	Glyceryl trinitrate
K_{OA}	Partition coefficient between octanol and air
K_{OW}	Partition coefficient between octanol and water
LPO	Lipid peroxidation
MDA	Malondialdehyde
MT	Microtubules
OTC	Oxytocycline
PCBs	Polychlorinated biphenyls
PCDD/Fs	Polychlorinated dibenzofurans
POPs	Persistent organic pollutants
TBZ	Thiabendazole
TCE	Trichloroethylene
TM	Tail moment

B. S. Zeb (✉) · M. T. Hayat · H. Z. Abbasi
Department of Environmental Sciences, COMSATS University Islamabad- Abbottabad Campus, Abbottabad, Pakistan
e-mail: saimazeb@cuiatd.edu.pk

T. Zeb · I. Nawaz
Bafa Agricultural Research Station, Mansehra, Pakistan

F. Y. Khan
Department of Water Management, District Battagram, Pakistan

A. Ebadi
Department of Agriculture, Jouybar Branch, Islamic Azad University, Jouybar, Iran

Q. Mahmood (ed.), *Sustainable Plant Nutrition under Contaminated Environments*, Sustainable Plant Nutrition in a Changing World,
https://doi.org/10.1007/978-3-030-91499-8_11

TNT	Trinitrotoluene
TSCF	Transpiration stream cofactor

11.1 Introduction

The mother environment receives large amounts of synthetic chemicals that are referred to as anthropogenic or xenobiotic because they are new to nature. Plant productivity and its normal functions are influenced by various internal and external factors. These chemicals acts as vital external factors that affect plants adversely. Hormones and persistent organic pollutants (POPs), including antibiotics, herbicides, polychlorinated biphenyls (PCBs), bisphenol A (BPA), polychlorinated dibenzo-*p*-dioxins, and polychlorinated dibenzofurans (PCDD/Fs), have raised substantial questions of this issue in environmental research. These chemicals have severe effects on plants, and hundreds and thousands of synthetically produced organic compounds used in agricultural, industrial, and domestic applications (Boxall et al. 2006).

These chemicals can transport for short or long distances and can transform into various compounds, thus undergoing unlimited competition. Thus, the fate of such pollutants is very difficult to forecast. The transport and buildup of some anthropogenic chemical compounds from environment to organism is known as the bioconcentration factor (Bernes 1998), the ratio between the concentration of a compound in an organism and its environment. The increase of concentration caused by transition of a pollutants from one trophic level to another higher level is referred to as biomagnification. Aquatic systems have been studied for the presence of hydrophobic chemicals in the food chain (McLachlan 1995; Kelly and Gobas 2001).

Bioremediation is facilitated by various plants and microbes in the rhizosphere for the treatment of contaminated soils. Facilitation of this process in the rhizosphere is possible it can change its environment (Anderson et al. 1993). Bioaccumulation also has a role in enhancing bioremediation in the rhizosphere. For this purpose, poplar trees (*Populus*) were grown in a bioreactor to study the transport and digestion of an atrazine herbicide (Burken and Schnoor 1997).

The study of utilization of pollutant concentrations in plant life as a possible pointer of atmospheric and soil contamination levels is referred to as biomonitoring (Paterson et al. 1990; Kylin 1994; Simonich and Hites 1994; Smith and Jones 2000). Many species are being used for biomonitoring of organic pollutants, for example, Scots pine (*Pinus sylvestris*), spruce (*Picea abies*), and saxifrage (*Saxifraga oppositofolia*) (Eriksson et al. 1989; Calamari et al. 1991; Jensen et al. 1992, Tremolada et al. 1996; Safe et al. 1992; Strachan et al. 1994; Juuti et al. 1995; Reischl 1988; Weiss 1998; France et al. 1997). Moreover, conifers are also widely used for detecting different patterns of organic pollutants geographically (Eriksson et al. 1989; Jensen et al. 1992; Tremolada et al. 1996; Weiss 1998) and for the investigation of polycyclic aromatic hydrocarbons (PAHs) in plants (Kylin 1994; Simonich and Hites 1994; Wagrowski and Hites 1996; Bakker 2000).

Table 11.1 Organic pollutants: classification and sources

Organic pollutants	Classification	Sources
Chlorinated solvents Polychlorinated biphenyls	Trichloroethylene Perchloroethylene 4-chlorobiphenyl 4,4-Dichlorobiphenyl	Drycleaners Chemical manufacture Electrical manufacturing Power station Railway yards
Chlorinated phenol	Pentachlorophenol	Timber treatment landfills
"BTEX"	Benzene Toluene Ethyl benzene Xylene	Oil production and storage Gas work sites Paint manufacture Port facilities Railway yards Chemical manufacture
Polyaromatic hydrocarbons (PAHs)	Napthalene Antracene Flourene	Oil production and storage Gas work sites Coke plants
	Benzopyrene	Engine works Landfills Tar production and storage Boiler and dump sites
Pesticides	Antrazine Carbaryl Carbofuran Coumphos Diazinon Glycophosphate Parathion Propham	Agriculture Timber treatment plants Pesticide manufacture Recreational areas Landfill

Organic pollutants can also be changed into less harmful products by plants, which contribute by plant growth and other supporting processes through the process of phytoremediation. Various classes of organic pollutants are shown in Table 11.1.

Many organic pollutants that are endogenous, xenobiotic, and toxic, such as herbicides (trichloroethylene, TCE; trinitrotoluene, TNT; glyceryl trinitrate, GTN) and PCBs are harmful for plants and have a certain process of degradation in which plant roots have an important role in many cases. The foremost system for that is the glutathione-*S*-conjugate pump or ATP-binding cassette (ABC) transporters, which identify different conjugates such as organic conjugates of glutathione, high molecular mass toxic organics, and other metal peptides such as phytochelatins (Tommasini et al. 1998).

POPs are made from manmade chemicals and are considered as one of the most hazardous compounds ever synthesized. POPs include various chemicals such as dioxins, organochlorine pesticides, and PCBs, all of which are very resistant to biodegradation. These pollutants accumulate in the body over a long time and sequester in adipose tissue and liver, affecting the reproductive functions and also causing cancer (Ruzzin et al. 2012). The various reproductive problems and diseases include disorders in the normal functioning of both males and females, including birth disorders, premature births, low fetal weight, impotency, decrease in fertility, and menstrual problems. POP contamination has been linked with reproductive system dysfunctions (Alharbi et al. 2018).

Organic pollutants are a unique group of compounds that differ widely from one another in chemical and biological functions. Organic pollutants are considered as environmental pollutants as they have a very toxic effect on animals, plants, and humans. Soil becomes contaminated with organic pollutants by natural as well as human activities, which include smelting operations, mining, and agriculture, causing overall increases in the level of heavy metals in soil to an alarming point. The effect of organic pollutants on the environment is clearly shown in Fig. 11.1, the organic pollutant cycle in the environment. These organic pollutants have a very persistent nature, and thus accumulate in the soil and in the plants growing in it. They interfere with the physiological functions of plants such as gaseous exchange, photosynthesis, and nutrient absorption, and also cause reductions in dry matter accumulation, plant growth, and overall yield. The heavy metals also impact the concentration of antioxidants in plants so that the overall nutritive level of the produce is lost. Dietary intake of any plant with a heavy metal accumulation may cause harmful impacts on human health (Kacálková and Tlustoš 2011). Phytoremediation is a very inexpensive and environmentally friendly approach that which can be carried out for remediation of polluted water on site (in situ), and the products of remediation do not need further processing. Many plant species grow well in contaminated soil or water. The organic pollutant cycle in the environment is shown in Fig. 11.1.

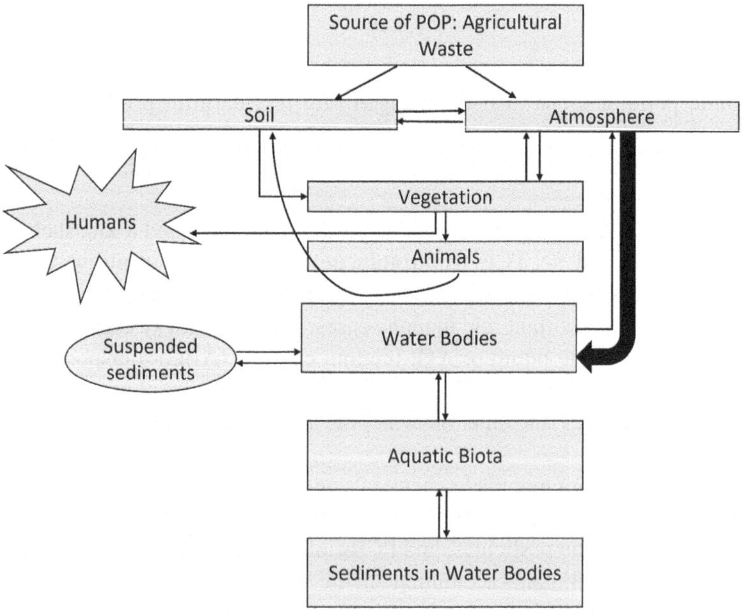

Fig. 11.1 Organic pollutant cycle in the environment

These plants clean the water either by absorbing the contaminants into their roots or by decomposing the contaminants into less harmful forms or accumulation in the plant tissues. Water plants such as duckweed, water hyacinth, and pennywort are becoming very well known in this regard (Mahajan and Kaushal 2018). Increasing levels of POPs) in various environmental matrices are the matter of a growing concern (Holoubek et al. 2007). The Stockholm Convention on POPs (UNEP 2001) was established in 2001 as a global treaty protecting human health and the environment against these chemicals. POPs tend to persist for long periods in the environment, become globally dispersed, and accumulate in human and wildlife lipids along with binding to the sediments (Kang et al. 2000; Lee et al. 2001).

Exposure to POP residues in drinking water continues to cause a health threat to many populations, particularly in areas of developing countries with inadequate water treatment facilities (Zhang et al. 2015). Samples from the rural area Farrukabad in the vicinity of the Ganges River in Utter Pradesh, northern India, were found to be polluted with residues of HCH, DDT, and aldrin at values exceeding the WHO guidelines. Herman et al. (2005) reported contamination by endosulfan, atrazine, aldrin, dieldrin, and chlorpyrifos in canals of South Florida with average amounts of 16, 11, 9, 2.6, and 6 ng/L, respectively. Contamination by endosulfan and other pesticides was maximum during high levels of agricultural activity. Fatoki et al. (2004) reported that the organochlorines range from 5.5 ng/L (2,4 D) to 160 ng/g (HCB) in water samples in South Africa. In addition, the agriculture of Pakistan is the main contributor to its GDP, and important increases in fertilizer use have been observed in previous years. The extensive and indiscriminate use of pesticides and fertilizers to promote agricultural activities in surrounding areas results in the leaching of such chemicals, which are POPs, through runoff to groundwater as well as surface water.

11.2 Effects of Pesticides on Plants

When pesticides increase in the soil, effects on the plants including seed germination, inhibition in food crops, and biochemical, histological, and cytological changes occur (Dalvi and Salunhe, 1975). Excess pesticide use inhibits the growth of crops and also reduces crop yields. The effects of pesticides on seeds depend on the following:

(a) Distribution, persistence, and metabolism in the soil
(b) Active form
(c) Concentration
(d) Type of soil
(e) Interaction between contaminated soil and seeds

Pesticides affect different processes: some pesticides affect the metabolic pathway of plants and some interfere with the photosynthesis process (Igbedioh 1991). Herbicides also disrupt many of the lipid-including processes. Metribuzin has increased the fatty acid content of different soya bean oil varieties (Hardcastle et al. 1974). The amount of linoleic acid increase appeared to tend to correlate with decrease in the content of stearic acid, which shows the potential stimulation to convert stearic acid to oleic acid to linoleic acid. Fungicides inhibit the production of protein and nucleic acid (Sisler 1969). The content of nitrate (as a percent of dry matter) in corn crops is increased by less than 1.7 ppm of linuron in the solution. Unfortunately, some of the findings obtained in pesticide effect research on plant nutrition are difficult to understand (Hance 1981). In different crops, some pesticides can decrease ascorbic levels by 30% (Igbedioh 1991). The effect of pesticides on different crops are as follows.

11.2.1 In Rice Seedlings

Imidacloprid and chlorpyrifos have been reported to cause reduction of root length, shooting time, fresh weight, and protein content in rice seedlings. Such pesticides also degraded the chlorophyll pigment (Sharma et al. 2013). These pesticides induce oxidative stress to rice seedling by producing reactive oxygen species (ROS) such as hydrogen peroxide (H_2O_2) and superoxide anions (Sharma et al. 2015a, b).

11.2.2 In Tomato

Zhou et al. (2015) reported the increased levels of H_2O_2 in tomato plants following the application of chlorothalonil pesticides. In addition, NADPH oxidases participate in the formation of reactive oxygen species (Kaur et al. 2014).

Additionally, it was reported that RBOH1 (respiratory burst oxidase homolog 1) regulates H_2O_2 levels in pesticide-stressed tomato plants (Zhou et al. 2015).

11.2.3 In Mustard

Sharma et al. (2016) noticed a reduction in photosynthesis in mustard plants after the application of imidacloprid insecticide. Application of diuron pesticide in *Saccharina japonica* has caused growth retardation, decreased carotenoid and chlorophyll content, optimal quantum yield, and maximum electron transport rate (Kumar et al. 2010).

11.2.4 *In* Brassica

Sharma et al. (2015b) reported reduced content of different phytochemicals following the application of imidacloprid pesticide to juncea plants in *Brassica*. Pesticides are considered an efficient way to control crop pests, but they have serious adverse effects on various species of plants and ecosystems (Aktar et al. 2009).

Pesticides also cause a number of biochemical changes, molecular and morpho-physiological, in plants that adversely affect growth and productivity along with the production of pest resistance (Table 11.2) (Cheng et al. 2012).

Table 11.2 Harmful impacts of selected organic pollutants

Pesticides	Harmful effects of pesticides	Toxic effects reduction from EB	Reference
Polychlorinated biphenyl	Photosynthesis and chlorophyll contents Decrease biomass accumulation Inhibits represses stomatal conductance	Reduces lipid peroxidation Alleviates photo-inhibition Improves growth rate of plant Modulates oxidation stress	Ahammed et al. (2013a)
Chlorpyrifos	Reduces net photosynthetic rate (Pn) and quantum yield of PS2	Improves the Pn and quantum yield of PS2 Activates fenese antioxidant scheme Improves the metabolism of pesticides by activation of key gene like P450	Xia et al. (2009)
Phenanthrene and pyrene	Photosynthetic inhibition Production of H_2O_2, OH^-, O^{2-} Xenobiotic effects at cellular level Inhibit seed germination and pigment accumulation	Production of H_2O_2, OH^-, O^{2-} Xenobiotic effects at cellular level Inhibit seed germination and pigment accumulation Enhances germination of seed Improves morphological characteristics Enhances biomass for plants Decreases inhibition of photosynthesis	Ahammed et al. (2013b)
Imidachloprid	Reduces plant growth Reduces chlorophyll content Impairs photosynthesis and stomatal conductance Reduces the concentration of active biomolecules	Decreases IMI toxicity Increases plant growth Enhances pigments accumulation Increases photosynthetic efficiency and stomatal conductance	Sharma et al. (2017)
Chlorothlonil	Adversely affected soluble protein and free proline Increases malondialdehyde contents	Improves osmoregulation Decreases lipid peroxidation Significantly stimulates the expression of antioxidant genes	Wang et al. (2017)

11.3 Detoxification Process

However, plants can detoxify the pesticide absorbed using a detoxification system (Cherian and Oliveira 2005). According to this system, in the presence of enzymes such as P_{450} monooxygenase, peroxidases, and carboxylesterases, the absorbed pesticide is metabolically activated by "phase 1." Conjugation of glutathione (GSH) and glucose occur with the help of glutathione, *S*-transferase (GST) and UDP-glucosyltransferase (UGT) during the second detoxification process. In the third phase the soluble molecules are deposited in vacuole and apoplast.

An overview of pesticide metabolism in plants is shown in Fig. 11.2. The conjugated glutathione molecules are transported with the help of ATP-dependent membrane pumps (Martinoia et al. 1993). Although the exact mechanism of EBL application and pesticide stress relief is not yet clear, studies have shown that EBL alleviation effects are specific to pesticides. EBL in various ways soothes the detrimental effects of pesticides on plants.

Figure 11.2. Mechanism of pesticide metabolism by GST in plants (revised from Cherian and Oliveira 2005)

Plants are exposed to polyaromatic hydrocarbons (PAHs) from both air and soil in the environment. PAH pollution in the soil affects the growth and survival of different plant species associated with fungi. Clover, ryegrass, and leek were affected by PAHs, but a maize crop was not affected (Bossert and Bartha 1985). When PAHs occur in soil, various plants respond in different ways. In the study it was observed that the low concentration of PAHs in soil stimulated the development of bean and sunflower plant although a high concentration of PAHs showed adverse effects for all other plants. Increased concentration of PAHs reduces the root length growth of wheat, tomato, bean, and sunflower.

If molecules of hydrocarbon penetrate in different plant tissues, they can damage the cell membrane of the plants, causing leakage of cell material and possibly blocking the intercellular space, which reduces transportation and respiration (Xu and Johnson 1995). This action may not affect the whole plant, but it can cause genetic mutation, slow the growth rate, and weaken the plants, thereby increasing their

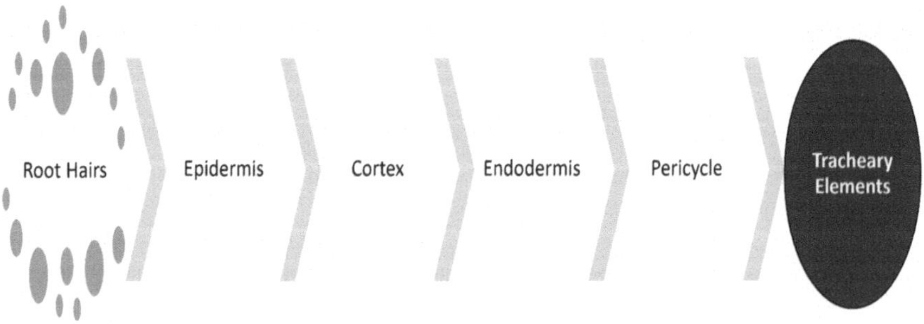

Fig. 11.2 Pathways for the flow of soil water through the elements of tracery

sensitivity to other types of stress (Ren et al. 1994). The PAHs effect on growth and survival of plants depends on different plant species associated with fungi.

The PAH pathway of exposure (i.e., direct or indirect) highlights similar plant response patterns. Significant leaf injuries, disturbance of plant carbon allocation, reduced production of biomass, and photosynthetic efficiency are all-important biological and physiological effects of contamination with PAHs.

11.3.1 Visible Symptoms in Leaves

The leaves can be affected by direct exposure to atmospheric PAHs by high mesophyll breakdown, chlorosis, and necrosis, silvering, and bronzing of the leaf surface (Oguntimehin et al. 2010a, b). Combined UV and PAH effects can also be responsible for damage to the leaves. Photoinduced toxicity of PAHs takes place in two pathways: photo-modification to more harmful photoproducts, and reaction to photosensitization with reactive oxygen species (ROS) formation (Wang et al. 2009). Indirect exposure to ambient PAHs (i.e., in the soil) causes the same adverse effects on the leaf as those from contaminated air, that is, necrotic lesions evident on the leaves as white spots (Liu et al. 2009a, b). Indirect exposure to atmospheric PAHs often decreases the number and size of stems and leaves. (Alkio et al. 2005).

11.3.2 Plant Biomass Production

The presence of PAHs usually affects the biomass of plants directly and indirectly. For example, the foliar use of PAHs greatly decreases the production of whole plant biomass in *Raphanus sativus*, *Cucumis sativus*, *Brassica rapa*, *Solanum lycopersicum*, *Brassica campestris*, *Lactuca sativa*, and *Brassica napus* (Ahammed et al. 2012a, b, c). Nevertheless, stimulation has also been reported for the production of biomass PAHs effects on growth that are usually related to the concentration of exposure. For example, in Petri dishes, the growth of *Pisum sativum* shoots exposed to fluoranthene was stimulated at 0.1 mg L^{-1} but inhibited at higher concentrations of 1, 5, and 7 mg L^{-1} (Váňová et al. 2009).

Contrasting effects have also been described for biomass partitioning, with either a decrease or an increase in the root-to shoot ratio: direct exposure to atmospheric phenanthrene greatly reduces the shoot biomass of *Trifolium pratense* but not root biomass, and has no observable impact on biomass production in *Lolium perenne* (Desalme et al. 2011). Indirect exposure to PAHs in polluted soil modifies both root and shoot elongation (Reynoso-Cuevas et al. 2008). Therefore, the impact of PAHs on plant biomass remains difficult to generalize because responses vary with the PAH in question, plant species, PAH application method, and exposure concentrations.

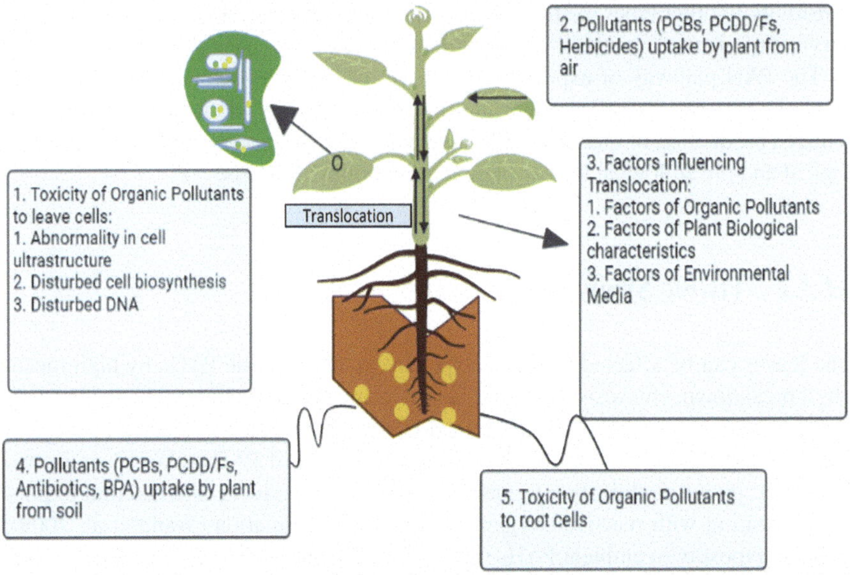

Fig. 11.3 Organic pollutant translocation and uptake in plants, uptake influencing factors, and the harmful effects of organic contaminants in plants. BPA bisphenol A, PCDD/Fs polychlorinated dibenzofurans, PCBs polychlorinated biphenyls

11.3.3 Photosynthesis

The phytotoxic effects of atmospheric PAHs in vascular plants have demonstrated negative impacts of photosynthesis in most studies (Fig. 11.3). With exposure to anthracene (Wieczorek and Wieczorek 2007), fluoranthene (Oguntimehin et al. 2008), and phenanthrene (Ahammed et al. 2012a, b, c), carbon assimilation, as determined by the photosynthetic rate (μmol CO_2/m^2 s), is decreased. Fluorescence of chlorophyll shows disruptions in photosynthesis photochemistry from the disruption of the electron transport chain in thylakoid membranes and oxygen levels (Ahammed et al. 2012a, b, c). The effect of PAHs on plant photosynthesis on atmospheric emission depends on the concentration and length of exposure. For the production of biomass, PAHs can promote (Wild et al. 1992) (short-term, small quantities) or inhibit (Mallakin et al. 2002) (environmental exposure) the electron transport capacity in PSII.

Different malfunctions from PAH effects on the leaves may explain the decrease in carbon assimilation by plants subjected directly or indirectly to environmental PAH. Reduction of stomatal conductance was reported in *R. sativus*, *L. esculentum*, *L. sativa*, and *P. densiflora* (Oguntimehin et al. 2010) and this participates in reduced carbon assimilation in plants exposed to atmospheric PAH. Photosynthetic process disruption is associated with PAH accumulation in thylakoid membranes, causing conformation changes in chloroplastic structures and reversible inactivation of PSII (Mallakin et al. 2002). However, the exact reason for this concentration decrease, or whether it represents the degradation of existing molecules or de novo synthesis inhibition, remains uncertain.

11.3.4 Oxidative Stress

The biochemical composition of leaves is one significant negative effect of PAHs. The oxidation of PAHs produces reactive oxygen species (ROS) in leaf cells.

The accumulation of reactive oxygen species in leaf cells results in cell organ deformities, loss of membrane integrity, lipid peroxidation of membranes, and eventually collapse of the cellular structures including the photosynthetic apparatus (Liu et al. 2009a, b).

The production of ROS and the problem they cause in chloroplasts identified the effect on photosynthetic methods as they potentially reduce the function of enzymes as well as the protein and pigment content (Oguntimehin et al. 2008).

11.4 Modes of Plant Uptake and Transport of Organic Contaminants

11.4.1 Uptake From Soil

In the soil, organic contaminants may be present in wet or dry deposition from transport from point sources such as industrial discharge, deposition of waste, and spraying of pesticides. In soil these pollutants are divided between soil particles, water, and air, and uptake is possible through the air or water phase (Hellström 2004). Some microbial species may even degrade plastics in the soil (Falah et al. 2020).

A compound follows three probable pathways when entering plants roots in the water phase: (i) through the cell wall or the apoplastic pathway, (ii) in the protoplasm or the symplastic pathway, or (iii) from vacuole to vacuole or the transcellular pathway (Wild et al. 1992). A root section shown in Fig. 11.2 illustrates the possible pathways.

The compounds avoid entrance into cells so they move in the apoplast until reaching the endodermis, a tubular connection of cells that are tightly packed by the casparian band, which is waxy and impermeable to water. There all the molecules pass over through the plasma membranes and the protoplasts of the tightly connected endodermal cells and enter the symplast to further stretch to the xylem to be transferred. This is a huge blockade for hydrophobic chemicals that need active transport, but unfortunately there is no proof of active transport of these compounds other than hormone-like chemicals (Trapp and McFarlane 1994). Different researchers have studied soil plant uptake by different hydrophobic compounds such as

PCBs, dioxins, furanes, and PAHs. Most of the studies in fields (Fries 1996) and in greenhouses (Waltke et al. 1990) on uptake of polychlorinated biphenyls, dioxins, dichloro-diphenyl-trichloroethane, and other various hydrophobic soil compounds have shown that only root absorption can be observed in the accumulation of lipid in the peel of carrot, beet, and turnip.

However, there is one interesting coincidence. Active transport was found in dioxins of two members of the genus *Cucurbita*, zucchini (*Cucurbita pepo* L. cv. Giromontiina) and pumpkin (*Cucurbita pepo* L. cv. Gelber Zentner), but not in cucumber (*Cucumis sativus* L. cv. Delikatess), which is also a member of the Cucurbitaceae family (White et al. 2003).

Uptake of organic contaminants from soil is the main issue for the health of humans regarding the risk of contamination of agricultural fields and the accumulation in crops. A few studies show uptake of environmental pollutants that are present in sludge or compost (Fonagy et al. 1996). In the studies this point is already discussed: there is no translocation through the edible parts of plants that are above the ground whereas the edible section of roots and tubers such as carrot can be contaminated. After volatilization and transport through air, the volatile compounds can contaminate edible leaves and fruits. There is the risk that the shoots and fruits near the surface could be directly polluted by soil. Undergoing both chemical and physical interaction with the soil, the roots of plants provide anchorage and energy storage.

11.4.2 Uptake From Air

A waxy cuticle layer on a leaf supports as a blockade for loss of water and entrance of xenobiotics. The cuticle membrane consists of large quantities of different cell-wall polysaccharides on the internal layer, and soluble and polymerized aliphatic lipids on the outer side of the layer. This cuticle is attached by a pectin-rich layer to the cell wall. The arrangements of fibrillae and lamellae are combined by cutin (Kirkwood 1999).

Organic compounds may influence the plant surface in two forms:

(i) Dry deposition (gaseous form or combined with particles)
(ii) Wet deposition (dissolved in water)

These settled compounds are consolidated in cuticular lipids that disperse by lipid layer and sooner or later are transferred by the phloem, or exchange through stomata may happen as a gas. There may be an aqueous path that contains pores, carbohydrate strands, and lipid bands, besides lipid content of cuticle (Paterson et al. 1990). For C14–C16 dioxins and furanes (PCDD/F), dry gaseous accumulation is the standard path (Welsch-Pausch et al. 1995). Modes of plant uptake and transport of organic contaminants are shown in Table 11.3.

Table 11.3 Modes of plant uptake and transport of organic contaminants

Organic pollutant	Tissue tested	Species	Reference
PCBs (PCB3, PCB15, PCB28, PCB52, PCB73)	Root Foliage	Hybrid poplar Bluegrass, *Luzula* and *Betula*	Liu and Schnoor (2008) Pier et al. (2002)
PCDD/fs	Root Foliage	*Cucurbita*, rouzi grass (Thylacospermum caespitosum), lettuce Potato, apple, pear, rice, pea and oil seed rape Rice and radish	Engwall and Hjelm (2000) Zhang and Zhu (2009) Li et al. (2014) Stefan and Michael (1997); Wu et al. (2002)
Antibiotics (tetracycline) Polyether (semisynthetic) Macrolides, Aminoglycosides, sulfa and beta lactams antibiotics	Root	Cucumber, spinach, lettuce Carrot, radish Potatoes, onion and garlic Wheat, carrot and pinto beans	Migliore et al. (2003) Alistair et al. (2006) Grote et al. (2007) Bassil et al. (2013) Kang et al. (2013) David et al. (2016)
BPA	Root	Rice, tobacco	Nakajima et al. (2002) Noureddin et al. (2004)
Herbicides (sulfonylurea, imidazolinone, triazines, phenylureas, uracils, and sulfonamide families	Root and foliage	Pea, coffee plant, orange plant	Sterling (1994); Jurado et al. (1999); Papiernik et al. (2012); Goncalves (2016)

11.5 Factors Affecting Uptake of Organic Pollutants By Plants

To understand the uptake mechanism of these organic contaminants by plants, it is important that we also understand that there are three major factors involved: (i) the physical and chemical factors or processes of the plant, (ii) the environmental medium, and (iii) plant species (Zhang et al. 2017). Physical processes determine the uptake of pollutants while biological and chemical processes define its degree, which are discussed next.

11.5.1 Molecular Mass

Roots and foliage absorb organic pollutants with low molecular mass and volatility. However, only roots absorb organic pollutants with high molecular mass and no volatility (Fismes et al. 2002). Organic contaminant molecular mass is one of the prominent factors affecting the process of uptake by plants. A pollutant that has less than 1000 molecular mass is easily absorbed and transported (Kvesitadze et al. 2015).

11.5.2 Hydrophobicity (K_{ow})

High hydrophobicity K_{OW} (octanol and water partition coefficient) values of PCBs have a strong affinity to bind with plant roots, and transpiration stream entry is prohibited (Liu and Schnoor 2008). The amount of organic contaminants in barley and K_{OW} of PCB represents a positive correlation (Briggs et al. 1982; Moeckel et al. 2008).

11.5.3 K_{OA}

PCB with less than 60 K_{OA} (the partition coefficient between octanol and air) value absorbs mainly through air rather than soil (Cousins and Mackay 2001).

Semivolatile organic pollutants with values of K_{OA} less than 85 penetrates leaves mainly in the form of gas, contaminants with $85 < K_{OA} < 110$ absorbed primarily from air in plant leaves, and contaminants with $K_{OA} > 110$ absorbed primarily as sediment particles in plant leaves (McLachlan 1999).

11.5.4 Plant Lipid Content

Positive correlation was observed between lipid and organic content of roots in plants (Gao et al. 2005). Lipophilic adsorption dominated PCDD/Fs from nutrient solution in root uptake. The amount of lipid content that is extractable in roots of plant may tell us about the root buildup of PCDD/Fs from soil solutions (Zhang and Zhu 2009).

11.5.5 Carbohydrate-Containing Plant Cell Walls

Carbohydrate-containing plant cell walls were detected mostly with polycyclic aromatic hydroxyls (PAHs) contamination, because the carbohydrate content of plant cells is responsible for plants PAH uptake (Chen et al. 2009; Zhang and Zhu 2009).

11.5.6 Transpiration Stream Concentration Factor (TSCF)

Transpiration stream concentration factor (TSCF) actually represents the transport of organic pollutants from roots to shoots (Dettenmaier et al. 2009).

11.5.7 Plant Membrane

The plant membrane characteristics such as permeability, pH, and electrical gradients are used to find the organic pollutant concentration of absorption (Sitte et al. 1991; Sterling 1994).

11.5.8 Epicuticular Waxes

PCB plant-uptake from air is greatly dependent on the quantity and composition of epicuticular waxes (Moeckel et al. 2008).

11.5.9 Electrolytes

Amplified uptake of plant's background electrolyte like chlor-tetracycline (CTC) and tylosin in the existence of Ca^{2+} ion related with Na^{+} ion (Lee et al. 2014).

11.5.10 Dissolved Organic Carbon (DOC)

Amplified plant CTC absorption through dissolved organic carbon (DOC) concentrations is primarily used as a competition between the two for soil surface assimilation sites (Tao et al. 2010; Lee et al. 2014).

11.5.11 pH

Acidic or alkaline media show a good relationship with plant uptake of organic pollutants rather than intermediate pH media, which shows a poor relationship (Sithole and Guy 1987). Solution that have pH values of 6 have been found to contain more antibiotics than solutions with pH 4 and 8 (Min 2012).

11.5.12 Organic Matter Content of Soil

Organic pollutant uptake of plants is decreased by the increase in the organic matter content of soil because of the binding of a strong anion/cation exchanger with soil organic matter (Trapp and McFarlane 1994).

11.5.13 Temperature

It is assumed that passive absorption, enzyme activity, and transpiration stream rate in plants are enhanced by a high temperature coefficient for organic pollutant dispersion procedures (Korte et al. 2000).

11.6 Toxic Effects Caused By Organic Pollutants in Plants

There are many possible threats caused by organic pollutants in different parts of the plants and alter its various functions as mentioned in Fig. 11.3.

11.6.1 Effects on Plant Cell Structure and Membrane Strength

Vital functions of the cell are controlled by microstructures within the plant cell: ribosomes, endoplasmic reticulum, cell membranes, retinal membranes, microtubules (MT) or microfilaments, and nuclear membrane. The cell functions will cease if these microstructures are damaged or altered (Gunning and Steer 1975; Agarwal 2006).

The cell structure becomes compromised at entry of organic pollutants into plant cells; for instance, when tobacco was exposed at the concentrations of 10,000 mg kg^{-1} to WHO98-TEQ of PCDD/Fs, the tobacco cells were destroyed completely. The exposure to certain organic pollutants induced irreversible damage to cell wall which became uneven and crumpled, leaking cytoplasm and change in the shape of chloroplasts to hexagonal from round (Zhang et al. 2012).

Organic pollutants such as bipyridylium and diphenylether cause disruption of the cell membrane, which disrupts plant tissues in result (Ramadass et al. 2015).

α- and β-Tubulins and MT-associated proteins assemble to form microtubes that may be dented by organic contaminants (Nancy et al. 2014). By exposure of microtubes to fungicides, such as thiabendazole (TBZ) or griseofulvin (GF) to the amount of 100 μg mL^{-1} to *Allium cepa*, in meristematic root cells microtubular damage occurs in the form of abnormal arrays. Griseofulvin exposure can cause anomalous pre-prophasic bands, mitotic spindles, phragmoplasts, and irregular tripolar spindles. When pea (*Pisum sativum* L.) meristematic root-tip cells are exposed to BPA, it causes microtubule array disruption and increase in diameter from 23 ± 0.70 nm to 32 ± 0.14 nm in the meristematic root-tip cells and complete depolymerization is found on further exposure (Adamakis et al. 2013).

Lipid peroxidation (LPO) refers to unsaturated lipids; when reacted with the oxygen molecule, they become noxious from oxidative deterioration (Zwart et al. 1999; Chen et al. 2016). Free radicals (nonenzymatic lipid peroxidation) or enzymes (enzymatic lipid peroxidation) use oxygen molecule as a catalyst upon entrance. It has been observed that LPO produces various disturbances in membrane-like damage of microstructures, alteration of its integrity, fluidity, permeability, and loss of efficiency, and which probably produce harmful products, which may be in proatherogenic and inflammation-inducing forms when low-density lipoprotein changes (Greenberg et al. 2008).

It is now possible to identify the analytical index of LPO using malondialdehyde (MDA); the stability of the membrane decreases with increase in MDA content. Thus, increase in MDA content can indicate a harmful effect and oxidative stress caused by PCDD/Fs (Zhang et al. 2012).

Reluctance rates of antioxidative enzyme activities in MDA content always show positive correlations; for example, the MDA content may increase when the concentrations of tetracycline exceed 5 mg L^{-1} (Xie et al. 2011b).

11.6.2 Effects on Cell Biogenesis

Impacts on cell biogenesis from organic pollutants include photosynthesis, protein synthesis, amino acids, nucleic acids, lipids, and hormones; for instance, cell biogenesis is altered by various PCBs and herbicides (Moore Jr. & Harriss 1974; Moore and Roberts 1998; Brain et al. 2010).

Commonly, inhibition of photosynthesis is caused by PCBs, antibiotics, and herbicides, which supports growth, nutrient regulation, and resistance against abiotic or biotic stresses (Xia et al. 2004; Yang et al. 2008; Bassil et al. 2013). Inhibited photosynthetic carbon assimilation was found in phytoplankton communities and unialgal cultures from exposure of PCBs at the concentration of 10.0 μg L^{-1} (Engwall and Hjelm 2000). PCBs are known to disrupt light reaction as by decreasing photosynthesis-irradiance curves (Harding and Phillips 1978; Liu et al. 2010).

Oxytocycline (OTC) stresses wheat and other plants with sensitive cultivars by decrease in photosynthesis and transpiration rates and stomatal ability, which bear a resemblance to stresses of heavy metals, alkali, and salt (Li et al. 2011). Some herbicide families contain photosynthetic inhibitors such as benzothiadiazoles, triazines, uracils, phenylureas, and nitriles, which bind with chloroplasts and close down the photosynthetic process in vulnerable plant parts (Bunce 2008). Leaves and other plant parts become white and transparent as herbicides prevent plants from making photosynthetic pigments (Bellé et al. 2014).

Long exposures to hexazinone and chlorsulfuron are known to delay RNA, protein, and lipid production (Fisher et al. 1975; Yang et al. 2010). The effects of some herbicides like imidazolinone, ulfonylurea, and sulfonamide may disrupt one key enzyme by suppressing the production of vital branched-chain amino acids, and some herbicides are known to inhibit three essential aromatic amino acids having amino acid derivatives (Azqueta and Collins 2013). All these organic contaminants are believed to compromise the activities of lock-and-key enzymes; for example, the biosynthesis ability of aromatic amino acids and branched-chain amino acids, such as the ubiquitin-26S protein system and papain-like cysteine proteases, is affected by glyphosate and imazamox, respectively (Zobiole et al. 2009). Although these contaminants disrupt different paths and processes in plants, they cause the same physiological damage, such as decreasing soluble protein content by increasing free amino acid content (Amaia et al. 2013).

Many studies have shown that complex organic pollutants like herbicides have shown some complex processes in plants, but antibiotic effects on cell functions is are unknown. Their systems may be more complex than other chemical contaminants.

11.6.3 Effects on DNA Materials, Processes, and Its Responses

PCBs exposure damaged DNA considerably in the leaf nuclei of tobacco plant, upon measurement; the tail moment (TM) value of leaf nuclei was 10.0 ± 1.4 μm, which is much higher than that for the equivalent value in control plant nuclei (2.3 ± 0.4) μm) (Mathur et al. 2007). Glyphosate exposure in soybean root tip cells that were cultivated for 8 h showed a clear difference among the indicators in control plants such as total DNA content in tail, comet rate, comet tail length, tails and olive-tailed margins, and some other indicators, and there was a clear dose–response relationship among all indicators (Li et al. 2012).

Many research studies have shown that polycyclic aromatic hydrocarbons (PAHs) pose great threats to DNA. Plants were exposed with phenanthrene at 50 mg kg^{-1}; the TM values from the comet assay increased to 122.04 μm from 46.41 μm in control plants. When pyrene concentrations increase (0–50 mg kg^{-1}), the extent of DNA damage also increases; for example, *Vicia faba* root tip cells treated with pyrene at 50 mg kg^{-1} showed substantial DNA damage, as TM values increased 44.3 μm in the control plants to 110.36 μm in treated plants, which was examined using the comet assay (Gao et al. 2014). Plastic wastes cause toxicity in the soil and hence should be properly monitored (Ahmed et al. 2020).

Moreover, PCBs are known to disrupt cell division rates in some algal strains such as *Thalassiosira pseudonana* strain 3H, *Monochrysis lutheri*, *Chaetoceros socialis*, *Thalassiosira pseudonana* 13-I, *Skeletonema costatum*, and *Isochrysis galbana* at low concentrations (Fisher et al. 1975).

Porphyra haitanensis cell division rates decreased with increased concentrations of PCBs (Wang et al. 2006). Chlortetracycline exposure may prevent cells from undergoing cell division in higher plants, because protein synthesis and enzyme activities might become low and cause disruption in normal organelle functions, as suppressing mitotic division in root cells (Luciana 2003).

The findings of the toxicity of chlortetracycline and tetracycline indicate that its exposure caused an insignificant increase in the occurrence of aberration of micronucleus and chromosomes, and also the exchange between sister chromatids at the levels of 0.0625–250 mg L^{-1} and 0.25–10 mg L^{-1} of chlortetracycline and tetracycline, respectively. By the rise of concentration of tetracycline (0.5–300 mg L^{-1}), chromosomal aberration was also increased. At one point, the aberration of micronuclei and chromosomes, and the sister chromatids exchange, decreased from acute cell damage and toxicity. However, these antibiotics are genotoxic to plant cell and DNA functions (Xie et al. 2010, 2011a, b).

Another disruptor in mitosis in *Allium cepa* meristematic root cells is griseofulvin (GF). GF causes change in mitotic division and inhibits the mitotic index, causing genotoxic effects such as increased fragmentation of chromosomes, tripolar cell division, disordered anaphases, bridging and lagging, C-metaphases, and anomalies in interphase (Nancy et al. 2014). Additionally, a study of the genotoxicity of BPA on onion root meristematic cells showed that contact to 50, 100, 150, and 200 mg L^{-1} BPA for 5 days increased the fraction of CA (break) from 0.4% to 7.2% and CA (fragments) from 0.3% to 3.2%, which caused chromosome segregation, incomplete cytokinesis, and disruption of the interphase, which suggests that BPA has acute toxicity for the meristematic cells of *P. sativum* (Adamakis et al. 2013).

11.7 Conclusion

Chemicals can alter both plant and microbial activities when these enter plants and soil and possibly contaminate groundwater. Besides these known effects, plants can be used to remediate these chemicals and can be an indicator, such as evergreen conifers, which can be used to monitor organic pollutants in the air.

Soil and air are exposed to thousands of chemicals, so it is too difficult to monitor every combination and known permutation of every known concentration of chemical and environmental factors. However, with some prior chemical research experience, predictions can be made on the basis of physical and chemical properties.

Research in the field of soil uptake has shown many interesting discoveries in many plant species, such as the uptake of hydrophobic organic pollutants in zucchini and pumpkin and dioxins in the genus *Cucurbita*. Many plants have been studied to find more plants that can be used as biomonitors of the organic pollutants in air. However, this objective needs further study and technological advances, as well as research and experimental experience to study the complex mechanisms within the interactions between air and plants.

Possible impacts on plant parts and functions such as cell membranes, DNA functions, mitotic division, and protein synthesis, and including influencing factors such as pH, lipid content, hydrophobicity, molecular mass, root extractable lipid content, dissolved organic carbon, and background electrolyte type, can directly affect cell ultrastructure, membrane structure, nucleic and genetic structure, and other organelle functions, resulting in the possible collapse of plant life.

Moreover, the release of new organic compounds cannot be denied, and efforts should be made to control those which may prove to be alarming, such as flame retardants. Some hydrophobic plants do not interact with these substances and may be ignored, but in some plants these compounds are translocated to edible plant parts, which are large urgent areas of research that need to be addressed.

References

Adamakis IDS, Panteris E, Cherianidou A, Eleftheriou EP (2013) Effects of bisphenol A on the microtubule arrays in root meristematic cells of *Pisum sativum* L. Mutation Res/Genet Toxicol Environ Mutagenesis 750(1-2):111–120

Agarwal UP (2006) Raman imaging to investigate ultrastructure and composition of plant cell walls: distribution of lignin and cellulose in black spruce wood (*Picea mariana*). Planta (Berl) 224(5):1141–1153

Ahammed GJ, Wang MM, Zhou YH, Xia XJ, Mao WH, Shi K, Yu JQ (2012a) The growth, photosynthesis and antioxidant defense responses of five vegetable crops to phenanthrene stress. Ecotoxicol Environ Saf 80:132–139

Ahammed GJ, Wang MM, Zhou YH, Xia XJ, Mao WH, Shi K, Yu JQ (2012b) The growth, photosynthesis and antioxidant defense responses of five vegetable crops to phenanthrene stress. Ecotoxicol Environ Saf 80:132–139

Ahammed GJ, Yuan HL, Ogweno JO, Zhou YH, Xia XJ, Mao WH et al (2012c) Brassinosteroid alleviates phenanthrene and pyrene phytotoxicity by increasing detoxification activity and photosynthesis in tomato. Chemosphere 86(5):546–555

Ahammed GJ, Choudhary SP, Chen S, Xia X, Shi K, Zhou Y, Yu J (2013a) Role of brassinosteroids in alleviation of phenanthrene–cadmium co-contamination-induced photosynthetic inhibition and oxidative stress in tomato. J Exp Bot 64(1):199–213

Ahammed GJ, Ruan YP, Zhou J, Xia XJ, Shi K, Zhou YH, Yu JQ (2013b) Brassinosteroid alleviates polychlorinated biphenyls-induced oxidative stress by enhancing antioxidant enzymes activity in tomato. Chemosphere 90(11):2645–2653

Ahmed S, Mahmood. Q, Elahi N, Nawab B (2020) Current practices and futuristic options in plastic waste management in Pakistan. Central Asian J Environ Sci Technol Innov 1(4):237–244. https://doi.org/10.22034/CAJESTI.2020.04.06

Aktar MW, Sengupta D, Chowdhury A (2009) Impact of pesticides use in agriculture: their benefits and hazards. Interdiscipl Toxicol 2(1):1–12

Alharbi OM, Khattab RA, Ali I (2018) Health and environmental effects of persistent organic pollutants. J Mol Liq 263:442–453

Alistair B, Paul J, Edward JS, Chris JS, Edward S, Len SL (2006) Uptake of veterinary medicines from soils into plants. J Agric Food Chem 54:2288–2297

Alkio M, Tabuchi TM, Wang X, Colon-Carmona A (2005) Stress responses to polycyclic aromatic hydrocarbons in *Arabidopsis* include growth inhibition and hypersensitive response-like symptoms. J Exp Bot 56(421):2983–2994

Amaia Z, Miriam GM, Villamor JG, Zabalza A, Hoorn RA, Royuela M (2013) Proteolytic pathways induced by herbicides that inhibit amino acid biosynthesis. PLOS ONE 8:e73847

Anderson TA, Guthrie EA, Walton BT (1993) Bioremediation in the rhizosphere. Environ Sci Technol 27(13):2630–2636

Azqueta A, Collins AR (2013) The essential comet assay: a comprehensive guide to measuring DNA damage and repair. Arch Toxicol 87(6):949–968

Bakker MI (2000) Atmospheric deposition of semivolatile organic compounds to plants (Doctoral dissertation)

Bassil RJ, Bashour II, Sleiman FT, Abou-Jawdeh YA (2013) Antibiotic uptake by plants from manure-amended soils. J Environ Sci Health (Part B: Pesticides Food Contaminants and Agricultural Wastes) 48:570–574

Bellé C, Kulczynski SM, Basso CJ, Edu Kaspary T, Lamego FP, Pinto MAB (2014) Yield and quality of wheat seeds as a function of desiccation stages and herbicides. J Seed Sci 36(1):63–70

Bernes C (1998) Persistent organic pollutants. Swedish Environmental Protection Agency, Stockholm

Bossert I, Bartha R (1985) Plant growth in soils with a history of oily sludge disposal. Soil Sci 140(1):75–77

Boxall AB, Johnson P, Smith EJ, Sinclair CJ, Stutt E, Levy LS (2006) Uptake of veterinary medicines from soils into plants. J Agric Food Chem 54(6):2288–2297

Brain RA, Solomon KR, Brooks BW (2010) Targets, effects and risks in aquatic plants exposed to veterinary antibiotics. ACS Symposium Series 1018:169–189

Briggs GG, Bromilow RH, Evans AA (1982) Relationships between lipophilicity and root uptake and translocation of non-ionised chemicals by barley. Pestic Sci 13(5):495–504

Bunce JA (2008) Acclimation of photosynthesis to temperature in *Arabidopsis thaliana* and Brassica oleracea. Photosynthetica 46:517–524

Burken JG, Schnoor JL (1997) Uptake and metabolism of atrazine by poplar trees. Environ Sci Technol 31(5):1399–1406

Calamari D, Bacci E, Focardi S, Gaggi C, Morosini M, Vighi M (1991) Role of plant biomass in the global environmental partitioning of chlorinated hydrocarbons. Environ Sci Technol 25(8):1489–1495

Chen L, Zhang S, Huang H, Wen B, Christie P (2009) Partitioning of phenanthrene by root cell walls and cell wall fractions of wheat (*Triticum aestivum* L.). Environ Sci Technol 43(24):9136–9141

Chen XN, Qiu DY, Lin HM (2016) Botanical characteristics and medicinal value of five *Glycyrrhiza* species cultivated in the Hexi region of Gansu. Acta Pratacul Sin 25:246–253

Cheng F, Li W, Zhou Y, Shen J, Wu Z, Liu G, ... Tang Y (2012) admetSAR: a comprehensive source and free tool for assessment of chemical ADMET properties

Cherian S, Oliveira MM (2005) Transgenic plants in phytoremediation: recent advances and new possibilities. Environ Sci Technol 39(24):9377–9390

Cousins IT, Mackay D (2001) Strategies for including vegetation compartments in multimedia models. Chemosphere 44(4):643–654

Dalvi RR, Salunkhe DK (1975) Toxicological implications of pesticides: their toxic effects on seeds of food plants. Toxicology 3(3):269–285

David A, Christiana M, Godfred D, Johan JW, Bjame S, Robert CA (2016) Uptake of antibiotics from irrigation water by plants. Chemosphere 157:107–114

Desalme D, Binet P, Bernard N, Gilbert D, Toussaint ML, Chiapusio G (2011) Atmospheric phenanthrene transfer and effects on two grassland species and their root symbionts: a microcosm study. Environ Exp Bot 71(2):146–151

Dettenmaier EM, Doucette WJ, Bugbee B (2009) Chemical hydrophobicity and uptake by plant roots. Environ Sci Technol 43(2):324–329

Engwall M, Hjelm K (2000) Uptake of dioxin-like compounds from sewage sludge into various plant species–assessment of levels using a sensitive bioassay. Chemosphere 40(9–11):1189–1195

Eriksson G, Jensen S, Kylin H, Strachan W (1989) The pine needle as a monitor of atmospheric pollution. Nature 341(6237):42–44

Falah W, Chen F, Zeb BS, Hayat MT, Mahmood Q, Ebadi A, Toughani M, Li EZ (2020) Polyethylene terephthalate degradation by microalga *Chlorella vulgaris* along with pretreatment. Mater Plastice 57(3):260–270

Fatoki OS, Awofolu OR, Genthe B (2004) Cadmium in the Umtata River and the associated health impact of on rural communities who are primary users of water from the river. Water SA 30(4):507–513

Fisher ER, Gregorio RM, Fisher B, with the assistance of Carol Redmond, ScD, Vellios F, Sommers SC, Cooperating Investigators (1975) The pathology of invasive breast cancer: a syllabus derived from findings of the National Surgical Adjuvant Breast Project (Protocol no. 4). Cancer (Phila) 36(1):1–85

Fismes J, Perrin-Ganier C, Empereur-Bissonnet P, Morel JL (2002) Soil-to-root transfer and translocation of polycyclic aromatic hydrocarbons by vegetables grown on industrial contaminated soils. J Environ Qual 31(5):1649–1656

Fonagy P, Leigh T, Steele M, Steele H, Kennedy R, Mattoon G, Gerber A (1996) The relation of attachment status, psychiatric classification, and response to psychotherapy. J Consult Clin Psychol 64(1):22

France RL, Muir DGC, Segstro MD (1997) Regional patterns in contamination of saxifrage from Ellesmere Island in the High Arctic (77–81° N). Environ Sci Technol 31(7):1879–1882
Fries GF (1996) Ingestion of sludge applied organic chemicals by animals. Sci Total Environ 185(1–3):93–108
Gao Y, Zhu L, Ling W (2005) Application of the partition-limited model for plant uptake of organic chemicals from soil and water. Sci Total Environ 336(1–3):171–182
Gao J, Sun SP, Zhu WP, Chung TS (2014) Chelating polymer modified P84 nanofiltration (NF) hollow fiber membranes for high efficient heavy metal removal. Water Res 63:252–261
Goncalves CG (2016) Selectivity of saflufenacil applied singly and in combination, with glyphosate on coffee and citrus crops. Revista Caatinga 29:45–53. (in Portuguese)
Greenberg ME, Li XM, Gugiu BG, Gu X, Qin J, Salomon RG, Hazen SL (2008) The lipid whisker model of the structure of oxidized cell membranes. J Biol Chem 283(4):2385–2396
Grote M, Schwake AC, Michel R, Stevens H, Heyser W, Langenkämper G, Betsche T, Freitag M (2007) Incorporation of veterinary antibiotics into crops from manured soil. Land Bauforschung Volkenrode 57:25–32
Gunning BE, Steer MW (1975) Ultrastructure and the biology of plant cells. Edward Arnold, London, pp 312–395
Hance RJ (1981) Effects of pesticides on plant nutrition. Residue reviews. Springer, New York, pp 13–41
Hardcastle WS, Wilkinson RE, Young CT (1974) Metribuzin effects on seed constituents of soybean varieties. Weed Sci 22(6):575–577
Harding LW, Phillips JH (1978) Polychlorinated biphenyl (PCB) effects on marine phytoplankton photosynthesis and cell division. Mar Biol 49(2):93–101
Hellström A (2004) Uptake of organic pollutants in plants. Department of Environment and Assessments, Swedish University of Agricultural Sciences, Uppsala
Herman JP, Ostrander MM, Mueller NK, Figueiredo H (2005) Limbic system mechanisms of stress regulation: hypothalamo-pituitary-adrenocortical axis. Prog Neuropsychopharmacol Biol Psychiatry 29(8):1201–1213
Holoubek I, Klanova J, Jarkovský J, Kohoutek J (2007) Trends in background levels of persistent organic pollutants at Kosetice Observatory, Czech Republic. Part I. Ambient air and wet deposition 1996–2005. J Environ Monit 9(6):557–563
Igbedioh SO (1991) Effects of agricultural pesticides on humans, animals, and higher plants in developing countries. Arch Environ Health 46(4):218–224
Jensen S, Eriksson G, Kylin H, Strachan WMJ (1992) Atmospheric pollution by persistent organic compounds: monitoring with pine needles. Chemosphere 24(2):229–245
Jurado Exposito M, Munoz Castejon M, Garcia-Torres L (1999) Uptake and translocation of imazethapyr in peas as affected by parasitism of Orobanche crenata and herbicide application methods. Weed Res 39:129–136
Juuti S, Norokorpi Y, Ruuskanen J (1995) Trichloroacetic acid (TCA) in pine needles caused by atmospheric emissions of kraft pulp mills. Chemosphere 30(3):439–448
Kacálková L, Tlustoš P (2011) The uptake of persistent organic pollutants by plants. Open Life Sciences 6(2):223–235
Kang Y, Sheng G, Fu J, Mai B, Zhang G (2000) Polychlorinated biphenyls in surface sediments from the Pearl River Delta and Macau. Mar Pollut Bull 40(9):794–797
Kang D H, Gupta S, Rosen C, Fritz V, Singh A, Chander Y, Murray H, Rohwer C (2013) Antibiotic uptake by vegetable crops from manure-applied soils. J Agric Food Chem 61:9992–10001
Kaur G, Sharma A, Guruprasad K, Pati PK (2014) Versatile roles of plant NADPH oxidases and emerging concepts. Biotechnol Adv 32(3):551–563
Kelly BC, Gobas FA (2001) Bioaccumulation of persistent organic pollutants in lichen−caribou−wolf food chains of Canada's central and western Arctic. Environ Sci Technol 35(2):325–334
Kirkwood RC (1999) Recent developments in our understanding of the plant cuticle as a barrier to the foliar uptake of pesticides. Pestic Sci 55(1):69–77
Korte F, Kvesitadze G, Ugrekhelidze D, Gordeziani M, Khatisashvili G, Buadze O, Coulston F (2000) Organic toxicants and plants. Ecotoxicol Environ Saf 47(1):1–26

Kumar KS, Choo KS, Yea SS, Seo Y, Han T (2010) Effects of the phenylurea herbicide diuron on the physiology of *Saccharina japonica* Aresch. Toxicol Environ Health Sci 2(3):188–199

Kvesitadze G, Khatisashvili G, Sadunishvili T, Kvesitadze E (2015) Plants for remediation: uptake, translocation and transformation of organic pollutants. In: Plants, pollutants and remediation. Springer, Dordrecht, pp 241–308

Kylin H (1994) Airborne lipophilic air pollutants in pine needles. Doctoral dissertation, Thesis Stockholm University

Lee KT, Tanabe S, Koh CH (2001) Distribution of organochlorine pesticides in sediments from Kyeonggi Bay and nearby areas. Korea Environ Pollution 114(2):207–213

Lee J, Seo Y, Essington ME (2014) Sorption and transport of veterinary pharmaceuticals in soil—a laboratory study. Soil Sci Soc Am J 78(5):1531–1543

Li GL, Liu YJ, Guan WF, Chen XL (2012) Study on the comet assay detects glyphosate-induced DNA damages in soybean root tip cells. J Henan Inst Sci and Technol 40:18–21

Li L, Wang Q, Qiu XH, Dong Y, Jia SL, Hu JX (2014) Field determination and QSPR prediction of equilibrium-status soil/vegetation partition coeffcient of PCDD/Fs. J Hazard Mater 276:278–286

Li ZJ, Xie XY, Zhang SQ, Liang YC (2011) Wheat growth and photosynthesis as affected by oxytetracycline as a soil contaminant. Pedosphere 21:244–250

Liu J, Schnoor JL (2008) Uptake and translocation of lesser-chlorinated polychlorinated biphenyls (PCBs) in whole hybrid poplar plants after hydroponic exposure. Chemosphere 73(10):1608–1616

Liu H, Weisman D, Ye YB, Cui B, Huang YH, Colón-Carmona A, Wang ZH (2009a) An oxidative stress response to polycyclic aromatic hydrocarbon exposure is rapid and complex in *Arabidopsis thaliana*. Plant Sci 176(3):375–382

Liu H, Weisman D, Ye YB, Cui B, Huang YH, Colón-Carmona A, Wang ZH (2009b) An oxidative stress response to polycyclic aromatic hydrocarbon exposure is rapid and complex in *Arabidopsis thaliana*. Plant Sci 176(3):375–382

Luciana M (2003) Practitioner review: computerized assessment of neuropsychological function in children: clinical and research applications of the Cambridge Neuropsychological Testing Automated Battery (CANTAB). J Child Psychol Psychiatry 44(5):649–663

Mahajan P, Kaushal J (2018) Role of phytoremediation in reducing cadmium toxicity in soil and water. J Toxicol 4864365:1–16

Mallakin A, Babu TS, Dixon DG, Greenberg BM (2002) Sites of toxicity of specific photooxidation products of anthracene to higher plants: inhibition of photosynthetic activity and electron transport in *Lemna gibba* L. G-3 (Duckweed). Environ Toxicol. 17(5):462–471

Martinoia E, Grill E, Tommasini R, Kreuz K, Amrhein N (1993) ATP-dependent glutathione *S*-conjugate 'export' pump in the vacuolar membrane of plants. Nature 364(6434):247–249

Mathur A, Cao M, Bhattacharya S, Dilger A, Tomas A, Vivier L (2007) The new ext4 filesystem: current status and future plans. In Proceedings of the Linux Symposium, vol. 2, pp. 21–33

McLachlan MS (1995) Bioaccumulation of hydrophobic chemicals in agricultural food chains. Environ Sci Technol 30(1):252–259

McLachlan MS (1999) Framework for the interpretation of measurements of SOCs in plants. Environ Sci Technol 33(11):1799–1804

Migliore L, Cozzolino S, Fiori M (2003) Phytotoxicity to and uptake of enrofloxacin in crop plants. Chemosphere 52:1233–1244

Min L (2012) Study on uptake and accumulation of three sulfonamides in plants. MSc thesis. Anhui Agricultural University, China (in Chinese)

Moeckel C, Thomas GO, Barber JL, Jones KC (2008) Uptake and storage of PCBs by plant cuticles. Environ Sci Technol 42(1):100–105

Moore SA Jr, Harriss RC (1974) Differential sensitivity to PCB by phytoplankton. Mar Pollut Bull 5(11):174–176

Moore K, Roberts LJ (1998) Measurement of lipid peroxidation. Free Radic Res 28(6):659–671

Nakajima N, Ohshima Y, Serizawa S, Kouda T, Edmonds JS, Shiraishi F (2002) Processing of bisphenol a by plant tissues: glucosylation by cultured BY-2 cells and glucosylation/translocation by plants of nicotiana tabacum. Plant and Cell Physiol 43:1036–42

Nancy BA, Sonia S, Marcelo LL, Marta DM (2014) Induction of microtubule damage in Allium cepa meristematic cells by pharmaceutical formulations of thiabendazole and griseofulvin. Mutat Res Genet Toxicol Environ Mutagen 772:1–5

Noureddin MI, Furumoto T, Yamagishi M (2004) Absorption, translocation and metabolism of bisphenol A, a possible endocrine disruptor, in rice seedlings. Environ Control Biol 42:31–40

Oguntimehin I, Nakatani N, Sakugawa H (2008) Phytotoxicities of fluoranthene and phenanthrene deposited on needle surfaces of the evergreen conifer, Japanese red pine (*Pinus densiflora* Sieb. et Zucc.). Environ Pollut 154(2):264–271

Oguntimehin I, Eissa F, Sakugawa H (2010a) Negative effects of fluoranthene on the ecophysiology of tomato plants (*Lycopersicon esculentum* Mill.): fluoranthene mists negatively affected tomato plants. Chemosphere 78(7):877–884

Oguntimehin I, Eissa F, Sakugawa H (2010b) Negative effects of fluoranthene on the ecophysiology of tomato plants (Lycopersicon esculentum Mill): Fluoranthene mists negatively affected tomato plants. Chemosphere 78(7):877–884

Papiernik SK, Koskinen WC, Barber BL (2012) Low sorption and fast dissipation of the herbicide saflufenacil in surface soils and subsoils of an eroded prairie landscape. J Agric Food Chem 60:10936–10941

Paterson S, Mackay D, Tam D, Shiu WY (1990) Uptake of organic chemicals by plants: a review of processes, correlations and models. Chemosphere 21(3):297–331

Pier MD, Zeeb BA, Reimer KJ (2002) Patterns of contamination among vascular plants exposed to local sources of polychlorinated biphenyls in the Canadian Arctic and Subarctic. Sci Total Environ 297:215–227

Ramadass K, Megharaj M, Venkateswarlu K, Naidu R (2015) Toxicity and oxidative stress induced by used and unused motor oil on freshwater microalga, *Pseudokirchneriella subcapitata*. Environ Sci Pollut Res 22(12):8890–8901

Reischl A (1988) Akkumulation luftbürtiger chlorierten Kohlenwasserstoffe in Koniferennadeln. Thesis. University of Bayreuth, Germany

Ren L, Huang XD, McConkey BJ, Dixon DG, Greenberg BM (1994) Photoinduced toxicity of three polycyclic aromatic hydrocarbons (fluoranthene, pyrene, and naphthalene) to the duckweed *Lemna gibba* L. G-3. Ecotoxicol Environ Saf 28(2):160–171

Reynoso-Cuevas L, Gallegos-Martínez ME, Cruz-Sosa F, Gutiérrez-Rojas M (2008) In vitro evaluation of germination and growth of five plant species on medium supplemented with hydrocarbons associated with contaminated soils. Bioresour Technol 99(14):6379–6385

Ruzzin J, Lee DH, Carpenter DO, Jacobs DR (2012) Reconsidering metabolic diseases: the impacts of persistent organic pollutants. Atherosclerosis 224(1):1–3

Safe S, Brown KW, Donnelly KC, Anderson CS, Markiewicz KV, McLachlan MS, Hutzinger O (1992) Polychlorinated dibenzo-*p*-dioxins and dibenzofurans associated with wood-preserving chemical sites: biomonitoring with pine needles. Environ Sci Technol 26(2):394–396

Sharma I, Bhardwaj R, Pati PK (2013) Stress modulation response of 24-epibrassinolide against imidacloprid in an elite indica rice variety *Pusa* Basmati-1. Pestic Biochem Physiol 105(2):144–153

Sharma A, Kumar V, Singh R, Thukral AK, Bhardwaj R (2015a) 24-Epibrassinolide induces the synthesis of phytochemicals effected by imidacloprid pesticide stress in *Brassica juncea* L. J Pharmacognosy Phytochem 4(3)

Sharma I, Bhardwaj R, Pati PK (2015b) Exogenous application of 28-homobrassinolide modulates the dynamics of salt and pesticides induced stress responses in an elite rice variety *Pusa* Basmati-1. J Plant Growth Regul 34(3):509–518

Sharma A, Kumar VINOD, Kanwar MK, Thukral AK, Bhardwaj R (2016) Multivariate analysis reveals the role of 24-epibrassinolide in elemental uptake by *Brassica juncea* L. under imidacloprid toxicity. Pollut Res 35:749–756

Sharma A, Thakur S, Kumar V, Kesavan AK, Thukral AK, Bhardwaj R (2017) 24-Epibrassinolide stimulates imidacloprid detoxification by modulating the gene expression of *Brassica juncea* L. BMC Plant Biol 17(1):1–10

Simonich SL, Hites RA (1994) Vegetation-atmosphere partitioning of polycyclic aromatic hydrocarbons. Environ Sci Technol 28(5):939–943

Sisler HD (1969) Effect of fungicides on protein and nucleic acid synthesis. Annu Rev Phytopathol 7(1):311–330

Sithole BB, Guy RD (1987) Models for tetracycline in aquatic environments. II: Interaction with humic substances. Water Air Soil Pollut 32(3–4):315–321

Sitte P, Ziegler H, Ehrendorfer F, Bresinsky A (1991) Lehrbuch der Botanik für Hochschulen, vol. 33. Fischer

Smith KE, Jones KC (2000) Particles and vegetation: implications for the transfer of particle-bound organic contaminants to vegetation. Sci Total Environ 246(2–3):207–236

Stefan T, Michael M (1997) Modeling volatilization of PCDD/F from soil and uptake into vegetation. Environ Sci Technol 31:71–74

Sterling TM (1994) Mechanisms of herbicide absorption across plant membranes and accumulation in plant-cells. Weed Sci 42:263–276

Strachan WM, Kylin H, Eriksson G, Jensen S (1994) Organochlorine compounds in pine needles: methods and trends. EnvironToxicol Chem 13(3):443–451

Tao YQ, Zhang SZ, Xue B (2010) Bioavailability of polycyclic aromatic hydrocarbons in soils and its evaluation method. J Graduate School Chinese Acad Sci 4:568–576

Trapp M, Mc Farlane C (1994) Modeling and simulation of organic chemical processes. In: Trapp S, Mc arlane JC, eds., Plant Contamination. CRC, USA

Tommasini R, Vogt E, Fromenteau M, Hörtensteiner S, Matile P, Amrhein N, Martinoia E (1998) An ABC-transporter of *Arabidopsis thaliana* has both glutathione-conjugate and chlorophyll catabolite transport activity. The Plant Journal 13(6):773–780

Tremolada P, Burnett V, Calamari D, Jones KC (1996) A study of the spatial distribution of PCBs in the UK atmosphere using pine needles. Chemosphere 32(11):2189–2203

UNEP (United Nations Environment Programme) (2001) Stockholm convention on persistent organic pollutants, as amended 2009. Switzerland, Geneva

Váňová L, Kummerová M, Klemš M, Zezulka Š (2009) Fluoranthene influences endogenous abscisic acid level and primary photosynthetic processes in pea (*Pisum sativum* L.) plants in vitro. Plant Growth Regul 57(1):39-47

Wagrowski DM, Hites RA (1996) Polycyclic aromatic hydrocarbon accumulation in urban, suburban, and rural vegetation. Environ Sci Technol 31(1):279–282

Waltke BK, O'Connor MP, O'Connor C (1990) An introduction to biblical Hebrew syntax. Eisenbrauns

Wang Y, Chen J, Li F, Qin H, Qiao X, Hao C (2009) Modeling photoinduced toxicity of PAHs based on DFT-calculated descriptors. Chemosphere 76(7):999–1005

Wang J, Dai JX, Yang KF (2006) The status and prospect of meiosis in Porphyra. Periodical of Ocean University of China 36:377–380. (in Chinese)

Wang Z, Jiang Y, Peng X, Xu S, Zhang H, Gao J, Xi Z (2017) Exogenous 24-epibrassinolide regulates antioxidant and pesticide detoxification systems in grapevine after chlorothalonil treatment. Plant Growth Regul 81(3):455–466

Weiss P (1998) Umweltbundesamt, Bundesministerium für Umwelt, Jugend und Familie. Vienna, Austria

Welsch-Pausch K, McLachlan MS, Umlauf G (1995) Determination of the principal pathways of polychlorinated dibenzo-*p*-dioxins and dibenzofurans to *Lolium multiflorum* (Welsh Ray Grass). Environ Sci Technol 29(4):1090–1098

White JC, Mattina MI, Lee WY, Eitzer BD, Iannucci-Berger W (2003) Role of organic acids in enhancing the desorption and uptake of weathered *p*, *p*′-DDE by *Cucurbita pepo*. Environ Pollut 124(1):71–80

Wieczorek JK, Wieczorek ZJ (2007) Phytotoxicity and accumulation of anthracene applied to the foliage and sandy substrate in lettuce and radish plants. Ecotoxicol Environ Saf 66(3):369–377

Wild SR, Berrow ML, McGrath SP, Jones KC (1992) Polynuclear aromatic hydrocarbons in crops from long-term field experiments amended with sewage sludge. Environ Pollut 76(1):25–32

Wu WZ, Schramm KW, Xu Y, Kettrup A (2002) Contamination and distribution of polychlorinated dibenzo-*p*-dioxins and dibenzofurans (PCDD/F) in agriculture felds in Ya-Er lake area, China. Ecotoxicol Environ Saf 53:141–147

Xia JR, Li YJ, Lu J, Chen B (2004) Effects of copper and cadmium on growth, photosynthesis, and pigment content in Gracilaria lemaneiformis. Bull Environ Contam Toxicol 73:979–986

Xia XJ, Zhang Y, Wu JX, Wang JT, Zhou YH, Shi K, Yu JQ (2009) Brassinosteroids promote metabolism of pesticides in cucumber. J Agric Food Chem 57(18):8406–8413

Xie X, Zhou Q, Bao Q, He Z, Bao Y (2011a) Genotoxicity of tetracycline as an emerging pollutant on root meristem cells of wheat (*Triticum aestivum* L.). Environ Toxicol 26(4):417–423

Xie X, Zhou Q, Lin D, Guo J, Bao Y (2011b) Toxic effect of tetracycline exposure on growth, antioxidative and genetic indices of wheat (*Triticum aestivum* L.). Environ Sci Pollut Res 18(4):566–575

Xu JG, Johnson RL (1995) Root growth, microbial activity and phosphatase activity in oil-contaminated, remediated and uncontaminated soils planted to barley and field pea. Plant and Soil 173(1):3–10

Yang X, Guschina IA, Hurst S, Wood S, Langford M, Hawkes T (2010) The action of herbicides on fatty acid biosynthesis and elongation in barley and cucumber. Pest Manag Sci 66:794–800

Yang CW, Jianaer A, Li CY, Shi DC, Wang DL (2008) Comparison of the effects of salt-stress and alkalistress on photosynthesis and energy storage of an alkaliresistant halophyte Chloris virgate. Photosynthetica 46:273–278

Zhang B, Zhang H, Jin J, Ni Y, Jiping C (2012) PCDD/Fs-induced oxidative damage and antioxidant system responses in tobacco cell suspension cultures. Chemosphere 88:798–805

Zhang M, Zhu L (2009) Sorption of polycyclic aromatic hydrocarbons to carbohydrates and lipids of ryegrass root and implications for a sorption prediction model. Environ Sci Technol 43(8):2740–2745

Zhang C, Yao FENG, Liu YW, Chang HQ, Li ZJ, Xue JM (2017) Uptake and translocation of organic pollutants in plants: A review. J Integr Agric 16(8):1659–1668

Zhang X, Yuan X, Shi H, Wu L, Qian H, Xu W (2015) Exosomes in cancer: small particle, big player. J Hematol Oncol 8(1):1–13

Zhou Y, Xia X, Yu G, Wang J, Wu J, Wang M et al (2015) Brassinosteroids play a critical role in the regulation of pesticide metabolism in crop plants. Sci Rep 5(1):1–7

Zobiole LHS, de Oliveira Junior RS, Bonato CM, Muniz AS, Castro CD, de Oliveira FA et al (2009) Effect of increasing doses of glyphosate on water use efficiency and photosynthesis in glyphosate-resistant soybeans. In: Embrapa Soja-Artigo em anais de congresso (ALICE)., WORLD SOYBEAN RESEARCH CONFERENCE, 8, 2009, Beijing. Developing a global soy blueprint for a safe secure and sustainable supply: proceedings. Beijing: Chinese Academy of Agricultural Sciences: Institute of Crop Science, 2009. Poster. WSRC 2009. 1 CD-ROM. Editado por Lijuan Qiu, Rongxia Guan, Jian Jin, Qijan Song, Shuntang Guo, Wenbin Li, Yuanchao Wang, Tianfu Han, Xiaobing Liu, Deyue Yu, Lianzhou Jiang, Deliang Peng

Zwart LL, Meerman JH, Commandeur JN, Vermeulen NP (1999) Biomarkers of free radical damage: applications in experimental animals and in humans. Free Radic Biol Med 26(1–2):202–226

Chapter 12
Transcription Factors That Scavenge Reactive Oxygen Species in Rhizobacteria

Amir Miraj Ul Hussain Shah, Allah Ditta, Abida Parveen, Sumaira Thind, and Abdol Ghaffar Ebadi

12.1 Introduction

The rhizobacteria, especially plant growth-promoting rhizobacteria (PGPR), have an important role in the growth and productivity of various crop plants under normal as well as under biotic and abiotic stress conditions through various mechanisms (Ditta et al. 2015; Ditta and Khalid 2016; Ditta et al. 2018a, b; Zeb et al. 2018; Sarfraz et al. 2019; Ullah et al. 2020; Hussain et al. 2020). Under biotic and abiotic stress conditions, there is an increased production of reactive oxygen species (ROS), that is, hydrogen peroxide (H_2O_2), superoxide anion ($O_2{\bullet}^-$), and hydroxyl radical (•OH) (Ditta 2018; Imtiaz et al. 2018; Shahzad et al. 2019; Ahmad et al. 2020; Sabir et al. 2020), which have deleterious effects on various metabolic processes, body structures (nucleic acids, lipids, and protein) of PGPR and ultimately may lead to the death of the microbes. The major harmful impacts of oxygen are intervened by its reduced structures, hydrogen peroxide and superoxide (Huan et al. 2016).

Amir Miraj Ul Hussain Shah
Department of Botany, University of Science & Technology Bannu, Khyber Pakhtunkhwa, Pakistan

A. Ditta (✉)
Department of Environmental Sciences, Shaheed Benazir Bhutto University Sheringal, Khyber Pakhtunkhwa, Pakistan

School of Biological Sciences, The University of Western Australia, Perth, WA, Australia
e-mail: allah.ditta@sbbu.edu.pk

A. Parveen · S. Thind
Department of Botany, Government College University, Faisalabad, Pakistan

A. G. Ebadi
Department of Agriculture, Islamic Azad University, Jouybar Branch, Jouybar, Iran

Q. Mahmood (ed.), *Sustainable Plant Nutrition under Contaminated Environments*, Sustainable Plant Nutrition in a Changing World,
https://doi.org/10.1007/978-3-030-91499-8_12

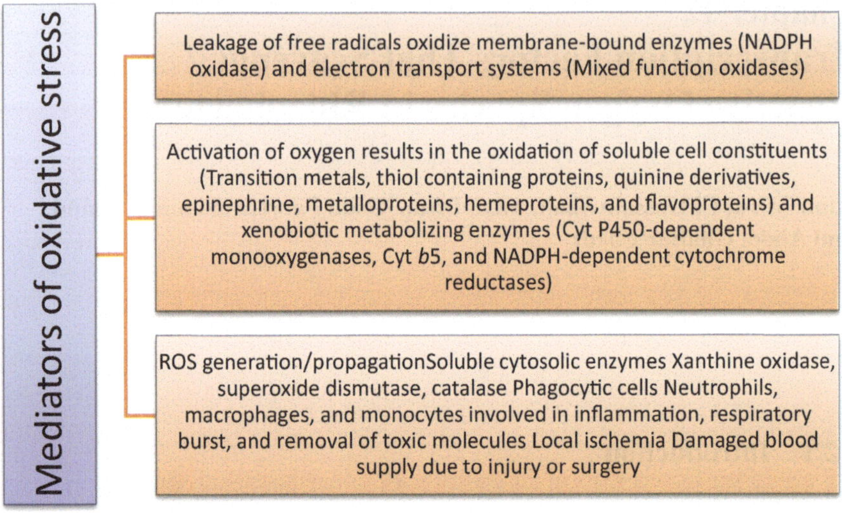

Fig. 12.1 Endogenous mediators of oxidative stress (Rahal et al. 2014)

The generation of ROS species persistently creates intimate cells because of the oxidation of redox enzymes (Bhattacharyya et al. 2014; Mehmood et al. 2018; Niamat et al. 2019; Liu et al. 2020). Various cellular components have been involved in the initiation and generation of reactive oxygen species in the cell (Fig. 12.1).

In response to ROS, microbes have sifting enzymes for their survival, which hold the oxidants from the onset of harmfulness (Shah et al. 2014; Zou et al. 2016). In the most recent decade, it was found that an assortment of conditions can lift these species over that limit and that microorganisms normally maintain an inducible defense system to balance such events (Ishikado et al. 2013; Pisoschi and Pop 2015). This system includes certain regulons that are controlled by the OxyR, PerR, or PerR-like homolog and SoxR transcription factors (Sun et al. 2012; Fillat 2014; Shah et al. 2014; Imlay 2015; Kim et al. 2016, 2017a, b; Wang et al. 2017). The present chapter aims to comprehensively describe the production of hydrogen peroxide (H_2O_2) and the role of certain regulons actively involved in coping oxidative stress. A special focus has been given to the OxyR, PerR, or PerR-like homolog and SoxR transcription factors that control these regulatory mechanisms.

12.2 Production of Hydrogen Peroxide (H_2O_2)

H_2O_2 is produced from redox cycling of pigments, sulfur, and oxygen reaction under aerobic and anaerobic conditions and by oxygen via chromophores during photochemical reduction processes. It is also excreted by plants, animals, and bacteria as a defensive chemical against native microorganisms (Bedard et al. 2007; Gu and Imlay 2013; Anjum et al. 2014; Imran et al. 2016; Ozyigit et al. 2016) and fauna

(Lee et al. 2015; Schiaffino et al. 2015; Zhao and Pluth 2016). Various researchers consider H_2O_2 as a defensive chemical as it is smaller in size with the ability to cross certain membranes such as water, thereby hindering its removal from the target cells (Seaver and Imlay 2001). In contrast, after entering into the cell H_2O_2 disturbs the metabolism of certain metals such as iron, which is essential for life (Uhl and Dukan 2016).

The mechanisms of how H_2O_2 harms cells have been studied in the model bacteria *Escherichia coli* and *Streptococcus pneumonia* (Ferrandiz et al. 2015; Rai et al. 2015). The studies have shown that exogenous H_2O_2 is harmful to DNA, therefore is mutagenic (Ni et al. 2016; Rima et al. 2016). As revealed through Fenton chemistry, it has been found that H_2O_2 reacts with freely available iron inside the cell and results in the production of the hydroxyl radical (HO•) (Imlay 2013; Yesilkaya et al. 2013).

$$[4\text{Fe}\quad 4\text{S}]^{2+} + \text{O}_2\bullet^- + 2\text{H}^+ \rightarrow [4\text{Fe}\quad 4\text{S}]^{2+} + \text{H}_2\text{O}_2$$

$$[4\text{Fe}-4\text{S}]^{3+} \rightarrow [4\text{Fe}-4\text{S}]^{+} + \text{Fe}^{2+}$$

$$\text{Fe}^{2+} + \text{H}_2\text{O}_2 \rightarrow \text{Fe}(\text{III}) + \text{OH}^- + \text{HO}\bullet$$

The HO• radical produced reacts with different biomolecules such as DNA, proteins, and lipids at diffusion-limited rates (Chatgilialoglu et al. 2011; Terzidis and Chatgilialoglu 2015; Terzidis et al. 2016). Some of the metal pool is insecurely connected with DNA (Rai et al. 2001), and the Fenton reaction on its surface produces lesions that ultimately result in mutagenesis.

Figure 12.2 shows various free radicals present inside the cell. The major impact of H_2O_2 stress is to deactivate two families of iron-containing enzymes, that is, non-redox uninuclear enzymes and [4Fe-4S]-containing dehydratases (Jang and Imlay 2007; Gu and Imlay 2013; Sobota et al. 2014). A solvent-exposed reduced Fe is used to bind and activate substrates in both types of enzymes. Moreover, H_2O_2 is capable of oxidizing Fe in both enzymes. After oxidation, the metal atom

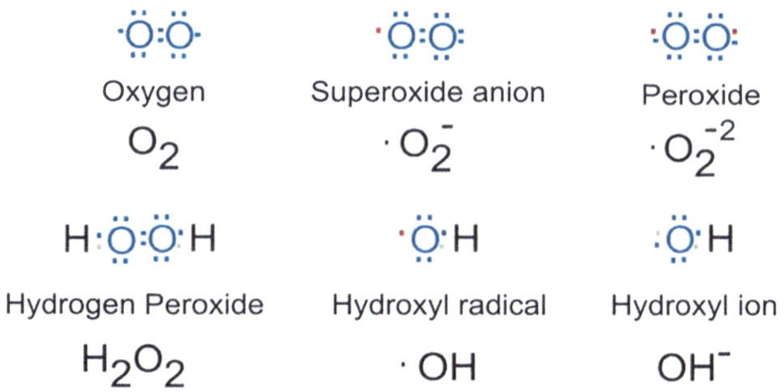

Fig. 12.2 Free radicals that may cause oxidative stress in plants

dissociates from the enzyme under consideration and its activity is lost. The aforementioned enzyme system is an important component of growth-related metabolic reactions (TCA cycle, pentose phosphate pathway, other biosynthetic pathways); loss of their activity results in reduced growth (Patra and Hay 2014; Kuehne et al. 2015; Lisher et al. 2017). The minimum inhibitory concentration (MIC) of H_2O_2 for most of the cells is 0.5 μM (Gardner et al. 2015). The scavenging enzymes maintain a steady-state H_2O_2 concentration below this limit. However, maintenance of this limit becomes challenging if the cell already has extracellular H_2O_2, having the ability to flow into the cell cytoplasm via cellular membranes. Once H_2O_2 has entered the cell, certain protective regulons controlled by the OxyR or the PerR or PerR-like homolog transcription factors are activated (Marinho et al. 2014; Netto and Antunes 2016).

12.3 Defensive Role of the OxyR Regulons

The OxyR regulons are widespread in living organisms. These regulons have a sensory cysteine residue that is oxidized after exposure to H_2O_2, resulting in exposure of the disulfide bond and its alteration to connect with DNA (Flores-Cruz and Allen 2011; Ahmad et al. 2016; Zhao et al. 2016; Wang et al. 2017). OxyR regulons are also reported as repressors in some bacteria, and the oxidation of OxyR regulons results in the loss of its activity while increasing its expression (Teramoto et al. 2013; Yu et al. 2016). Generally, OxyR regulons have a stimulatory effect on the expression of certain proteins that are involved in the recruitment of RNA polymerases (Luo et al. 2005). To scavenge the problems caused by H_2O_2, about two dozen genes are activated via oxidized protein in *E. coli* as is clear from Table 12.1 (Zheng et al. 2001).

Similarly, OxyR regulons drive down the level of H_2O_2 by inducing the expression of a catalase and NADP peroxidase (Pannala and Dash 2015; Passaia and Margis-Pinheiro 2015). To sequester unincorporated iron, a mini-ferritin, Dps, is synthesized. Mini-ferritin with induced Fur and Yaa proteins reduces the rate of DNA damage (Chiancone and Ceci 2010; Shi et al. 2015). At the same time, manganese importer (MntH) is induced to replace Fe with Mn to make available Fe as a cofactor of mononuclear enzymes (Kehres et al. 2002; Sun et al. 2012; Shah et al. 2014; Sobota et al. 2014).

Initiation of the Suf iron-sulfur-meeting framework empowers the overhaul of broken Fe-S clusters (Outten et al. 2004; Fontecave et al. 2005). Despite the fact that the cell uses an Isc framework, when the iron level in the cell is reduced, Isc functionality is severely impaired as observed in Dps activity (Yeo et al. 2006; Lee et al. 2008). Likewise, the Fe-dependent enzyme in the heme biosynthesis is replaced as HemF or incited as HemH, to maintain the smooth running of the system as intracellular iron concentration is scarce. Moreover, glutathione reductase, glutaredoxin-1, and thioredoxin-2 have important roles in reducing the disulfide (-S-S-) bond (Eser

Table 12.1 OxyR regulon in *Escherichia coli*

Genes	Activity	Role
*kat*G	Catalase	Scavenge H_2O_2
*ahp*CF	NADH peroxidase	
dps	Mini-ferritin	Reduce free iron
fur	Iron-import repressor	
*yaa*A	Unknown biochemistry	
*mnt*H	Manganese importer	Activates mononuclear enzymes
*suf*A-E	Iron-sulfur assembly	Activates Fe/S enzymes
*hem*F	Coproporphyrinogen III oxidase	Sustain heme synthesis
*hem*H	Ferrochelatase	
gor	Glutathione reductase	Thiol maintenance
*trx*C	Thioredoxin	
*grx*A	Glutaredoxin	
*dsb*G	Protein sulfonate reductase	

et al. 2009). Once the level of H_2O_2 has declined, glutaredoxin deactivates OxyR (Imlay 2015).

12.4 Sensitivity and Specificity of OxyR for H_2O_2

Various studies have focused on revealing the sensitivity and specificity of OxyR toward H_2O_2 (Choi et al. 2001; Lee et al. 2004; Jaroensuk et al. 2016; Tanner et al. 2017). The studies have found that there is a hyperreactive thiol (Cys199) component of OxyR in *E. coli* that is oxidized to sulfenic acid (–SOH) after reaction with H_2O_2 (Rahal et al. 2014; Hillion and Antelmann 2015; Imlay 2015). In response, the adducted residue is displaced from the residue cleft in which it is normally buried. It condenses with Cys208 to form a disulfide bond after moving into its proximity. This change results in a locked domain with a certain conformation that can activate the transcription proteins. These proteins are dimer in functionality and help in preventing the activation of OxyR under the reduced concentration of H_2O_2 (Lee et al. 2004). Conclusively, OxyR utilized cysteine residues for the detection of H_2O_2 proximity and acts as a rheostat as disulfide bond formation is among most of the amino acids that are only with reversible functionality (Choi et al. 2001; Antelmann and Helmann 2011).

Surprisingly, the cysteine reaction is very sluggish because of the very low rate constant, 2 M^{-1} s^{-1}. Notably, a concentration of 0.5 μM H_2O_2 is sufficient to hinder the growth of *E. coli* (Sobota et al. 2014). At this rate, the half time for the oxidation of 0.5 μM H_2O_2 is 1 week. In clear comparison, OxyR senses cysteine residues at the rate of 10^5 M^{-1} s^{-1} and detects the presence of H_2O_2 within a few seconds. The premise behind this dynamic enhancement might come from the elucidation of the role of OxyR C199. It has been found that hyperreactive behaviors of OxyR C199 resemble the catalytic cysteine residues on peroxiredoxins, which are common peroxidases involved in the degradation of H_2O_2 via cyclic redox reactions with cysteine (Hugo et al. 2009; Ferrer-Sueta et al. 2011).

In OxyR, the dioxygen bond in H_2O_2 is cleaved with a reaction between H_2O_2 and cysteine thiolate during a nucleophilic attack (Flohe et al. 2011). To make this reaction favorable, a cationic residue ensures the deprotonation of cysteine. However, this reaction only helps in improving the order of magnitude of the reaction (Nelson et al. 2008). During the rest of the stimulation process, a peroxidation cysteine residue of peroxiredoxins takes part in the H-bond network. This network protects its charge like a shield in a water-repellent cleft (Poole 2007). With the entry of H_2O_2, these bonds are shifted toward H_2O_2 itself, and this shift makes thiolate free to take part in further reactions. These conditions are favorable for many catabolic reactions (Peskin et al. 2013). A nucleophile whose reaction potential is increased under a low dielectric environment is generated with the release of a cysteine residue. The shifted H-bond toward H_2O_2 polarizes its O_2-bond and makes itself more conducive to attack other molecules. The forward reaction occurs with the protonation of one of these residues with the hydroxide leaving group. A similar role has been suggested for the arginine and histidine residues that invariably flank the cysteine residue of OxyR.

It has been suggested that the H-bond network configured to fit hydrogen peroxide (H_2O_2) should also enhance the rate constant by disfavoring water (H_2O) in the active site as a water molecule is much smaller to pass through this configured network. The reason is that the thiol-based peroxidases are appropriate for these physiological functions as these achieve a reaction rate of 10^7 M^{-1} s^{-1}. The case with OxyR, which performs similar functions, is similar. Recently, it has been found that the H-bond network might be engaged with H_2O_2 binding as revealed from the OxyR structure in *Pseudomonas aeruginosa*. This engagement ultimately leads toward the protonation of the product OH group and deprotonation of cysteine (Jo et al. 2015, 2016).

OxyR in *E. coli* rapidly responds to very minute concentrations (<1 μM) of H_2O_2 that threaten the normal metabolic activities of the organism under consideration (Mosel et al. 2013). In some microbes, such as *Vibrio vulnificus*, there are double OxyR proteins that help scavenge the deleterious effect of H_2O_2 on metabolic activities of the organism (Kim et al. 2014; Bang et al. 2016). Overall, the more sensitive OxyR proteins can detect low concentrations of H_2O_2 and are synthesized endogenously under lower concentrations of oxygen, whereas the less sensitive OxyR proteins are capable of scavenging H_2O_2, which is transported inside the cell from the external environment via extracellular transport.

12.5 The Response of OxyR to Stresses Other than H_2O_2

Since OxyR responds to the abduction of its sensory SH^-, it is highly likely to be triggered by electrophiles (Sainsbury et al. 2010; Svintradze et al. 2013). Scientists around the world have typically used artificial organic HN-NH as a proxy for such complexes. In this regard, whether OxyR is capable of sensing molecules other than H_2O_2 would be very interesting to know (Seth et al. 2012; Fang et al. 2016; Yu et al. 2016). It has been found that nitric oxide escaping from the system has an important role in nitrosylation of OxyR under anoxic conditions (Seth et al. 2012; Fang et al. 2016; Crack et al. 2016; Munnoch et al. 2016). It was found that some elements present in the regulon have an important function in tolerating this stress as revealed through experimentation on mutant OxyR bacteria (Crack et al. 2016).

In many bacteria, some other transcription factors are responsible for their responses against electrophiles. The attack of electrophiles on thiols results in the generation of disulfide bonds as commonly observed in many species of bacteria. To avert the effects of disulfide bond formation and thiol adduction, Spx and SigR/RsrA systems in *Bacillus subtilis* and *Streptomyces coelicolor*, respectively, induce a mixture enzyme system that includes proteases, chaperones, glutaredoxins, and thioredoxins (Zuber 2009; Kim et al. 2012). The mentioned bacteria contain another system, PerR, in addition to OxyR to regulate the negative effects of H_2O_2 on cell growth. In comparison to the OxyR system discussed earlier, the PerR system does not include genes present in OxyR regulon in *E. coli* and involved in disulfide reduction. In this regard, *E. coli* seems robust and novel as it contains only the OxyR system which helps it in relieving both the stresses, that is, H_2O_2 (a threat for iron-containing enzymes) and the electrophile attack, which otherwise react with cysteine residues more rapidly compared to H_2O_2.

12.6 PerR or PerR-homologue: An Alternative to OxyR

In microorganisms, especially those belonging to gram-positive bacteria, PerR, or PerR-homologue is used as a sensor of H_2O_2 instead of OxyR (Marinho et al. 2014). The PerR or PerR-homologue belongs to the Fur family of transcription factors that can sense different metal atoms, such as Fe, Mn, Zn, Ni, and H_2O_2. The transcription factor that senses Fe is called Fur proper. For Mn, Zn, Ni, and H_2O_2, the transcription factors involved in their sensing are called Mur, Zur, Nur, and PerR, respectively as clear in Fig. 12.3 (Mulrooney and Hausinger 2003; Bellini and Hemmings 2006; Li et al. 2009; Fillat 2014; Helmann 2014; Shah et al. 2014).

Each subunit of PerR is composed of a structural and regulatory site (Hohle and O'Brian 2016). The former attaches Zn in an irreversible reaction whereas the latter is involved in the reversible binding of activating metal. During binding with DNA, only metal-bound forms of PerR are capable and act as repressors. The regulatory metal, which is ferrous iron in the case of PerR in *Bacillus subtilis*, is affixed in the

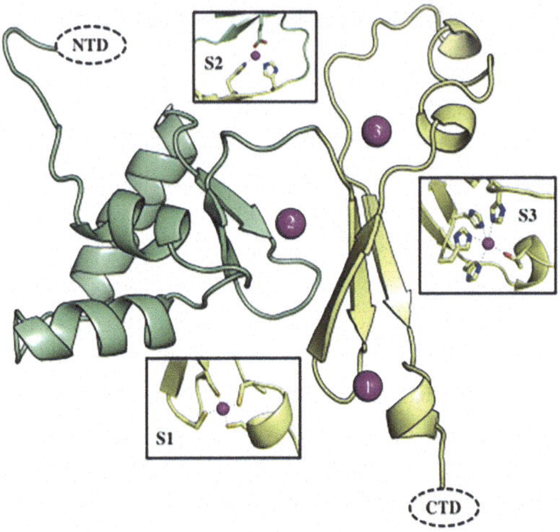

Fig. 12.3 Structure of PerR homolog Mur in its holo-forms. Structures were derived from the Protein Data Bank (PDB 3F8N and 2FE3) (Shah et al. 2014). Orange circles represent Zn, Fe, or Mn

form of a pyramid with four ligands along one plane and a single axial ligand (Jacquamet et al. 2009; Huang et al. 2017). During H_2O_2 stress, an iron atom present at the sixth coordination site is oxidized, resulting in the formation of the ferryl or hydroxyl radical. These radicals subsequently oxidize histidine ligands to 2-oxohistidine (Herbig and Helmann 2001; Lee and Helmann 2006). In this way, the modified residue is unable to bind another metal atom and undergoes conformational changes that result in its inability to bind with DNA (Dian et al. 2011; Kim et al. 2016). As mentioned earlier, holo-PerR acts as a repressor as it represses regulon members during H_2O_2 stress.

In *Bacillus subtilis*, the reaction rate of PerR is similar to OxyR in *E. coli*, 105 M^{-1} S^{-1} (Herbig and Helmann 2001). The rate constant is similar because the rate of the Fenton reaction is similar, as is clear from the similar inactivation rates of mononuclear Fe proteins such as peptide deformylase in *E. coli* (Anjem and Imlay 2012). To resist oxidation under H_2O_2 stress, there are additional Fur proteins in *B. subtilis*. While elucidating the mechanisms behind this process, it has been suggested that coordinating glutamate residue (Fur protein) may act as a bidentate ligand for the Fe atom and occlude its sixth coordination site (Parent et al. 2013). Given that H_2O_2 should unswervingly contact iron or move metal with a specific end goal to oxidize it, H_2O_2 oxidizes PerR, however, not the fur-like protein (Anjem and Imlay 2012).

The PerR regulon of *Bacillus subtilis* and *Staphylococcus aureus* contains a large portion of similar genes that OxyR reins in other bacteria, such as genes encoding a catalase, a Dps homolog (MrgA), and Fur protein (Mielecki and Grzesiuk 2014; Kim et al. 2016). However, PerR compromises a few alluring takeoffs from the OxyR system of H_2O_2 detecting. In the first stance, which is very rare, the reaction might be irreversible and act as an on/off switch, as the exact mechanism behind these processes has not yet been elucidated. It has been found that there is dynamic equilibrium during activation of its subsequent deactivation in bacterial regulators, with some exceptions (Mielecki and Grzesiuk 2014). Moreover, it is also still unclear that the irreversibility of PerR deactivation is beneficial or not in any way and needs to be elucidated in the future (Imlay 2015). Second, the PerR PerR-homolog (Fur, Mur, Zur or Nur) is capable to bind with Mn in place of Fe, as observed in many mononuclear proteins under in vitro and in vivo conditions (Zuber 2009; Sun et al. 2012; Shah et al. 2014). The activation of PerR against H2O2 stress starts only under the condition that Mn is oxidized with H_2O_2 when Mn is present in the cytoplasm in sufficient amounts. It has also been suggested that PerR and OxyR might be involved in increasing the concentration of Mn, thereby elevating the Mn/Fe ratio in the cytoplasm as Mn is directly involved in H_2O_2 scavenging (Yamamoto et al. 2011; Shah et al. 2014). This PerR and OxyR system remains shut down until the level of Mn in the cytoplasm is high (Huang et al. 2017). Furthermore, PerR is not involved in controlling disulfide-reducing redoxins because the Spx response regulator, which is not activated in H_2O_2 sensing, is involved in their synthesis. Conclusively, PerR acts as a switch to turn the Fenton reaction on or off.

12.7 Superoxide Stress and SoxR

12.7.1 Toxicity of Redox-Cycling Compounds

It is understood that superoxide (O_2^-) is internally induced in microbes during stress conditions. The O_2^- is produced when redox enzymes are self-oxidized under any biotic or abiotic stress (Korshunov and Imlay 2002). Superoxide dismutase activity scavenges O_2^- stress and keeps it at a steady-state level about 10^{-10} M (Diaz et al. 2016). The mutant allele of superoxide dismutase (SOD) in *E. coli* lacks the branched amino acid in their protein sequence that shapes them to be unable to dissociate tricarboxylic acid cycle substrates. Because of the SOD mutant allele, *E. coli* shows growth deficiency similar in function to that caused by H_2O_2 that inactivates the non-redox mononuclear enzyme and [4Fe-4S] dehydratases. Superoxide oxidizes the cofactor Fe of any of these two families of the iron-comprised enzymes (Gu and Imlay 2013; Sobota et al. 2014).

It is a fact that superoxide (O_2^-) is a charged atom that cannot enter into microscopic organisms from the outside (Korshunov and Imlay 2002); in this way, any cytoplasmic superoxide (O_2^-) stress is caused because of the superoxide (O_2^-) that is created internally (Diaz et al. 2016). The O_2^- that is shaped by catalyst autoxidation is saved exceptionally uncommonly (~10^{-10} M) by the activity of SODs (Imlay 2013; Esteve-Gassent et al. 2015). *E. coli* mutants that non-attend cytoplasmic SODs show development flaws that have been moved back to needs in similar proteins which H_2O_2 inactivates: [4Fe-4S] dehydratases and mononuclear iron enzymes (Gu and Imlay 2013; Sobota et al. 2014). Going deep into the literature has uncovered that total overarticulation of superoxide reductase (SOR) ably rewards for the absence of SOD in *E. coli* (Dietrich et al. 2006). Along these lines, mutants cannot incorporate stretched chain or sweet-smelling amino acids, and they cannot catabolize TCA cycle substrates.

$$O_2 \bullet^- + O_2 \bullet^- + 2H^+ \rightarrow H_2O_2 + O_2$$

The second framework, found only in prokaryotic cells, is the recently considered non-heme superoxide reductase (SOR) that catalyzes the diminishment of $O_2 \bullet^-$ into H_2O_2 (Kurtz 2004; Niviere and Fontecave 2004):

$$O_2 \bullet^- + 1e^- + 2H+ \rightarrow H_2O_2 eq\,2(SOR)$$

The comparative phenotypes are level in wild-type *E. coli* if it has viologens, quinones, or phenazines (Guttenberger et al. 2017). These cyclic natural chemicals can go into cells, where they acquire electrons from redox catalysts and exchange them to oxygen (Wang and Newman 2008). Deadly measures of superoxide (O_2^-) are along these lines delivered. Quinones and phenazines are positively unconfined by many microorganisms and plants (Song et al. 2013; Milackova et al. 2015), as regularly as possible to harm smaller-scale flora (de Paiva et al. 2003; Inbaraj and Chignell 2004). A few years earlier it was seen that *E. coli* answers to this worry by creating higher titers of SOD. This adjustment is mediated by two proteins named SoxR and SoxS. SoxR is a [2Fe-2S] having the interpretation of satisfying the misfit; it then prompts the translation of SoxS. SoxS is a moment translation figure that enacts scores of guarded qualities scattered around the chromosome (Blanchard et al. 2007). Some of these qualities encode proteins that help stifle the noxiousness of O_2^{-1}:superoxide dismutase, aconitase, and fumarase C, which are superoxide-safe isozymes of defenseless [4Fe-4S] enzymes (Varghese et al. 2003); and YggX, which facilitates group repair processes (Gralnick and Downs 2003; Pomposiello et al. 2003; Gu and Imlay 2011). Similar to OxyR, the SoxRS framework is not activated all through normal oxygen-consuming development and does not switch off the basal level of self-legitimizing frameworks; in its place, it becomes noticeably dynamic only when constraining conditions emerge (Yang et al. 2016; Aquino et al. 2017; Kim et al. 2017a, b).

12.7.2 *The Role of SoxR*

SoxR is a homodimer that ties one [2Fe-2S] bunch for every monomer that contains a DNA-restricting HTH theme (Watanabe et al. 2008; Kobayashi et al. 2015). Electron paramagnetic reverberation considers that when cells exposed to redox cycling deepen, the clusters are rapidly oxidized to the $[2Fe\text{-}2S]^{2+}$ state (Landry and Ding 2014; Landry et al. 2015; Shimberg et al. 2016). The endless supply of the oxidant groups returns inside to reduced structures. In *E. coli*, SoxS is consists of a bizarrely long 20-bp separation between its −10 and −35 promoter themes. The present model is that group oxidation twists the SoxR dimer and along these lines changes the compliance of the bound DNA, moving the promoter themes to a similar side of the helix so that RNA polymerase can successfully tie. Some researchers have suggested that changes in SoxR redox state distort bound DNA (Fujikawa et al. 2015). Established reviews specify that the SoxR reduction resulting in conformational changes is catalyzed by Rsx/Rse proteins that probably utilize NADPH as a source of electrons (Griffith et al. 2004). The combined information recommends that translation of the soxS relies on the dynamic harmony between the oxidation and reduction of the soxR press sulfur groups. At the point when SoxR returns to its reduced state, the translation of the full regulon rapidly stops because of the fast proteolysis of surviving SoxS proteins (Griffith et al. 2004).

12.7.3 How Does SoxR Oxidize?

In parallel with the detection of H_2O_2 via OxyR, the possibility has been suggested that SoxR is initiated when O_2^- specifically oxidizes its groups, which are arranged close to the protein surface and seem to be attacked by small atoms like O_2^- (Watanabe et al. 2008). Surprisingly, the findings of two researchers to straightforwardly recognize bunch oxidation by O_2^- under in vitro conditions suggested conflicting results (Gu and Imlay 2011; Fujikawa et al. 2012), implying that one of these confounded tests was somehow coordinated. The in vivo findings are clearer and to some degree shocking. In superoxide dismutase (SOD) mutants, which show overpowering levels of O_2^-, the *soxS* gene is actuated only two- to four fold (Gu and Imlay 2011). This activation is reliable with the possibility that O_2^- can oxidize much; however, it is not exactly the approximately 40-fold enlistment that happens when wild-type cells are treated with redox-cycling mixtures. Similarly, unobtrusive effects were found for individuals from other regulons (Gu and Imlay 2011). This enlistment is too minor to be in any way viable: the SoxRS regulon gives no assurance to SOD mutants unless the *soxS* gene is persuasively initiated through a heterologous promoter. Consequently, high O_2^- focuses are not adequate to initiate SoxR. Moreover, overproduction of SOD does not seem sensible relative to the fitness of redox specialists to initiate SoxR in wild-type cells (Gu and Imlay 2011; Mahavihakanont et al. 2012), driving the workers to finish these medications that oxidize SoxR independent of O_2^-. Unquestionably, contact with redox combinations can enact SoxR in cells under anaerobic conditions (Gu and Imlay 2011; Krapp et al. 2011).

An abnormal elucidation for SoxR oxidation by redox response is that they reduce NADPH pools and, in this manner, intrude the aptitude of Rse/Rsx to keep SoxR in its reduced frame. Most likely, a few qualities inside the *E. coli* SoxR regulon give a system to restore NADPH pools (Krapp et al. 2002; Nakayama et al. 2013). However, limits determine that the move in NADPH redox status ascertained (−0.314 V) is insufficient to clarify the close aggregate oxidation of SoxR, that is, $Eo' = -0.293$ V (Kobayashi et al. 2015), and changes that harm the cell NADPH arrangement do not enact SoxR (Gu and Imlay 2011; Zheng et al. 2017). Along these lines, although NADPH decrease may extend the prompting impacts of redox operators, without anything else it is most likely not adequate to activate the framework.

In this way, it looks probable that the redox mixes straightforwardly oxidize the groups of SoxR that are reasonable as they do with the metal and flavin moieties of commonplace redox catalysts. This response has been set up in vitro, with rate constants that follow the rates of SoxR oxidation seen in vivo (Gu and Imlay 2011). The position of the iron molecule close to the SoxR protein surface affirms that it can be oxidized by various univalent oxidants without structures. The constituency of the regulon supports the fact that SoxRS ensure inconsistency of redox-cycling compounds in microscopic organisms (Kobayashi et al. 2015). A noteworthy component of the SoxRS reaction in *E. coli* is the acceptance of proteins that together stop the passage. The gradual addition of redox-dynamic mixes inside the cell, lipopolysaccharide-adjusting catalysts, decreases the bent of such mixes to penetrate the envelope. Expansive specificity spreads the frameworks that energetically draw them out of the cytoplasm and bringing out chemicals that kill them completely through covalent modification (Rau and Stolz 2003; Lee et al. 2009; Lin et al. 2010). Fascinatingly, SoxR is noteworthy insurance against redox-dynamic mediators even under anaerobic conditions and delegate that these specialists can apply O_2^- autonomous components of toxicity. Their principal activity under these conditions is vague; however, such mixes can frame adducts to proteins by going as Michael acceptors, and these can deactivate [4Fe-4S] dehydratases by straightforwardly oxidizing the groups (Anjem and Imlay 2012), and can exhaust cell NADPH pools by oxidizing NADPH-reducible enzymes and subsequent expulsion of the electrons into the anaerobic respiratory chain (Gu and Imlay 2011).

Along these lines, it is significant to note that H_2O_2 is an unsuccessful activator of SoxR (Gu and Imlay 2011). Nitric oxide (NO) can enact it by framing dinitrosyl press edifices with the iotas of the iron-sulfur group (Fujikawa et al. 2014). The ramifications of this result are reluctant, in part because dedicated NO rummaging frameworks are controlled by free administrative proteins (NsrR and NorR) that are a great deal more open to NO. NO dosages that activated the oxic and anaerobic NO-searching frameworks by 30- to 180 fold just prompted *soxS* gene 5-overlay (Pullan et al. 2007). Similarly, this slight SoxS enlistment had no apparent outcome upon *soxS*-controlled genes, driving the creators to suggest that this level of SoxS acceptance was lacking for the enactment of the full regulon. That derivation converges with the failure of three-fold SoxS acceptance to convey any guard to SOD cells (Gu and Imlay 2011).

12.8 A More Expansive View: SoxR Beyond *E. coli*

Investigations of SoxR and SoxS in other microorganisms have supported a significant reevaluation of their natural function. Some homologs of SoxR are expressed among α-, β-, γ-, and δ-proteobacteria and actinobacteria, yet not elsewhere

(Dietrich et al. 2008). SoxR straightforwardly controls the statement of all regulon individuals in the various microscopic organisms. Incredibly, only enterics utilize SoxS. Similarly, in an exceptional way in these non-enterics, the SoxR regulon is genuinely unique from that of *E. coli*: it distinctively incorporates just a modest bunch of qualities of poorly defined function (Naseer et al. 2014). These qualities generally do not involve sodA or different qualities that make them able to encounter O_2^- (Dietrich et al. 2006; Dietrich et al. 2008; Shin et al. 2011; Kim and Park 2014). The intimation is that the physiological drive of SoxR in non-enterics is unique with its motivation in *E. coli*.

A sign concerning this energy has originated from investigations of *Pseudomonas aeruginosa* and *Streptomyces coelicolor* (Kim et al. 2017a, b). Both microbes combine and produce redox-cycling mixes: pocyanin for *P. aeruginosa* and actinorhodin for *S. coelicolor*. Unique parts have been discretionary for these mixes, numbering notifying competitors, flagging, supplementary electron exchange to detached respiratory substrates, and solubilizing iron (Dietrich et al. 2006; Wang et al. 2010, 2011). The compounds that enter are incited in both microbes when SoxR is oxidized by the redox specialist itself and a precursor of it (Dietrich et al. 2008; Shin et al. 2011). The SoxR protein can likewise respond to exogenous engineered redox mixes, for example, paraquat, yet this is unusual, as SoxR enactment does not enhance showdown to such compounds (Palma et al. 2005; Sakhtah et al. 2016). These microorganisms utilize SoxR to control the endogenous redox stress.

It is still under discussion whether it is gainful for the creation of living beings; these specialists can enter and toxify different organisms. It is conceivable that the least transversely learned SoxR is a sensor that could control these substances and incite strongholds. In *E. coli*, the internal SoxR quality may have attached to a prior framework that reacts with other harmful chemicals. SoxS is a paralog of the MarA and Ro translation elements (Duval and Lister 2013), which react and give insurance against exogenous non-redox operators such as salicylate and bile salts. The three proteins are sufficiently comparable (~50% personality) that they perceive basic administrative destinations, thus their regulons cover extensively (Martin and Rosner 2011). The straightforward model is that *E. coli* procured SoxR and after duplication of MarA or Rob, molded SoxS to interface SoxR to a previous regulon that serves to exclude change or fare risky mixes (Duval and Lister 2013). SoxR provided the way to enact this framework in light of redox–dynamic mixes. The lessening capability of *E. coli* SoxR is quite minor compared to those of *P. aeruginosa* and *S. coelicolor* SoxR proteins, and in like manner it reacts to a larger scope of redox operators to which it responds (Sheplock et al. 2013; Singh et al. 2013).

12.9 About Other "Antioxidant" Regulons

RpoE/ChrR, natural hydro-peroxides OhrR, disulfide push (Spx), SigR/RsrA, or hypochlorous corrosive agents (HypT) have also been perceived to have their role in scavenging reactive oxygen species such as H_2O_2 in a diverse group of bacterial species (Hong et al. 2005; Zuber 2009; Dubbs and Mongkolsuk 2012; Kim et al. 2012; Drazic et al. 2013; Drazic et al. 2014; Lemke et al. 2014; Drazic et al. 2015). Improperly spaced edges discourage a discourse of these frameworks here, and in this way, specialists are encouraged to examine more about them with later writing. From this exchange, the oxidative anxiety assortment is implicit in the way of an overly broad term that incorporates oxidants. In contrast to their substance properties, different targets are assaulted altogether and detect and guard against ROS stress in unique ways.

12.10 Conclusions and Future Perspectives

It has been found that Fe-intensive biochemistry inherited by present organisms is undergoing some functional problems resulting from the evolution of life under anaerobic conditions in earlier times. Seen is a problem of susceptibility for most bacteria and an opportunity for antagonists (reactive oxygen species) to inhibit/reduce the growth of bacteria under consideration. The present book chapter provided an insight into the transcription factors that are involved in helping bacteria to survive under dynamic concentrations of oxygen. It also focused on the conditions under which these regulons are activated and have optimal activity in scavenging reactive oxygen species. Studies have found that these regulons are under tight control, which makes them work under dynamic concentrations of oxygen and allows bacteria to survive and grow optimally under these circumstances. More studies should be conducted in non-enteric bacteria to clarify these research findings. In this way, the use of genetic engineering will be justified for gene manipulation and creating robust bacteria with the ability to scavenge reactive oxygen species at a better rate with maximal growth.

References

Ahmad S, Lee SY, Kong HG et al (2016) Genetic determinants for pyomelanin production and its protective effect against oxidative stress in *Ralstonia solanacearum*. PLoS One 11(8):e0160845

Ahmad I, Tahir M, Daraz U et al (2020) Responses and tolerance of cereal crops to metals and metalloids toxicity. In: Hassanuzzaman M (ed) Agronomic crops. Springer, Singapore, pp 235–264. https://doi.org/10.1007/978-981-15-0025-1_14

Anjem A, Imlay JA (2012) Mononuclear iron enzymes are primary targets of hydrogen peroxide stress. J Biol Chem 287:15544–15556

Anjum NA, Gill SS, Gill R et al (2014) Metal/metalloid stress tolerance in plants: role of ascorbate, its redox couple, and associated enzymes. Protoplasma 251(6):1265–1283

Antelmann H, Helmann JD (2011) Thiol-based redox switches and gene regulation. Antioxid Redox Signal 14:1049–1063

Aquino P, Honda B, Jaini S et al (2017) Coordinated regulation of acid resistance in *Escherichia coli*. BMC Syst Biol 11(1):1–5

Bang YJ, Lee ZW, Kim D et al (2016) OxyR2 functions as a three-state redox switch to tightly regulate production of Prx2, a peroxiredoxin of *Vibrio vulnificus*. J Biol Chem 291:16038–16047

Bedard K, Lardy B, Krause KH (2007) NOX family NADPH oxidases: not just in mammals. Biochimie 89:1107–1112

Bellini P, Hemmings AM (2006) In vitro characterization of a bacterial manganese uptake regulator of the fur superfamily. Biochemistry 45:2686–2698

Bhattacharyya A, Chattopadhyay R, Mitra S et al (2014) Oxidative stress: an essential factor in the pathogenesis of gastrointestinal mucosal diseases. Physiol Rev 94:329–354

Blanchard JL, Wholey WY, Conlon EM et al (2007) Rapid changes in gene expression dynamics in response to superoxide reveal SoxRS-dependent and independent transcriptional networks. PLoS One 2(11):e1186

Chatgilialoglu C, Ferreri C, Terzidis MA (2011) Purine 5′,8-cyclonucleoside lesions: chemistry and biology. Chem Soc Rev 40:1368–1382

Chiancone E, Ceci P (2010) The multifaceted capacity of Dps proteins to combat bacterial stress conditions: detoxification of iron and hydrogen peroxide and DNA binding. Biochim Biophys Acta 8:798–805

Choi HJ, Kim SJ, Mukhopadhyay P et al (2001) Structural basis of the redox switch in the OxyR transcription factor. Cell 105(1):103–113

Crack JC, Svistunenko DA, Munnoch J et al (2016) Differentiated, promoter-specific response of [4Fe-4S] NsrR DNA binding to reaction with nitric oxide. J Biol Chem 291:8663–8672

de Paiva SR, Figueiredo MR, Aragao TV et al (2003) Antimicrobial activity in vitro of plumbagin isolated from *Plumbago* species. Mem Inst Oswaldo Cruz 98:959–961

Dian C, Vitale S, Leonard GA et al (2011) The structure of the *Helicobacter pylori* ferric uptake regulator Fur reveals three functional metal binding sites. Mol Microbiol 79:1260–1275

Diaz JM, Hansel CM, Apprill A et al (2016) Species-specific control of external superoxide levels by the coral holobiont during a natural bleaching event. Nat Commun 7(1):13801

Dietrich LE, Price-Whelan A, Petersen A et al (2006) The phenazine pyocyanin is a terminal signaling factor in the quorum sensing network of *Pseudomonas aeruginosa*. Mol Microbiol 61:1308–1321

Dietrich LE, Teal TK, Price-Whelan A et al (2008) Redox-active antibiotics control gene expression and community behavior in divergent bacteria. Science 321:1203–1206

Ditta A (2018) Physiological and phenological responses of crop plants under heat stress. In: Ramakrishna A, Gill SS (eds) Metabolic adaptations in plants during abiotic stress. CRC Press, Taylor & Francis, Boca Raton, pp 55–63

Ditta A, Khalid A (2016) Bio-organo-phos: a sustainable approach for managing phosphorus deficiency in agricultural soils. In: Larramendy M, Soloneski S (eds) Organic fertilizers – from basic concepts to applied outcomes. InTech, Croatia, pp 109–136. https://doi.org/10.5772/62473

Ditta A, Arshad M, Zahir ZA et al (2015) Comparative efficacy of rock phosphate enriched organic fertilizer vs. mineral phosphatic fertilizer for nodulation, growth and yield of lentil. Int J Agric Biol 17:589–595

Ditta A, Imtiaz M, Mehmood S et al (2018a) Rock phosphate enriched organic fertilizer with phosphate solubilizing microorganisms improves nodulation, growth and yield of legumes. Commun Soil Sci Plant Anal 49(21):2715–2725. https://doi.org/10.1080/00103624.2018.1538374

Ditta A, Muhammad J, Imtiaz M et al (2018b) Application of rock phosphate enriched composts increases nodulation, growth and yield of chickpea. Int J Recycl Org Waste Agric 7(1):33–40. https://doi.org/10.1007/s40093-017-0187-1

Drazic A, Miura H, Peschek J et al (2013) Methionine oxidation activates a transcription factor in response to oxidative stress. Proc Natl Acad Sci U S A 110:9493–9498

Drazic A, Gebendorfer KM, Mak S et al (2014) Tetramers are the activation-competent species of the HOCl-specific transcription factor HypT. J Biol Chem 289:977–986

Drazic A, Kutzner E, Winter J et al (2015) Metabolic response of *Escherichia coli* upon treatment with hypochlorite at sub-lethal concentrations. PLoS One 10:e0125823

Dubbs JM, Mongkolsuk S (2012) Peroxide-sensing transcriptional regulators in bacteria. J Bacteriol 194:5495–5503

Duval V, Lister IM (2013) MarA, SoxS and Rob of *Escherichia coli*. Global regulators of multidrug resistance, virulence and stress response. Int J Biotechnol Wellness Ind 2:101–124

Eser M, Masip L, Kadokura H et al (2009) Disulfide bond formation by exported glutaredoxin indicates glutathione's presence in the *E. coli* periplasm. Proc Natl Acad Sci U S A 106:1572–1577

Esteve-Gassent MD, Smith TC II, Small CM et al (2015) Absence of sodA increases the levels of oxidation of key metabolic determinants of *Borrelia burgdorferi*. PLoS One 10:e0136707

Fang FC, Frawley ER, Tapscott T et al (2016) Discrimination and integration of stress signals by pathogenic bacteria. Cell Host Microbe 20:144–153

Ferrandiz MJ, Martin-Galiano AJ, Arnanz C et al (2015) Reactive oxygen species contribute to the bactericidal effects of the fluoroquinolone moxifloxacin in *Streptococcus pneumoniae*. Antimicrob Agents Chemother 60:409–417

Ferrer-Sueta G, Manta B, Botti H et al (2011) Factors affecting protein thiol reactivity and specificity in peroxide reduction. Chem Res Toxicol 24:434–450

Fillat MF (2014) The FUR (ferric uptake regulator) superfamily: diversity and versatility of key transcriptional regulators. Arch Biochem Biophys 546:41–52

Flohe L, Toppo S, Cozza G et al (2011) A comparison of thiol peroxidase mechanisms. Antioxid Redox Signal 15:763–780

Flores-Cruz Z, Allen C (2011) Necessity of OxyR for the hydrogen peroxide stress response and full virulence in *Ralstonia solanacearum*. Appl Environ Microbiol 77:6426–6432

Fontecave M, Choudens SO, Py B et al (2005) Mechanisms of iron-sulfur cluster assembly: the SUF machinery. J Biol Inorg Chem 10:713–721

Fujikawa M, Kobayashi K, Kozawa T (2012) Direct oxidation of the [2Fe-2S] cluster in SoxR protein by superoxide: distinct differential sensitivity to superoxide-mediated signal transduction. J Biol Chem 287:35702–35708

Fujikawa M, Kobayashi K, Kozawa T (2014) Mechanistic studies on formation of the dinitrosyl iron complex of the [2Fe-2S] cluster of SoxR protein. J Biochem 156:163–172

Fujikawa M, Kobayashi K, Kozawa T (2015) Redox-dependent DNA distortion in a SoxR protein-promoter complex studied using fluorescent probes. J Biochem 157:389–397

Gardner PR, Gardner DP, Gardner AP (2015) Globins scavenge sulfur trioxide anion radical. J Biol Chem 290:27204–27214

Gralnick JA, Downs DM (2003) The YggX protein of *Salmonella enterica* is involved in Fe(II) trafficking and minimizes the DNA damage caused by hydroxyl radicals: residue CYS-7 is essential for YggX function. J Biol Chem 278:20708–20715

Griffith KL, Shah IM, Wolf RE Jr (2004) Proteolytic degradation of *Escherichia coli* transcription activators SoxS and MarA as the mechanism for reversing the induction of the superoxide (SoxRS) and multiple antibiotic resistance (Mar) regulons. Mol Microbiol 51:1801–1816

Gu M, Imlay JA (2011) The SoxRS response of *Escherichia coli* is directly activated by redox-cycling drugs rather than by superoxide. Mol Microbiol 79:1136–1150

Gu M, Imlay JA (2013) Superoxide poisons mononuclear iron enzymes by causing mismetallation. Mol Microbiol 89:123–134

Guttenberger N, Blankenfeldt W, Breinbauer R (2017) Recent developments in the isolation, biological function, biosynthesis, and synthesis of phenazine natural products. Bioorg Med Chem 5:31180–31184

Helmann JD (2014) Specificity of metal sensing: iron and manganese homeostasis in Bacillus subtilis. J Biol Chem 289:28112–28120

Herbig AF, Helmann JD (2001) Roles of metal ions and hydrogen peroxide in modulating the interaction of the *Bacillus subtilis* PerR peroxide regulon repressor with operator DNA. Mol Microbiol 41:849–859

Hillion M, Antelmann H (2015) Thiol-based redox switches in prokaryotes. Biol Chem 396:415–444

Hohle TH, O'Brian MR (2016) Metal-specific control of gene expression mediated by *Bradyrhizobium japonicum* Mur and *Escherichia coli* Fur is determined by the cellular context. Mol Microbiol 101:152–166

Hong M, Fuangthong M, Helmann JD et al (2005) Structure of an OhrR-ohrA operator complex reveals the DNA binding mechanism of the MarR family. Mol Cell 20:131–141

Huan C, Jiang L, An X et al (2016) Potential role of reactive oxygen species and antioxidant genes in the regulation of peach fruit development and ripening. Plant Physiol Biochem 104:294–303

Huang X, Shin JH, Pinochet-Barros A et al (2017) *Bacillus subtilis* MntR coordinates the transcriptional regulation of manganese uptake and efflux systems. Mol Microbiol 103:253–268
Hugo M, Turell L, Manta B et al (2009) Thiol and sulfenic acid oxidation of AhpE, the one-cysteine peroxiredoxin from *Mycobacterium tuberculosis*: kinetics, acidity constants, and conformational dynamics. Biochemistry 48:9416–9426
Hussain A, Zahir ZA, Ditta A et al (2020) Production and implication of bio-activated organic fertilizer enriched with zinc-solubilizing bacteria to boost up maize (*Zea mays* L.) production and biofortification under two cropping seasons. MDPI-Agronomy 10(39). https://doi.org/10.3390/agronomy10010039
Imlay JA (2013) The molecular mechanisms and physiological consequences of oxidative stress: lessons from a model bacterium. Nat Rev Microbiol 11:443–454
Imlay JA (2015) Transcription factors that defend bacteria against reactive oxygen species. Annu Rev Microbiol 69:93–108
Imran QM, Falak N, Hussain A et al (2016) Nitric oxide responsive heavy metal-associated gene AtHMAD1 contributes to development and disease resistance in *Arabidopsis thaliana*. Front Plant Sci 7:1712
Imtiaz M, Ashraf M, Rizwan MS et al (2018) Vanadium toxicity in chickpea (*Cicer arietinum* L.) grown in red soil: effects on cell death, ROS and antioxidative systems. Ecotox Environ Saf 158:139–144. https://doi.org/10.1016/j.ecoenv.2018.04.022
Inbaraj JJ, Chignell CF (2004) Cytotoxic action of juglone and plumbagin: a mechanistic study using HaCaT keratinocytes. Chem Res Toxicol 17:55–62
Ishikado A, Sono Y, Matsumoto M et al (2013) Willow bark extract increases antioxidant enzymes and reduces oxidative stress through activation of Nrf2 in vascular endothelial cells and *Caenorhabditis elegans*. Free Radic Biol Med 65:1506–1515
Jacquamet L, Traoré DA, Ferrer JL et al (2009) Structural characterization of the active form of PerR: insights into the metal-induced activation of PerR and Fur proteins for DNA binding. Mol Microbiol 73:20–31
Jang S, Imlay JA (2007) Micromolar intracellular hydrogen peroxide disrupts metabolism by damaging iron-sulfur enzymes. J Biol Chem 282:929–937
Jaroensuk J, Atichartpongkul S, Chionh YH et al (2016) Methylation at position 32 of tRNA catalyzed by TrmJ alters oxidative stress response in *Pseudomonas aeruginosa*. Nucleic Acids Res 44:10834–10848
Jo I, Chung IY, Bae HW et al (2015) Structural details of the OxyR peroxide-sensing mechanism. Proc Natl Acad Sci U S A 112:6443–6448
Jo I, Park N, Chung IY et al (2016) Crystal structures of the disulfide reductase DsbM from *Pseudomonas aeruginosa*. Acta Crystallogr D Struct Biol 72:1100–1109
Kehres DG, Janakiraman A, Slauch JM et al (2002) Regulation of *Salmonella enterica* serovar Typhimurium mntH transcription by $H(_2)O(_2)$, $Fe(^{2+})$, and $Mn(^{2+})$. J Bacteriol 184:3151–3158
Kim J, Park W (2014) Oxidative stress response in *Pseudomonas putida*. Appl Microbiol Biotechnol 98:6933–6946
Kim MS, Dufour YS, Yoo JS et al (2012) Conservation of thiol-oxidative stress responses regulated by SigR orthologues in actinomycetes. Mol Microbiol 85:326–344
Kim S, Bang YJ, Kim D et al (2014) Distinct characteristics of OxyR2, a new OxyR-type regulator, ensuring expression of Peroxiredoxin 2 detoxifying low levels of hydrogen peroxide in *Vibrio vulnificus*. Mol Microbiol 93:992–1009
Kim JH, Ji CJ, Ju SY et al (2016) *Bacillus licheniformis* contains two more PerR-like proteins in addition to PerR, Fur, and Zur orthologues. PLoS One 11(5):e0155539
Kim JH, Won YB, Ji CJ et al (2017a) The difference in in vivo sensitivity between *Bacillus licheniformis* PerR and *Bacillus subtilis* PerR is due to the different cellular environments. Biochem Biophys Res Commun 16:30094–30093
Kim J, Park C, Imlay JA et al (2017b) Lineage-specific SoxR-mediated regulation of an endoribonuclease protects non-enteric bacteria from redox-active compounds. J Biol Chem 292:121–133
Kobayashi K, Fujikawa M, Kozawa T (2015) Binding of promoter DNA to SoxR protein decreases the reduction potential of the [2Fe-2S] cluster. Biochemistry 54:334–339

Korshunov SS, Imlay JA (2002) A potential role for periplasmic superoxide dismutase in blocking the penetration of external superoxide into the cytosol of gram-negative bacteria. Mol Microbiol 43:95–106

Krapp AR, Rodriguez RE, Poli HO et al (2002) The flavoenzyme ferredoxin (flavodoxin)-NADP(H) reductase modulates NADP(H) homeostasis during the soxRS response of *Escherichia coli*. J Bacteriol 184:1474–1480

Krapp AR, Humbert MV, Carrillo N (2011) The soxRS response of *Escherichia coli* can be induced in the absence of oxidative stress and oxygen by modulation of NADPH content. Microbiology 157:957–965

Kuehne A, Emmert H, Soehle J et al (2015) Acute activation of oxidative pentose phosphate pathway as first-line response to oxidative stress in human skin cells. Mol Cell 59:359–371

Kurtz DM Jr (2004) Microbial detoxification of superoxide: the non-heme iron reductive paradigm for combating oxidative stress. ACC Chem Res 37:902–908

Landry AP, Ding H (2014) Redox control of human mitochondrial outer membrane protein MitoNEET [2Fe-2S] clusters by biological thiols and hydrogen peroxide. J Biol Chem 289:4307–4315

Landry AP, Cheng Z, Ding H (2015) Reduction of mitochondrial protein mitoNEET [2Fe-2S] clusters by human glutathione reductase. Free Radic Biol Med 81:119–127

Lee JW, Helmann JD (2006) The PerR transcription factor senses H_2O_2 by metal-catalysed histidine oxidation. Nature 440:363–367

Lee C, Lee SM, Mukhopadhyay P et al (2004) Redox regulation of OxyR requires specific disulfide bond formation involving a rapid kinetic reaction path. Nat Struct Mol Biol 11:1179–1185

Lee KC, Yeo WS, Roe JH (2008) Oxidant-responsive induction of the suf operon, encoding a Fe-S assembly system, through Fur and IscR in *Escherichia coli*. J Bacteriol 190:8244–8247

Lee JH, Lee KL, Yeo WS et al (2009) SoxRS-mediated lipopolysaccharide modification enhances resistance against multiple drugs in *Escherichia coli*. J Bacteriol 191:4441–4450

Lee D, Park S, Bae S et al (2015) Hydrogen peroxide-activatable antioxidant prodrug as a targeted therapeutic agent for ischemia-reperfusion injury. Sci Rep 5:16592

Lemke RA, Peterson AC, Ziegelhoffer EC et al (2014) Synthesis and scavenging role of furan fatty acids. Proc Natl Acad Sci U S A 111(33):E3450–E3457

Li Y, Qiu Y, Gao H et al (2009) Characterization of Zur-dependent genes and direct Zur targets in *Yersinia pestis*. BMC Microbiol 9:1471–2180

Lin CN, Syu WJ, Sun WS et al (2010) A role of ygfZ in the *Escherichia coli* response to plumbagin challenge. J Biomed Sci 17(1):84

Lisher JP, Tsui HC, Ramos-Montañez S et al (2017) Biological and chemical adaptation to endogenous hydrogen peroxide production in *Streptococcus pneumoniae* D39. mSphere 2:00291–00216

Liu YZ, Imtiaz M, Ditta A et al (2020) Response of growth, antioxidant enzymes and root exudates production towards As stress in *Pteris vittata* and *Astragalus sinicus* colonized by arbuscular mycorrhizal fungi. Environ Sci Pollut Res 27:2340–2352. https://doi.org/10.1007/s11356-019-06785-5

Luo L, Qi MS, Yao SY et al (2005) Role of oxyR from *Sinorhizobium meliloti* in regulating the expression of catalases. Acta Biochim Biophys Sin 37:421–428

Mahavihakanont A, Charoenlap N, Namchaiw P et al (2012) Novel roles of SoxR, a transcriptional regulator from *Xanthomonas campestris*, in sensing redox-cycling drugs and regulating a protective gene that have overall implications for bacterial stress physiology and virulence on a host plant. J Bacteriol 194:209–217

Marinho HS, Real C, Cyrne L (2014) Hydrogen peroxide sensing, signaling and regulation of transcription factors. Redox Biol 2:535–562

Martin RG, Rosner JL (2011) Promoter discrimination at class I MarA regulon promoters mediated by glutamic acid 89 of the MarA transcriptional activator of *Escherichia coli*. J Bacteriol 193:506–515

Mehmood S, Saeed DA, Rizwan M et al (2018) Impact of different amendments on biochemical responses of sesame (*Sesamum indicum* L.) plants grown in lead-cadmium contaminated soil. Plant Physiol Biochem 132:345–355. https://doi.org/10.1016/j.plaphy.2018.09.019

Mielecki D, Grzesiuk E (2014) Ada response: a strategy for repair of alkylated DNA in bacteria. FEMS Microbiol Lett 355:1–11

Milackova I, Rackova L, Majekova M et al (2015) Protection or cytotoxicity mediated by a novel quinonoid-polyphenol compound? Gen Physiol Biophys 34:51–64

Mosel M, Li L, Drlica K et al (2013) Superoxide-mediated protection of *Escherichia coli* from antimicrobials. Antimicrob Agents Chemother 57:5755–5759

Mulrooney SB, Hausinger RP (2003) Nickel uptake and utilization by microorganisms. FEMS Microbiol Rev 27:239–261

Munnoch JT, Martinez MT, Svistunenko DA et al (2016) Characterization of a putative NsrR homologue in *Streptomyces venezuelae* reveals a new member of the Rrf2 superfamily. Sci Rep 6(1):1–4

Nakayama T, Yonekura S, Yonei S et al (2013) *Escherichia coli* pyruvate:flavodoxin oxidoreductase, YdbK: regulation of expression and biological roles in protection against oxidative stress. Genes Genet Syst 88:175–188

Naseer N, Shapiro JA, Chander M (2014) RNA-Seq analysis reveals a six-gene SoxR regulon in *Streptomyces coelicolor*. PLoS One 9:e106181

Nelson KJ, Parsonage D, Hall A et al (2008) Cysteine pK(a) values for the bacterial peroxiredoxin AhpC. Biochemistry 47:12860–12868

Netto LE, Antunes F (2016) The roles of peroxiredoxin and thioredoxin in hydrogen peroxide sensing and in signal transduction. Mol Cells 39:65–71

Ni K, Steger K, Yang H et al (2016) A comprehensive investigation of sperm DNA damage and oxidative stress injury in infertile patients with subclinical, normozoospermic, and astheno/oligozoospermic clinical varicocoele. Andrology 4:816–824

Niamat B, Naveed M, Ahmad Z et al (2019) Calcium-enriched animal manure alleviates the adverse effects of salt stress on growth, physiology and nutrients homeostasis of *Zea mays* L. MDPI-Plants 8(11):480. https://doi.org/10.3390/plants8110480

Niviere V, Fontecave M (2004) Discovery of superoxide reductase: an historical perspective. J Biol Inorg Chem 9:119–123

Outten FW, Djaman O, Storz G (2004) A suf operon requirement for Fe-S cluster assembly during iron starvation in *Escherichia coli*. Mol Microbiol 52:861–872

Ozyigit II, Filiz E, Vatansever R et al (2016) Identification and comparative analysis of H_2O_2-scavenging enzymes (ascorbate peroxidase and glutathione peroxidase) in selected plants employing bioinformatics approaches. Front Plant Sci 7:301

Palma M, Zurita J, Ferreras JA et al (2005) *Pseudomonas aeruginosa* SoxR does not conform to the archetypal paradigm for SoxR-dependent regulation of the bacterial oxidative stress adaptive response. Infect Immun 73:2958–2966

Pannala VR, Dash RK (2015) Mechanistic characterization of the thioredoxin system in the removal of hydrogen peroxide. Free Radic Biol Med 78:42–55

Parent A, Caux-Thang C, Signor L et al (2013) Single glutamate to aspartate mutation makes ferric uptake regulator (Fur) as sensitive to H_2O_2 as peroxide resistance regulator (PerR). Angew Chem Int Ed Engl 52:10339–10343

Passaia G, Margis-Pinheiro M (2015) Glutathione peroxidases as redox sensor proteins in plant cells. Plant Sci 234:22–26

Patra KC, Hay N (2014) The pentose phosphate pathway and cancer. Trends Biochem Sci 39:347–354

Peskin AV, Dickerhof N, Poynton RA et al (2013) Hyperoxidation of peroxiredoxins 2 and 3: rate constants for the reactions of the sulfenic acid of the peroxidatic cysteine. J Biol Chem 288:14170–14177

Pisoschi AM, Pop A (2015) The role of antioxidants in the chemistry of oxidative stress: a review. Eur J Med Chem 97:55–74

Pomposiello PJ, Koutsolioutsou A, Carrasco D et al (2003) SoxRS-regulated expression and genetic analysis of the yggX gene of *Escherichia coli*. J Bacteriol 185:6624–6632
Poole LB (2007) The catalytic mechanism of peroxiredoxins. Subcell Biochem 44:61–81
Pullan ST, Gidley MD, Jones RA et al (2007) Nitric oxide in chemostat-cultured Escherichia coli is sensed by Fnr and other global regulators: unaltered methionine biosynthesis indicates lack of S nitrosation. J Bacteriol 189:1845–1855
Rahal A, Kumar A, Singh V et al (2014) Oxidative stress, prooxidants, and antioxidants: the interplay. Biomed Res Int 761264:23
Rai P, Cole TD, Wemmer DE et al (2001) Localization of Fe($^{2+}$) at an RTGR sequence within a DNA duplex explains preferential cleavage by Fe($^{2+}$) and H_2O_2. J Mol Biol 312:1089–1101
Rai P, Parrish M, Tay IJ et al (2015) *Streptococcus pneumoniae* secretes hydrogen peroxide leading to DNA damage and apoptosis in lung cells. Proc Natl Acad Sci U S A 112(26):E3421–E3430
Rau J, Stolz A (2003) Oxygen-insensitive nitroreductases NfsA and NfsB of *Escherichia coli* function under anaerobic conditions as lawsone-dependent Azo reductases. Appl Environ Microbiol 69:3448–3455
Rima D, Shiv BK, Bhavna C et al (2016) Oxidative stress induced damage to paternal genome and impact of meditation and yoga: can it reduce incidence of childhood cancer? Asian Pac J Cancer Prev 17:4517–4525
Sabir A, Naveed M, Bashir MA et al (2020) Cadmium mediated phytotoxic impacts in *Brassica napus*: managing growth, physiological and oxidative disturbances through combined use of biochar and *Enterobacter* sp. MN17. J Environ Manag 265:110522. https://doi.org/10.1016/j.jenvman.2020.110522
Sainsbury S, Ren J, Nettleship JE et al (2010) The structure of a reduced form of OxyR from *Neisseria meningitidis*. BMC Struct Biol 10:1472–6807
Sakhtah H, Koyama L, Zhang Y et al (2016) The *Pseudomonas aeruginosa* efflux pump MexGHI-OpmD transports a natural phenazine that controls gene expression and biofilm development. Proc Natl Acad Sci U S A 113(25):E3538–4E3547
Sarfraz R, Hussain A, Sabir A et al (2019) Role of biochar and plant growth-promoting rhizobacteria to enhance soil carbon sequestration: a review. Environ Monit Assess 191:251. https://doi.org/10.1007/s10661-019-7400-9
Schiaffino S, Reggiani C, Kostrominova TY et al (2015) Mitochondrial specialization revealed by single muscle fiber proteomics: focus on the Krebs cycle. Scand J Med Sci Sports 4:41–48
Seaver LC, Imlay JA (2001) Hydrogen peroxide fluxes and compartmentalization inside growing *Escherichia coli*. J Bacteriol 183:7182–7189
Seth D, Hausladen A, Wang YJ et al (2012) Endogenous protein S-nNitrosylation in *E. coli*: regulation by OxyR. Science 336:470–473
Shah AM, Zhao Y, Wang Y et al (2014) A Mur regulator protein in the extremophilic bacterium *Deinococcus radiodurans*. PLoS One 9(9):e106341
Shahzad H, Ullah S, Iqbal M et al (2019) Effects of salinity sources and levels on growth, physiology and nutrient contents of maize crop. Ital J Agron 14:199–207. https://doi.org/10.4081/ija.2019.1326
Sheplock R, Recinos DA, Mackow N et al (2013) Species-specific residues calibrate SoxR sensitivity to redox-active molecules. Mol Microbiol 87:368–381
Shi H, Huang G, Xia D et al (2015) Role of in situ resultant $H_{(2)}O_{(2)}$ in the visible-light-driven photocatalytic inactivation of *E. coli* using natural sphalerite: a genetic study. J Phys Chem B 119:3104–3111
Shimberg GD, Michalek JL, Oluyadi AA et al (2016) Cleavage and polyadenylation specificity factor 30: an RNA-binding zinc-finger protein with an unexpected 2Fe-2S cluster. Proc Natl Acad Sci U S A 113:4700–4705
Shin JH, Singh AK, Cheon DJ et al (2011) Activation of the SoxR regulon in *Streptomyces coelicolor* by the extracellular form of the pigmented antibiotic actinorhodin. J Bacteriol 193:75–81
Singh AK, Shin JH, Lee KL et al (2013) Comparative study of SoxR activation by redox-active compounds. Mol Microbiol 90:983–996

Sobota JM, Gu M, Imlay JA (2014) Intracellular hydrogen peroxide and superoxide poison 3-deoxy-D-arabinoheptulosonate 7-phosphate synthase, the first committed enzyme in the aromatic biosynthetic pathway of *Escherichia coli*. J Bacteriol 196:1980–1991

Song Y, Huang H, Chen Y et al (2013) Cytotoxic and antibacterial marfuraquinocins from the deep South China Sea-derived *Streptomyces niveus* SCSIO 3406. J Nat Prod 76:2263–2268

Sun H, Li M, Xu G et al (2012) Regulation of MntH by a dual Mn(II)- and Fe(II)-dependent transcriptional repressor (DR2539) in *Deinococcus radiodurans*. PLoS One 7(4):e35057

Svintradze DV, Peterson DL, Collazo-Santiago EA et al (2013) Structures of the *Porphyromonas gingivalis* OxyR regulatory domain explain differences in expression of the OxyR regulon in *Escherichia coli* and *P. gingivalis*. Acta Crystallogr D Biol Crystallogr 69:2091–2103

Tanner JR, Patel PG, Hellinga JR et al (2017) *Legionella pneumophila* OxyR is a redundant transcriptional regulator that contributes to expression control of the two-component CpxRA system. J Bacteriol 19:00690–00616

Teramoto H, Inui M, Yukawa H (2013) OxyR acts as a transcriptional repressor of hydrogen peroxide-inducible antioxidant genes in *Corynebacterium glutamicum* R. FEBS J 280:3298–3312

Terzidis MA, Chatgilialoglu C (2015) An ameliorative protocol for the quantification of purine 5′,8-cyclo-2′-deoxynucleosides in oxidized DNA. Front Chem 3:47

Terzidis MA, Prisecaru A, Molphy Z et al (2016) Radical-induced purine lesion formation is dependent on DNA helical topology. Free Radic Res 50:S91–S101

Uhl L, Dukan S (2016) Hydrogen peroxide induced cell death: the major defences relative roles and consequences in *E. coli*. PLoS One 11:e0159706

Ullah N, Ditta A, Khalid A et al (2020) Integrated effect of algal biochar and plant growth promoting rhizobacteria on physiology and growth of maize under deficit irrigations. J Soil Sci Plant Nutr 20:346–356. https://doi.org/10.1007/s42729-019-00112-0

Varghese S, Tang Y, Imlay JA (2003) Contrasting sensitivities of *Escherichia coli* aconitases A and B to oxidation and iron depletion. J Bacteriol 185:221–230

Wang Y, Newman DK (2008) Redox reactions of phenazine antibiotics with ferric (hydr)oxides and molecular oxygen. Environ Sci Technol 42:2380–2386

Wang Y, Kern SE, Newman DK (2010) Endogenous phenazine antibiotics promote anaerobic survival of *Pseudomonas aeruginosa* via extracellular electron transfer. J Bacteriol 192:365–369

Wang Y, Wilks JC, Danhorn T et al (2011) Phenazine-1-carboxylic acid promotes bacterial biofilm development via ferrous iron acquisition. J Bacteriol 193:3606–3617

Wang H, Naseer N, Chen Y et al (2017) OxyR2 modulates OxyR1 activity and *Vibrio cholerae* oxidative stress response. Infect Immun 30:00929–00916

Watanabe S, Kita A, Kobayashi K et al (2008) Crystal structure of the [2Fe-2S] oxidative-stress sensor SoxR bound to DNA. Proc Natl Acad Sci U S A 105:4121–4126

Yamamoto K, Ishihama A, Busby SJ et al (2011) The *Escherichia coli* K-12 MntR miniregulon includes dps, which encodes the major stationary-phase DNA-binding protein. J Bacteriol 193:1477–1480

Yang L, Yurkovich JT, Lloyd CJ et al (2016) Principles of proteome allocation are revealed using proteomic data and genome-scale models. Sci Rep 6:36734

Yeo WS, Lee JH, Lee KC et al (2006) IscR acts as an activator in response to oxidative stress for the suf operon encoding Fe-S assembly proteins. Mol Microbiol 61:206–218

Yesilkaya H, Andisi VF, Andrew PW et al (2013) *Streptococcus pneumoniae* and reactive oxygen species: an unusual approach to living with radicals. Trends Microbiol 21:187–195

Yu C, Wang N, Wu M et al (2016) OxyR-regulated catalase CatB promotes the virulence in rice via detoxifying hydrogen peroxide in *Xanthomonas oryzae* pv. *oryzae*. BMC Microbiol 16(1):269

Zeb H, Hussain A, Naveed M et al (2018) Compost enriched with ZnO and Zn-solubilizing bacteria improves yield and Zn-fortification in flooded rice. Ital J Agron 13(4):310–316. https://doi.org/10.4081/ija.2018.1295

Zhao Y, Pluth MD (2016) Hydrogen sulfide donors activated by reactive oxygen species. Angew Chem Int Ed Engl 55:14638–14642

Zhao Y, Nickels LM, Wang H et al (2016) OxyR-regulated catalase activity is critical for oxidative stress resistance, nodulation and nitrogen fixation in *Azorhizobium caulinodans*. FEMS Microbiol Lett 363:13

Zheng M, Wang X, Templeton LJ et al (2001) DNA microarray-mediated transcriptional profiling of the *Escherichia coli* response to hydrogen peroxide. J Bacteriol 183:4562–4570

Zheng JY, Tan HL, Matsudaira PT et al (2017) Excess reactive oxygen species production mediates monoclonal antibody-induced human embryonic stem cell death via oncosis. Cell Death Differ 20:164

Zou Y, Wang J, Peng J et al (2016) Oregano essential oil Induces SOD1 and GSH expression through Nrf2 activation and alleviates hydrogen peroxide-induced oxidative damage in IPEC-J2 cells. Oxidative Med Cell Longev 5987183:26

Zuber P (2009) Management of oxidative stress in *Bacillus*. Annu Rev Microbiol 63:575–597

Chapter 13
Na^+ Sensing, Transport, and Plant Salt Tolerance

Aniqah Akhter, Gulnaz Bibi, Nabgha Rasti, Hira Rasheed, Zainab Noor, and Jamshaid Hussain

13.1 Introduction

13.1.1 Na^+ Influx

A flux can be defined as the movement of substances through an area in unit time (Maathuis 2006). The movement of sodium ions along the membrane is known as sodium influx, which occurs in almost all conditions because the charge inside the cell is −150–200 mV. Na^+ concentration in soil increases during salinity that mediates Na^+ diffusion into the cytoplasm through generating an electrochemical difference (Blumwald 2000).

Sodium enters the plants through ion channels, i.e., voltage-dependent and voltage-independent channels (VICs) and carrier type transporters (Table 13.1). The former type includes inward and outward rectifying channels, while the latter type is the main route of Na^+ entry during higher NaCl concentrations. Voltage-dependent channels consist of AKT1 (inward rectifying channel) type transporters, which increase K^+ uptake on hyperpolarization of plasma membrane (Sentenac et al. 1992). Nonselective outward rectifying channels (NORCs) involved in Na^+ uptake through activation of internal Ca^{2+} concentration (Wegner and Boer 1997). VICs also known as NSCCs (nonselective cation channels) are selective for Na^+ than K^+ (Tyermann et al. 1997), which constitutes the main route of Na^+ entry during high salinity (Amtmann and Sanders 1998). NSCCs are Ca^{2+} sensitive channels, and genetically their nature remains unclear yet. NSCCs are further divided into CNGCs and GLRs. A number of transporters are involved in Na^+ influx and movement through plasma membrane including AKT, HKT, NSSC, PIP, NHA, LCT, and

A. Akhter · G. Bibi · N. Rasti · H. Rasheed · Z. Noor · J. Hussain (✉)
Department of Biotechnology, COMSATS University Islamabad, Abbottabad, Pakistan
e-mail: jamshaidhussain@cuiatd.edu.pk

Q. Mahmood (ed.), *Sustainable Plant Nutrition under Contaminated Environments*, Sustainable Plant Nutrition in a Changing World,
https://doi.org/10.1007/978-3-030-91499-8_13

Table 13.1 Function of some sodium transporters and ion channels in plants

Name	Species	Function	Reference
HKT	*Triticum turgidum, Arabidopsis thaliana, Oryza sativa*	Na+ uptake in roots and recovery in xylem	Mäser et al. (2002), Hori et al. (2007), Wang et al. (2015)
NHX	*Oryza sativa, Arabidopsis thaliana, Brassica napus, Nicotiana tabacum, Gossypium hirsutum, Cynodon dactylon, Triticum aestivum*	Na+ compartmentation in vacuole	Zhang et al. (2001), Roy and Chakraborty (2018), Yokoi et al. (2002)
CNGCs	*Oryza sativa, Arabidopsis thaliana*	Na+ influx into roots and unidirectional flow	Gobert et al. (2006), Essah et al. (2003)
SOS1	*Suaeda salsa, Cynodon dactylon, Arabidopsis thaliana*	Na extrusion from roots	Roy and Chakraborty (2018), Liu et al. (2018)
HAK	*Phragmites australis, Arabidopsis thaliana*	Na+ uptake from roots in low-affinity range and Na+ selective conductance	Apse et al. (1999), Wang et al. (2015)
Ouabain-sensitive Na-K ATPase	*Zea mays, Helianthusannuus, Dacus carota, Hordium vulgare*	Na+ extrusion from roots	Mukherjee and Bhatla (2014)
VI-NSCC (voltage-insensitive NSCC)	*Secale cereal, Lotus japonicus, Clivia, Pisum sativum, Triticum aestivum, Arabidopsis thaliana, Haemanthus, Hordeum vulgare,*	Na+ entry into roots and Na+ selective conductance	Maaathuis and Sanders (2001), Demidchik and Tester (2002), Essah et al. (2003), Tyerman et al. (1997)
GLRs	*Arabidopsis thaliana*	Na+ influx into roots and Na+ selective conductance	Tapken and Hollmann (2008), Plett and Moller (2010)

HAK. Byrt et al. (2017) reported recently the role of aquaporins in plants in Na^+ influx (AtPIP2;1).

Ion movement in plants occurs through apoplastic and symplastic pathway, and the former pathway is direct in xylem and its outside (Maathuis 2014; Krishnamurthy et al. 2009). As a result, this bypass flux can only contribute to less than 1% of transpirational stream volume (Yeo 1981). In *Arabidopsis* and many other plants as well, symplastic mechanism of Na^+ influx occurs through a system of transporters, but their physiological role and contribution remains uncertain (Ranathunge and Schriber 2011). In rice, 50% of the total Na^+ influx is created through apoplastic mechanism (Krishnamurthy et al. 2009). In other species, mostly Na^+ uptake is apoplastic in nature, and additionally 50% Cl^- movement takes place via this pathway (Petrsonet al. 1986; Shi et al. 2013). In monocots, especially in rice, Na^+ and Cl^- uptake is relevant with each other, whereas in *Arabidopsis* it is not much different in solute transport suggesting that apoplastic pathway needs to be studied further.

In the roots of higher plants, Na^+ influx occurs through HKTs, NSCCs/VICs, HAK/KUP/KT, CCC, and AKT members of transport family.

13.1.2 NSSCs/VICs

NSCCs not only function in Na^+ influx but also mediate K^+ efflux directly or through K^+ outward rectifying channels (KORGs). During salinity, NSCCs inhibit Na^+ entry into the cytoplasm by vacuolar compartmentation mechanism by activating KORG, which increases function of cell death enzymes (Demidchik et al. 2010). VICs are more selective for Na^+ than K^+, and plasma membrane permeability value is 1 for K^+ than Na^+. The order of selectivity for monovalent ions for NSSCs is NH > Rb^+ < C s^+ > Na^+ > Li^+ > TEA. Therefore, VICs are the prominent mechanism of Na^+ entry into roots under increased saline habitats, while VIC's blockage increases salt stress in some cases. Furthermore, it is studied that 30–50% of NSSCs mediated flux is reduced by 0.2–0.4 mM Ca^{2+} concentration in apoplastic region in root cells (Legué et al. 1997; Demidchick and Mathuis 2007; Essah et al. 2003). In monocots, they contain multiple HKT isoforms as compared to *Arabidopsis* which have AtHKT1 isoform; hence, monocot HKTs can mediate Na^+ uptake while AtHKT1 transport Na^+ through xylem and phloem (Berthomieu et al. 2003; Byrt et al. 2007; Sunrapi et al. 2005). In wheat TaHKT2:1, Na^+ accumulation is affected in low-affinity range (Laurie 2002). Na^+ entry does not depend on other inhibitors such as Cs, verapamil, flufenamate, TEA, and Quinine (Davenport and Tester 2000). Quinine partly blocked NSCCs but not all. Surprisingly in rice in low Na^+ concentration, i.e., less than 2 mM and decreased K^+ concentration, Na^+ is influxed by OsHKT2:1 type channels efficiently (Hori et al. 2007). Furthermore, in *Arabidopsis* (AtPIP2:1) intact root tissues, Na^+ uptake is 0.1–1 mol m^{-2} s^{-1} (Ahmed et al. 2015), while in oocytes PIP2:1 Na^+ fluxes were in the range of 0.2–0.5 μmol m^{-2} s^{-1}, which is six times smaller than AtPIP2:1 (Byrt et al. 2017). Therefore, Na^+ influx in intact tissues through aquaporins was not measurable significantly. Interestingly, in comparison of glycophytes with halophytes such as *Suaeda maritima*, studies suggested that 5–10 mM Ca^{2+} did not show considerable effect on Na^+ uptake, whereas at 150 mM NaCl, the inhibitors (Cs^+ and TEA) decrease Na^+ influx (Wang et al. 2007). Concludingly this difference in influx in glycophytes with that of halophytes was not due to NSCCs but may be salt resistant and evolutionary background of these in planta species are the main reasons hence, further studies may be required (Wang et al. 2007).

Adams (2014) reported about 20 GLRs and CNGCs in *Arabidopsis* in which AtCNGC 3 functions in the transport of Na^+ in root epidermal and cortex regions. However, Na^+ influx is decreased at initial salinity stages, while long-term salt stress has negligible effect on mutant and wild-type *Arabidopsis* plants that confirmed the role of AtCNGC3 in Na^+ influx in salinity conditions (Gobert et al. 2006).

Maathuis and Sander (2001) suggested that NSCCs were blocked by CNGCs (cGMP/cAMP) to make plant salt tolerant, i.e., AtCNGC1,2 decreases Na^+ entry, but this study was opposed by Lengin (2002) indicating that these channels were not included in NSSCs and have different selectivity described by Maathuis and Sander (selective for K^+ over Na^+). Therefore, VICs/NSSCs in some plants work for Na^+ influx, while Ca^{2+} signaling occurs via CNGCs.

GLRs are activated by Ca^{2+}-and Na^+-dependent glutamate ion channels in Na^+ uptake in *Arabidopsis* plant (AtGLR1.1 and AtGLR1.4) and Xenopus oocytes (Demidchik et al. 2004).

13.1.3 HKTs

High-affinity K^+ transporter type are present in fungi, bacteria, and plants and have two major subgroups on the basis of Na^+ and K^+ transport. Subfamily 1 of HKTS is specific for Na^+ influx when K^+ is not present in outside of the cell while HKT subgroup 2 involves in co-transport of Na^+–K^+ (Deinlein et al. 2014).

Rus et al. (2001) studied HKT1:1 in *Arabidopsis*, which regulates Na^+ influx in roots showing salt-tolerant mechanism in plants. Na^+ uptake also decreased in wheat tissues in saline conditions due to decreased expression of TaHKT2:1 (Laurie et al. 2002). Pharmacological studies of HKT transporters showed that they are resistant to Cs^+ and TEA while sensitive to Ba^+; therefore, under 25 mM NaCl, Na^+ influx is reduced by Ba^+, but Cs^+ and TEA have negligible effect on *S. maritima* (Liu et al. 2001; Garciadeblas et al. 2003). Therefore, under low salt conditions, HKTs' ability to transport sodium also decreases in *S. maritima.*

13.1.4 Na⁺ Efflux

Within the cell, maintenance of ions takes place through ion pumps including antiporters, carriers, and symporters. Cytoplasmic Na^+ extrusion also occurs through vacuolar Na^+/H^+ exchanger that removes harmful ions from cytoplasm into vacuole to maintain ion homeostasis during salinity stress condition. In high salt situations, Na^+ efflux is mediated by high energy proton motive forces, i.e., H-pump ATPases, high-affinity K^+-transporters, and Na^+/H^+ antiporters in higher plants. Cytosolic Na^+ can be excluded by ATP-dependent Na^+ channels in algae. In higher plants, proton motive force is provided by pyrophosphatase (PPiase), triphosphates, and translocating enzymes for Na^+ extrusion through vacuole and cytosol. Na^+ exclusion requires expenditure of energy in fungi. But such type of pumps (Na^+ ATPase pumps) could not be found in land plants as described in *Arabidopsis*. Salt tolerance may be elevated in fungal root cortex and epidermal cells by ATPase-dependent Na^+ efflux channels, but their selectivity for Na^+/K^+ is yet to be determined.

Rice and tobacco roots also showed increase activity of H^+-ATPases in cell membranes, which maintain the activity of Na^+/H^+ antiporters through maximum H^+-efflux (Niu et al. 1996; Zhang et al. 1999). It is described that at neutral pH, electroneutral Na^+/H^+ exchangers will be blocked in high salinity conditions. This will be determined by experimentally through stoichiometric calculations of Na^+ efflux at metabolic cost of 50%:50%, i.e., 1:1 exchange of Na^+ and H^+. For example, one Na^+ ion will be excluded from the plasma membrane at the expense of 1 molecule of ATP used (Briskin et al. 1991, 1995). As compared to experimental calculation, Na^+/H^+ exchange requires higher amount of energy than one Na^+ ion extrusion determined theoretically, which depends variably on intracellular transport, influx rate, and salt stress conditions. Therefore, Na^+ efflux from roots requires much expensive procedure; hence, Na^+ compartmentation into vacuole is an economical way like vacuolar H^+ pumps (Rea and Sanders 1987). Largely in plant tissues, Na^+ is removed into the apoplast through root-soil areas. This removal is mediated by SOS1 family of Na^+/H^+ antiporters. SOS1 gene family is exclusively important in xylem parenchyma and root tip cells (Wu et al. 1996). Therefore, these tissues completely depend on Na^+ extrusion from cytosol into apoplast through SOS1 because they lack vacuolar organization, and as a result compartmentation of Na^+ does not take place in such tissues. Additionally, most of these tissues yet not described which antiport isoform is mainly involved in this extrusion because of SOS1 gene expression lacking studies. Because they only contain one or two Na^+/H^+ antiport isoform, other transporters are also involved in extrusion.

Cation H^+ exchangers (CHXs) are also involved in K^+/H^+ exchange. Maathuis and Sander (2001) suggested Na^+ influx is 10 times higher in CHX mechanism than Na^+ efflux through Na^+/H^+ antiporter, i.e., 90% Na^+ is removed from cytoplasm to the outside of the cell membrane than Na^+ influx. In *Arabidopsis* 28 CHX members have been identified (Hall et al. 2006), whereas only AtCHX21 member mediates Na^+ flux. This member is present only in root endodermal cells; hence, it does not involve in Na^+ extrusion, while AtCHX17 involves in K^+/H^+ transport in outer root cells (Cellier et al. 2004). Remaining CHX family member identification may unhide the mechanism of Na^+ extrusion from root to the soil boundary.

The most important pathway for Na extrusion is a member of three protein system (SOS pathway) that provides salt stress tolerance and ion homeostasis in most of the plants (Ma et al. 2014; Kalschnig et al. 2015). This pathway works in the following sequence. First of all, a Ca^+-binding protein (SOS_3) on the cell membrane senses Ca^+ signals and binds to SOS_2 to make SOS_3-SOS_2 complex as a result phosphorylating SOS_1 on the surface of the cell membrane to exclude Na^+ with the exchange of H^+ (Gong et al. 2004).

SOS mechanism of Na^+ extrusion takes place both in root and shoot of the plant cell but in shoot SOS_3-like Ca^+-binding protein is CBL10, whereas in root it is CBL4 (Quan et al. 2007) in mutant plant with difference in ScaBP sensor differences. In effect of salt stress response, ScaBP81 CBL10 (SOS3-Homolog) recruits SOS2, which then binds to SOS1 to the cell membrane. A flowering gene (GIGANTEA) GI inactivates SOS2 binding with SOS3. GI is broken and facilitates SOS2-SOS3 interaction by activating SOS1 to increase salt tolerance mechanism in plants (Kim et al. 2013).

In *Arabidopsis*, root cap and xylem cells show increased SOS1 gene expression in NaCl stress, while AtSOS1 has increased Na^+ in root and shoot cells (Davenport et al. 2007). As a result, Na^+ concentration decreases in shoots and xylem due to increased gene expression of SOS1. Therefore, in mutants, increased Na^+ concentration in shoot and root confirmed SOS1 role in Na^+ efflux in plants (Shi et al. 2003).

13.2 Na^+ Loading on Xylem

Plants need adequate supply of water and nutrients for their normal growth and development. Both water and nutrients are provided to the plants from soil through xylem. Xylem takes nutrients and streams them up to the areal parts of the plant.

Among other nutrients, Na^+ is a major nutrient which is taken up through xylem. For this purpose, Na^+ enters the root hair, crosses epidermis, moves through cortical cells, crosses endodermic cells, and finally reaches to xylem for travelling toward its final destiny, that is, upper plant parts. This process is called xylem loading for Na^+ ions.

Xylem loading of Na^+ is a beneficial process especially in C4 plants to fulfill demands of plant for Na^+ in normal condition, but in saline state the concentration of Na^+ is very high in the soil so it can prove injurious for plant: as xylem will circulate large quantity of Na^+ to all parts of the plant, these concentrations may reach to toxic levels. Hence, it is very important to prevent bulk of Na^+ loading on xylem to make plant more tolerant against salinity stress.

Many transporters are known to play a role in loading of Na^+ on xylem such as SOS1 (salt overly sensitive 1), which transports Na^+/H^+ in opposite direction, CCC (chloride cation co-transporter), and ion transport channels such as NSCC (Zhu et al. 2017). Besides these, CHX transporters are also reported to be involved in loading of Na^+ on xylem (Fig. 13.1).

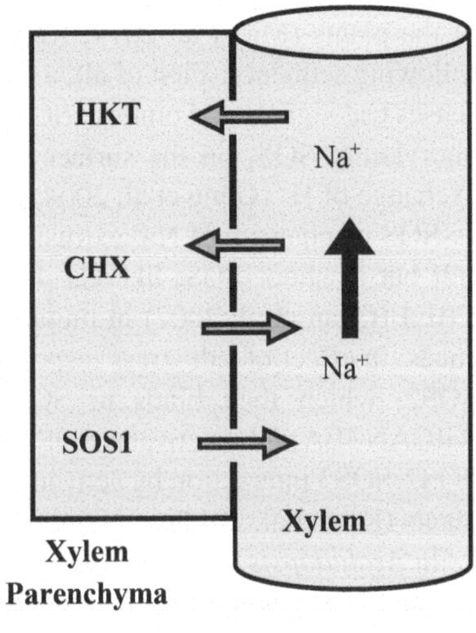

Fig. 13.1 Loading of Na^+ on xylem from parenchyma through various transporters. Xylem parenchyma are living cells which are involved in active and passive transport of Na^+ on apoplast of xylem and lumen of xylem (nonliving tissues) (Maathuis et al. 2014)

13.2.1 Loading of Na^+ on Xylem Through Active Transport

The active loading of xylem is furnished with the help of Na^+/H^+ antiporters. Active transport may be required when there is a decrease in transpiration (high Na^+ accumulation) or in old or mature roots where large quantity of Na^+ alongside of the pathway of xylem is present. SOS1 is presumed very crucial to accomplish the role of active Na^+ loading on xylem. Besides SOS1, many other transporters such as CHX21 transporters are also involved in loading of Na^+ on xylem through active transport.

13.2.2 Role of SOS1 in Na^+ Loading on Xylem

SOS1, which is a membrane-bound Na^+/H^+ exchanger, is expressed in various parts of the plant including xylem parenchyma. When stress is pretty high, it functions to unload Na^+ from xylem, but on the contrary, it helps to load Na^+ on xylem in a condition when plants face moderate level of salt stress. For instance, in *Arabidopsis* it was observed that when Sos1-mutant plants – in which Na^+/H^+ exchanger system was disabled – were subjected to a moderate level of salt stress (40 mM) showed less amount of Na^+ in areal parts, opposing to wild type that showed handsome loading of Na^+ on xylem (Munns et al. 2008; Zhu et al. 2016). This experiment depicts that SOS1 has a significant role in xylem loading of Na^+.

The activation of SOS1 takes place through systematic pathway. It was observed in *Arabidopsis thaliana* that at first the level of Ca^{++} increases in cytosol under salt stress. Then *At*SOS3, which is a calcium-binding protein, senses this intensification of Ca^{++} in the cytosol. Later a physical interaction occurs between this *At*SOS3 and AtSOS2 (serine/threonine protein kinase). The phosphorylation of this kinase complex formed by *At*SOS3 and *At*SOS2 results in the activation of Na^+/H^+ antiporter that is present on the plasma membrane and is coded by SOS1 gene (Ren et al. 2013).

13.2.3 Role of CHX21 in Na^+ Loading on Xylem

CHX21 is a cation exchanger, and it also helps in loading of Na^+ on xylem. The function of this gene was studied in *Arabidopsis thaliana* where its major expression was seen in root endodermis. It was observed that on the loss of function of AtCHX21 in the roots leads to decrease in level of Na^+ in xylem sap of *Arabidopsis thaliana*, but nothing happened to concentration of Na^+ in phloem. On the other hand, it was also observed to be involved in K^+ homeostasis (Evans et al. 2012).

Another experiment was conducted on *Arabidopsis,* in which it was observed that chx21 mutant plants showed substantial decrease in xylem Na^+ content. In plants that were not given salt stress treatment, the concentration of Na^+ on xylem decreased from 0.45 to 0.11 mM, while in plants that were subjected to 50 mM salt stress treatment, the concentration of Na^+ decreased from 2.45 to 1.45 mM in the xylem of the plant (Hall et al. 2006).

13.2.4 Passive Loading of Na^+ on Xylem

Passive loading of xylem is achieved through ion channels that are permeable for Na^+ and are located on the boundary between parenchyma and xylem.

The Na^+ travels through the symplast through endodermis and then enters xylem apoplast from parenchyma cells of xylem. The release of Na^+ into xylem is dependent upon electrochemical gradient, which can fluctuate and may entertain passive loading of Na^+ where its concentration may be less and the cells which are loading Na^+ become depolarized. Many transporters such as NSCC and some uniporters or symporters can help to passively load Na^+ on xylem (Davenport et al. 2007).

13.2.5 Role of CCC Transporters in Na^+ Loading on Xylem

CCC is a symporter which shows expression at xylem/symplast border and is also involved in loading of Na^+ on xylem (Colmenero-Flore et al. 2007). A study was conducted on *Oryza sativa* which revealed the function of OsCCC1 gene in Na^+ loading on xylem. In this approach, the difference between OsCCC1 mutant, wild type, and two complementation lines (in which cDNA of OsCCC1 was complemented in mutant to restore function of mutated gene) was observed. The concentration of Na^+ was low in mutant type as compared to wild type and two complementation lines. This observation clarifies that CCC co-transporter is involved in loading of Na^+ on xylem (Zhi et al. 2016).

13.2.6 Role of HKT Transporters on Na^+ Unloading from Xylem

In contrast to transporters which are involved in loading of Na^+ on xylem, HKT transporters are those which control Na^+ unloading by retrieving it from xylem. It prevents movement of Na^+ on shoots. Kobayashi et al. (2017) conducted an experiment on *Oryza sativa* in which they observed that disabling function of OsHKT1;5 gene resulted into substantial rise of Na^+ in shoots under salinity. It was also found that OsHKT1;5 is located near basal stem and in cells which are adjoining to xylem in roots. Owing to straight insertion of $^{22}Na^+$ tracer in lamella of leaves, this experiment justified the role of OsHKT1;5 in Na^+ unloading from xylem. Moreover, it was found that OsHKT1;5 is also present in plasma membrane and in phloem of associated vascular bundles.

Another study also emphasized similar trends. It was specified that HKT family transporters such as AtHKT1 in *Arabidopsis* along with OsHKT8 in rice, both are responsible for desalting of xylem (Ketehouli et al. 2019).

13.3 Na^+ Recirculation from Shoot to Root via Phloem

Salinity affects overall plant parts, but leaves of the plants are much more sensitive to it than other parts. Nature has bestowed plants with an important mechanism which helps them to keep less Na^+ level in leaves. This mechanism involves recirculation of Na^+ from shoot through phloem sap back to the root so that it may not accumulate in the leaves and hence providing a shield against damages of salt stress (Munns 2002).

As observed in *Arabidopsis*, SOS1 present in the root apex, may expel most of the Na^+ entering the roots, back to the soil through plasma membrane (Qiu et al. 2003). The other Na^+ present in the roots may be shipped to shoots or sequestered into compartments with the help of NHX transporters which are Na^+/H^+ exchangers present on tonoplast. Membrane shows transport in two directions and Na^+ leaks back into the cytosol through NSCC present on tonoplast. The Na^+ is loaded on the xylem from stele, and it happens through passive transport in specific conditions and also through active transport with the help of SOS1 transporters. Afterward the passive unloading of Na^+ takes place in leaves. This sodium can be recirculated back to the roots from leaves through phloem (Fig. 13.2).

13.3.1 Role of HKT Family Transporters in Na^+ Recirculation Through Phloem

Although we do not have sufficient information about loading and unloading of Na^+ from phloem, it is proposed that HKT transporters such as AtHKT1;1 are involved in both of these processes. An et al. (2017) suggested the role of AtHKT1 in recirculation of Na^+ through phloem. It is said to be involved in loading of Na^+ on phloem in shoot region and its unloading to the region of root. In the higher parts of the plant, it unloads Na^+ from stream of sap of xylem and sends it back to the roots through phloem. This phenomenon prevents accumulation of Na^+ in leaves and provides plant protection against salinity stress.

Another study conducted on *Oryza sativa*, using $^{22}Na^+$ tracer, identified that OsHKT1;5 mutants (lost function) were detected with the presence of $^{22}Na^+$ in leaf blades of immature leaves, which suggests the role of this gene in protecting leaf blades from accumulating Na^+ by transferring this Na^+ to phloem for recirculation.

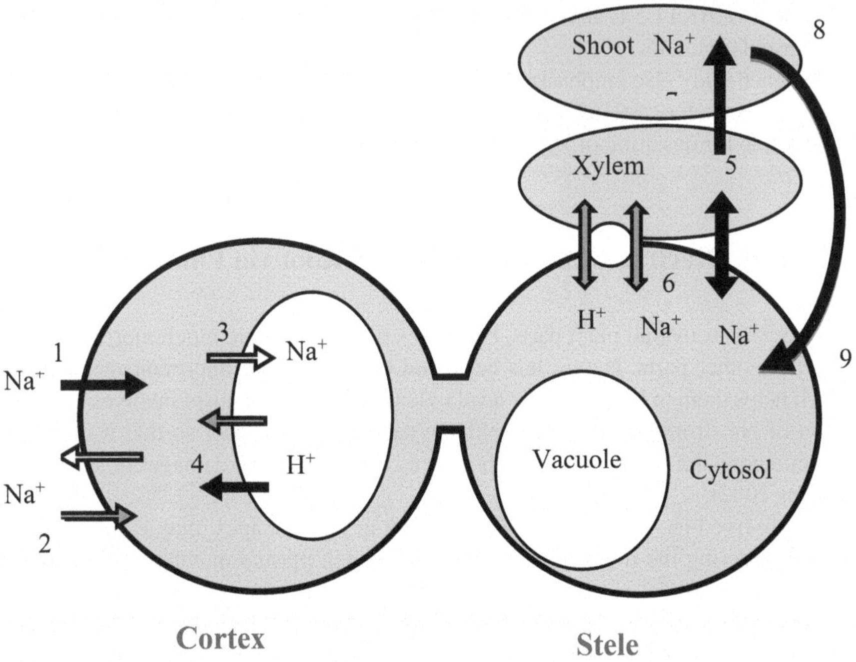

Fig. 13.2 A general model of Na^+ movement from roots to shoots and its recirculation through phloem under salt stress. Entrance of Na^+ into root cytosol chiefly via NSCC (1). Pumping out of the Na^+ from root to soil through SOS1 antiporter (2) or its vacuolization via NHX antiporters (3). The passive outflow of Na^+ from vacuole to cytosol (4). The passive loading of Na^+ in some conditions (5) and its active loading (via SOS1) from stele to xylem (6). Retrieval of Na^+ through similar processes in certain conditions such as decreased transpiration (5 and 6). Passive delivery of Na^+ into the cells of leaf (7), which can be subjected to recirculation back to roots via phloem (8 and 9)

This study also implied that OsHKT1;5 gene is also important to provide shield to next-generation seeds against salinity (Kobayashi et al. 2017).

13.4 Current Model of Sodium Transport

Meeting the challenge of saline soils can take the form of improvements to make our crops saline resistant as well as making improvements in water and land management as well (Tavakkoli et al. 2010). For both, clearly understanding the mechanisms involved in Na+ toxicity is of critical importance, and a major area of focus has been the of Na^+ transport regulation. For decades, researchers have been investigating the pathways of Na^+ transport, their regulatory mechanisms, and their distribution in plants, especially through root surface considered as their first point of contact. Na^+ uptake through the roots can be mediated through different pathways like (1) apoplastic pathway, by which Na^+ can diffuse through an intervening space of cell walls but without crossing the plasma membrane, (2) the symplastic pathway, by which Na^+ must cross the plasma membrane and to reach the cell cytoplasm, or (3) a combination of the apoplastic and symplastic (Gao et al. 2007; Negrão et al. 2011; Maathuis 2014). Recent detailed investigations of pathways leading to Na+ transport and its kinetics into barley (Hordeum vulgare L.) roots found that it had a biphasic character depending upon different sodium concentrations. When there is low Na^+ concentrations externally (1 mM), then Na+ is transported via a high-affinity transport system (HATS), a mechanism widely characterized by the saturable concentration dependence of influx (Britto and Kronzucker 2006). At higher concentrations (1 mM), Na+ transportation is mediated by a low-affinity transport system (LATS), which, unlike HATS, is characterized by a linear influx response to substrate concentration (Kochian et al. 1985; Britto and Kronzucker 2006). Because the problem of Na+ toxicity mainly occurs in the LATS range, it is important to understand how Na+ transport occurs in this phase. The unidirectional influx of Na^+ at high substrate concentrations has been reported to reach extraordinarily high rates in numerous studies and has been summarized in Fig. 13.3.

However, it remains unclear until now that how many Na+ transporters are involved in the LATS, and, after decades of intense investigation, there is still a lack of clarity. Although researchers still investigating the molecular mechanisms of Na+

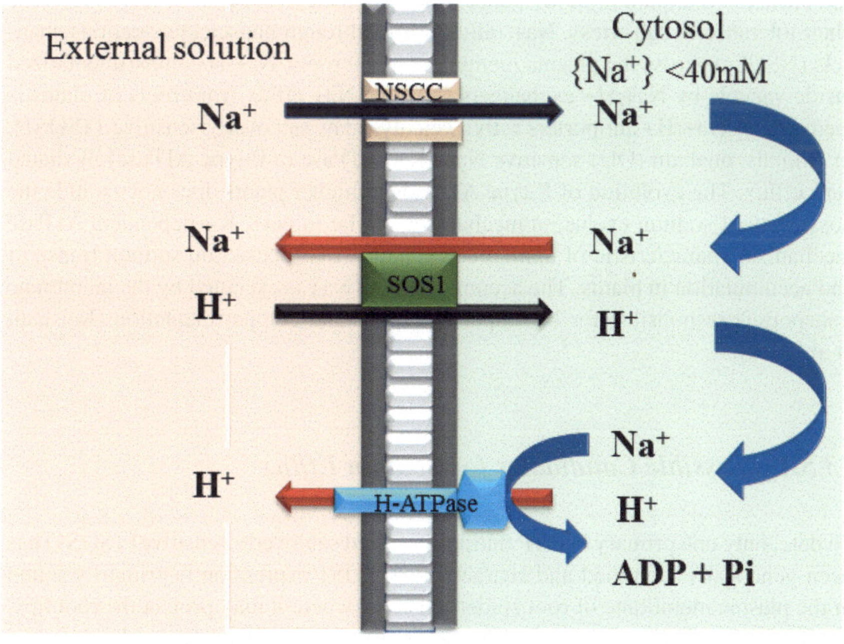

Fig. 13.3 The unidirectional influx of Na^+ at high substrate concentrations has been reported to reach extraordinarily high rates in numerous studies that rely on the use of very brief exposures of root material to radioactive Na^+

transport under salinity have produced conflicting results (Kronzucker and Britto 2011), a basic Na+ transport model via the LATS has been proposed, based mainly on the thermodynamics of unidirectional and net fluxes of Na+ as well as electrophysiological measurements. In this model, the Na+ enters across the plasma membrane of roots cells and into the cytoplasm passively down its electrochemical gradient (Cheeseman 1982), likely through ion channels such as nonselective cation channels (NSCCs), although other transporters may also be involved. The unidirectional influx of Na+ via HATS (high substrate concentrations) has been reported to reach extraordinarily high rates in numerous studies that rely on the use of very brief exposures of radioactive Na+ to root material (Davenport 1998; Essah et al. 2003; Malagoli et al. 2008; Møller et al. 2009; Wang et al. 2009; Jha et al. 2010).

13.5 Molecular Candidates for Transporters and Channels Catalyzing Na^+ Transport

To better understand Na+ toxicity, it is important to determine the transporters and channels at molecular level involved in Na+ transport (Horie and Schroeder 2004). Understanding the basic mechanism of sodium (Na+) influx, compartmentalization, and sodium efflux in higher plants. It is necessary to manipulate Na+ accumulation and ensure to maintain low Na+ concentration in the cytosol and, hence, to increase plant tolerance to salt stress. Na+ influx occurs through nonselective cations channels (NSCCs) across the plasma membrane in the roots. Na+ is compartmentalized inside vacuole by Na+/H+ exchangers (NHXs). Na+ efflux from roots of plants is mediated by Na+/H+ antiporters activity catalyzed by salt overly sensitive 1 (SOS1). In animals, ouabain (OU)-sensitive Na+, K+ -ATPase (a P-type ATPase) mediated Na+ efflux. The evolution of P-type ATPases in higher plants does not exclude the possibility of sodium exclusion mechanism similar to Na+, K+-dependent ATPase mechanism characteristic of animal cell. This part is focused on sodium transport and accumulation in plants. The accumulation of Na+ is executed by the membrane transporters responsible for Na+ uptake, export, and compartmentation (Keisham et al. 2018).

13.5.1 Possible Candidates for Sodium Efflux

To date, only one primary Na^+/H^+ antiporter called salt overly sensitive1 (SOS1) has been genetically identified and characterized. SOS1 expression is primarily found in the plasma membrane of root epidermal cells, where it may protect the root tips' meristematic region under salt stress, and in the xylem parenchyma, where it catalyzes the loading of Na^+ into the xylem stream.

13.5.2 The SOS Signaling Pathway and Na^+ Exclusion

The identification of salt overly sensitive (SOS) genes is an important progress in understanding of plant salt tolerance in the last two decades. Among these SOS genes, SOS1, SOS2, and SOS3 encode proteins that constitute the SOS signaling pathway (Zhu 2001). SOS1 encodes a plasma membrane Na^+/H^+ antiporter and thus reduces cytosolic Na^+ concentration (Shi et al. 2000, 2002, 2003), which uses the H^+ gradient to drive sodium efflux. While SOS2 is a protein kinase and can phosphorylate SOS1 (Liu et al. 2000; Quintero et al. 2002), under normal condition SOS2 is auto inhibitory kinase but its auto inhibition gets released when SOS3 binds with an auto inhibitory domain of SOS2 under salt stress condition (Guo et al. 2001). SOS3 is a Ca^{2+}-binding protein with three EF-hands (Liu and Zhu 1998). In the SOS signaling pathway, under salt stress the elevated Ca^{2+} level is sensed by SOS3, and the Ca^{2+}-bound SOS3 interacts with and activates SOS2. The N- terminus of SOS3 is myristoylated and associated with the plasma membrane, and thus the complex of SOS3-SOS2 is formed. This complex is recruited to the plasma membrane, where SOS2 leads to phosphorylation of SOS1. The Na^+ /H^+ antiporter activity of SOS1 is enhanced upon phosphorylation of SOS1. Thus, it reduces intracellular Na^+ concentration by Na^+ efflux transport through SOS1.

13.5.3 Role of SOS1 in Na Efflux

SOS1 ubiquitously exists in all sequenced genomes. Its unique feature is a SOS1 long cytosolic C-terminal region, which contains more than 700 amino acid residues in *Arabidopsis* (Shi et al. 2000). Another unique feature of SOS1 is that its C-terminal region possesses an auto inhibitory domain at the very end, which interacts with a putative cNMP-binding motif of the upstream sequences that inhibit the transport activity of SOS1 under normal conditions. SOS2 phosphorylates the serine residues in the auto inhibitory domain of SOS1, which causes activation of SOS1 upon salt stress (Quintero et al. 2011). The putative cNMP-binding motif in the SOS1 C-terminal region is necessary for SOS1 function, and mutations in this motif result in malfunctioning of SOS1 (Shi et al. 2000). This suggests that cyclic nucleotides might regulate SOS1 function; cyclic nucleotides are signaling molecules mediating environmental adaptation in plants in response to these stresses (Świeżawska et al. 2018). The SOS1 C-terminal region resembles its mammalian ortholog NHE1. The C-terminal part of NHE1 binds with many regulatory proteins, which regulate its antiport activity by phosphorylating them (Fliegel 2019). The SOS1 C-terminal part also binds with signaling protein RCD1 (Katiyar-Agarwal et al. 2006), which in turn interacts with many transcription factors and plays vital roles in plant stress responses (Kragelund et al. 2012). The cellular function of SOS1 is to remove Na^+ from the cytosol by transporting Na^+ into the extracellular space (Shi et al. 2002, 2003).

13.5.4 Role of SOS2 and SOS3 in Na Efflux

SOS2 belongs to a family of SNF1-related kinase 3 proteins (SnRK3s, also abbreviated as PKS or CIPKs) and contains 25 members. Similarly, SOS3 contains 10 members of the so-called SOS3-like calcium-binding proteins (SCaBPs or CBLs) (Yang and Guo 2018a; Zhu 2016). CBL-CIPK signaling network has been produced in plant responses to salt stress. Upon its release CBL-CIPK interaction with SOS3, SOS2 also interacts and phosphorylates SCaBP8 under salt stress conditions. Phosphorylation of SCaBP8 enhances the SOS2-SCaBP8 association and promotes the SOS2 recruitment to the plasma membrane, where SOS2 phosphorylates SOS1 and thus enhancing Na^+ efflux transport (Lin et al. 2009; Quan et al. 2007). The SOS2-SOS3 complex functions in the root, while the SOS2-SCaBP8 functions in the shoot, to confer salt tolerance and enhance SOS1 transport activity (Lin et al. 2009; Quan et al. 2007; Yang and Guo 2018a, b). Moreover, SOS2 interacts with flowering regulators ABI2, GIGANTEA (GI) and 14-3-3, and these interactions inhibit SOS2 kinase activity. The SOS2 interacts with ABI2 that indicates a possible interplay between ABA signaling and salt stress response, but how these interactions are regulated by ABA and thus modulate SOS2-mediated salt stress response are not known. Under salt stress, the flowering regulator GI is degraded mediates release of SOS2 from the GI-SOS2 inhibitory complex so that SOS2 is activated to function in the salt tolerance pathway (Kim et al. 2013). Under salt stress, elevated cytosolic Ca^{2+} binds the 14-3-3 protein, which promotes PKS5 with 14-3-3 binding that inhibits PKS5 kinase activity but in turn decreases the interaction between 14-3-3 with SOS2, thereby activating SOS2 for salt tolerance (Yang et al. 2019). Na^+ transporters and their regulators for Na^+ uptake, exclusion, and compartmentation have been shown in Fig. 13.4.

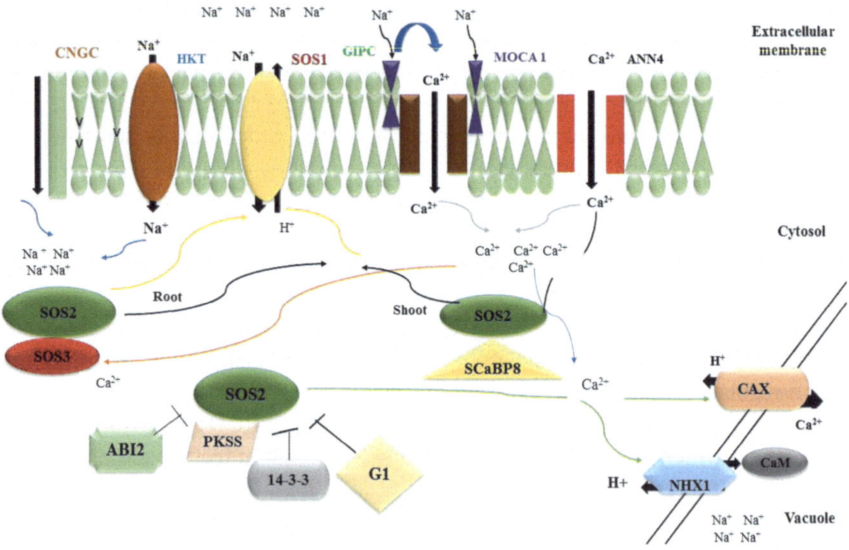

Fig. 13.4 Na^+ transporters and their regulators for Na^+ uptake, exclusion, and compartmentation

13.5.5 HKT Family of Transporters

Mostly very well-studied and well-defined Na^+-permeable plant proteins are in the HKT family of transporters. Presently, two classes of HKT transporters have been identified and characterized: (1) Class 1 (HKT1) transporters, of which AtHKT1 is the most well studied, are implicated in long distance Na^+ translocation between the roots and shoots. (2) Class 2 (HKT2) transporters are those trans porters that are known to mediate primary Na^+ entry into the roots. In *Arabidopsis*, there is only one gene encoding a (Class 1, AtHKT1) HKT transporter, whereas in rice, there are nine HKT genes and possibly even more in other cereal crops as well. High-affinity Na^+ uptake in rice is mediated by OsHKT2 but under low K^+ conditions and in barley roots, it was shown that low-affinity Na^+ uptake under salinity stress increases by the overexpression of HvHKT2.

13.5.6 The HKT-Type Transporters and Na^+ Uptake

AtHKT1 mediates Na^+ influx in roots. It was also observed that mutation in the AtHKT1 gene causes hyperaccumulation of Na^+ in shoots and hypersensitivity to sodium (Berthomieu et al. 2003). AtHKT1 is also involved in transportation of Na^+ from shoot to root via phloem sap. Further studies on AtHKT1 have established a consensus that AtHKT1 functions in root parenchyma cells to retrieve Na^+ from xylem by reducing Na^+ accumulation in the shoot. Notably, AtHKT1 is also expressed in leaf vascular system (Mäser et al. 2002; b). However, the molecular functions of leaf vasculature expressed in AtHKT1 is not determined yet. According to transport activity of HKT transporters, these transporters are classified into two classes (class I and II). The "class I" HTK transporters are Na^+ transporters usually mediate relatively Na^+ selective transport. However, the "class II" HKT transporters are mostly Na^+/K^+ symporters that co-transport Na^+ and K^+ (Ali et al. 2019; Hamamoto et al. 2015). The role of AtHKT1 transporter in salt stress adaptation for plant reproduction is vital, a most pivotal process for all living organisms. A recent study reveals that using large-scale genomic analysis further consolidates the role of AtHKT1 in plant adaptation to saline stress (Busoms et al. 2018).

13.5.7 The NHX-Type Transporters and Na^+ Compartmentation

It has long been known that the Na^+ accumulated in large central vacuoles of plant cells through a Na^+ /H^+ exchanger activity (Blumwald and Poole 1985). In principle, the central vacuole is an ideal location for Na^+ compartmentalization to reduce Na^+ accumulation in the cytoplasm. Na^+ sequestrated inside the vacuole can also serve

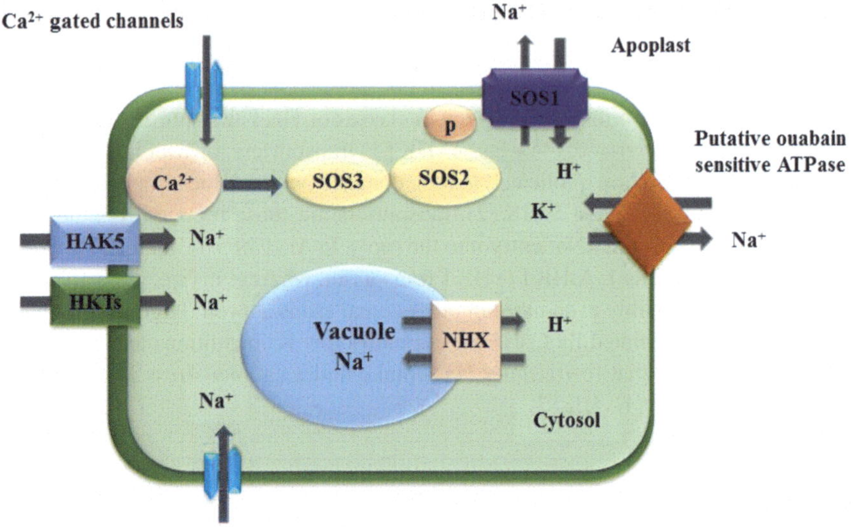

Fig. 13.5 An overview of the mechanisms of Na+ influx, compartmentalization of Na^+ in vacuoles, and Na^+ sensing followed by Na+ efflux in response to salt stress. Na^+ enters the cell via channels such as NSCCs and transporters such as HAK5 and HKTs. Na^+ is sensed by an unidentified sensory mechanism, and Ca2+ signaling cascade is activated. The salt stress-induced increase in cytosolic concentration of Ca^{2+} is sensed by SOS3 (a myristoylated Ca^{2+} sensor protein). SOS3 interacts with SOS2 and activates its kinase activity. The SOS2-SOS3 complex becomes localized to the plasma membrane. SOS2 then phosphorylates SOS1 and activates its Na^+/H^+ antiporter activity facilitating Na^+ efflux from the cell. Additionally, Na^+ is effluxed through a novel efflux mechanism involving a putative ouabain (OU)-sensitive ATPase. Na^+ is also compartmentalized inside vacuoles by NHX (a Na^+/H^+ exchanger) as a response to salt stress

as an osmolyte to lower the cellular water potential and thus promote water uptake under saline conditions. In roots, several cell layers with large vacuoles can accumulate Na^+ including cortex cells and parenchyma cells, which reduce the rate of Na^+ entering into the root xylem. Na^+ compartmentation into the vacuoles of root cells decreases Na^+ accumulation in leaves where essential processes like photosynthesis occur that is essential for plant growth. The Na^+ transported from root to shoot can also be stored in the vacuoles of stem cells and leaf cells, which prevents excessive Na^+ accumulation in their cytoplasm. Na^+ sequestration into the vacuoles is an effective cellular mechanism to reduce cytosolic toxicity of Na^+ under salt stress especially when plants are growing, as more vacuolar space is created during plant growth (Fig. 13.5).

13.6 Sensing of Na by Plant Cell Proteins that Perceive the Entry/Presence of Na in Plant

Earth is a salty planet. From the total land area of the world, 3.1% is covered by saline soil. Saline soil is a major threat to the agriculture globally. Due to salinization, there is great decrease in the yield of crop plants, which can be estimated around billions of dollars annually (Whu 2018). Salinization can be caused naturally such as geology or close proximity to coastal areas or by anthropogenic activities or by false irrigation practices (Maathuis 2013). Salinity contributes to reduction of crop yield by producing reactive oxygen species (ROS), causing hormonal imbalance and ionic and osmotic stress (Munns 2002).

Plants can be categorized in two types on the basis of their response to salt stress.

1. **Glycophytes:** Glycophytes are intolerant to salts, being inhibited by NaCl concentrations around 25–50 mmol L. Mostly crop species are glycophytic (Whu 2018).
2. **Halophytes:** Halophytes are types of plants which grow in high saline soil. Examples of halophytes include *Suaeda maritima* and *Salicornia spp*, and they show a tremendous increase in growth and development when salt is present in growth substrate (Maathuis 2013).

Sodium is the sixth most abundantly present element on the earth. Sodium salts are dominating throughout the saline areas of the world as compared to other salts. Sodium is not an essential element for plants except subgroup of C4 plants. For C4 plants, sodium is shown as an essential element (Whu 2018). Examples include *Atriplex spp., Kochia, childsii, mellets,* and a number of other C4 species (Maathuis 2013). Less amount of sodium is harmless. Excess of sodium into plant becomes toxic because Na competes with other ions for involvement in biological reactions (Kudla 2019).

High concentration of sodium can affect plants in two ways:

Osmotic Effect Due to the high concentrations of salts in the soil, there will be the deficiency of water, which is an initial effect of salinity (Monik et al. 2018). Rapid reduction in leaf elongation and root growth rates can be observed in early phases of salinity (Cramer 1991). A sudden increase in the extracellular Na^+ will cause dehydration, which in turn damage the cell membrane integrity (Munns 2002).

Ionic-Specific Effect Ionic stress specifically caused by Na^+ results in disruption of K^+ homeostasis in plants (Britto 2010). Alteration in the ratios of Na/K means that due to the increase in the amount of Na, K will be deficit (Monika et al. 2018). K^+ is considered as an essential and important component in many cellular processes such as pH homeostasis, ions stomatal functions, cytoplasmic charge balance, membrane potential, and growth, and all these processes are affected by Na^+ high concentration (Maathuis 2019). Due to accumulation of Na in the cytosol, not only K deficiency will occur but also there will be enzymatic inhibition. Because more than

50 enzymes are activated by potassium, Na^+ substitution in place of K^+ due to similarity between their ionic radii, Na^+ cannot activate them (Whu 2018).

It was not known for a long time whether plants can differentiate between osmotic and ionic perturbation (Kudla 2019). To cope with stress caused by Na^+ , a clear understanding of Na^+ sensing mechanisms and its transport within the plants have great importance. For the last two decades, researchers are investigating the sensing and transport system of Na^+ within the plant most specifically how it is sensed by the plants (Maathuis 2013).

Root Root can always perceive the changes that occur due to environmental factors such as nutrient level and salinity level. Root is first organ that encounters salinity in most plants. Na^+ uptake by the root can be can be mediated by two pathways: (1) Apoplastic pathway is the extracellular transport of Na^+ in which Na^+ diffuses through the cell wall with direct entry into the xylem by escaping the plasma membrane. (2) Symplastic pathways by which the sodium make entry into cytoplasm without escaping the plasma membrane (Maathuis 2013).

Another considerable part of root which is known to have a role in Na^+ sensing is root meristem. Root meristem is located at the tip of root. Removal of root meristem from salt-tolerant plants can disrupt the Na distribution in cell and a salt-sensitive type can result. These findings suggest two possible statements: one is that root meristem acts as a salt stress sensor or second would be that it harbors some components that are salt stress sensor (Whu 2018). Early investigation on transport mechanisms of Na^+ into barley (*Hordeum vulgare*) roots suggested that it has two different uptake mechanisms and their activation depended on the concentration of Na^+ and other ions such as K^+ , NH4, and NO3 (Britto and Kronzucker 2006). When the extracellular concentration of Na^+ is lower than 1 mM, then high-affinity transport systems (HATS) will be activated. In case of higher concentration of Na^+ externally, Na^+ will be transported by low-affinity transport system (LATS) (Britto and Kronzucker 2006). Stress caused by Na^+ will occur in the range of LATS. In this model, the entry of sodium occurs through the plasma membrane of root cells and is then transported into cytoplasm more likely through ion channels and transporters (Cheeseman 1982; Munns 2008).

The onset of salt stress is followed by rapid depolarization of cortical and epidermal root cells. By contrast, cell hyperpolarization is caused by hyperosmotic shock. Therefore, changes in membrane would act as the signals where salinity and osmotic pressure are concerned. Generally, cell depolarization is obtained when there is an increase in ion concentration in root medium (Maathuis 2013).

13.7 Molecular Candidates for Na Influx

Some putative salt stress sensors that are responsible for early signaling and symplastic Na transport are as follows:

NSCC NSCC or nonselective cation channels perform the fluxes of cations through the plasma membrane. A number of NSCC genes have been identified in different plant genomes. NSCCs are involved in multiple stress responses, growth, development, nutrient uptake, and calcium signal transduction. There are several areas of plant research where interest in NSCCs is substantial. NSCCs play a considerable role in plant salinity stress especially in Na entry. Mechanism of Na entry into root is similar as for Ca, no selection ion channels for sodium have been found in plant. NSCCs are proposed to provide a pathway for the entry of Na into plant (Maathuis 2007). For the majority of species, there is considerable evidence that NSCCs are the main pathways of sodium influx into plant cells (Maathuis 2019). Plasma membrane NSCCs can be categorized into several sub-classes: (1) voltage-independent NSCC VI-NSCCs, (2) hyperpolarization-activated NSCC HA-NSCCs, and (3) depolarization-activated NSCC DA-NSCCs (Demidchik 2002, 2007). Most of the electrophysiological studies on protoplast or membrane patches suggested that VI-NSCCs have a primary role in Na influx under salinity stress (Cheeseman 2013). Na influx through VI-NSCCs has been identified in *Arabidopsis thaliana, maize, barley, wheat,* and *Capsicum annum* (Kronzucker 2011).

HKT Transporters HKT transporter family is well defined in many plant systems. The most well-studied Na permeable proteins also belong to the HKT sub-classes (Kronzucker 2011). Two sub-classes of HKT transporters have been identified and classified. Among the class 1 (HKT1) transporters, AtHKT1 is well studied and is known to have role in Na distribution from the roots to the shoots (Kronzucker 2011; Maathuis 2013). Class 2 (HKT2) transporters are known to mediate the primary entry of Na^+ (Maathuis 2019; Britto and Kronzucker 2006). In *Arabidopsis*, there is only one gene that codes for HKT transporter (AtHKT1), in rice they are nine, and in cereal crops they may exceed in number (Platten 2006). OsHKT2;1 has been shown to have high-affinity uptake of Na in rice, but under low K concentration and overexpression of HvHKT 2:1 was shown to increase the low-affinity Na uptake in the roots of barley (Mian et al. 2011).

KUP/HAK/K Transporters The K transporters include KUP/HAK/KT, and K channel AtAKT1 is known to catalyze the K transport but also have some concerns with Na transport in salinity stress (Amtmann 1998; Blumwald 2000). In *Arabidopsis* AtHAK5 has been known to show low-affinity binding uptake of Na under the low K concentration. OsAKT1 was excessively present in the epidermal and endodermal regions of root and has very low amount in leaves of rice. In addition to this, salt-sensitive variety of rice shows a higher expression of OsAKT1 than the tolerant one (Golldack et al. 2003).

Low-Affinity Cation Transporter 1 (LCT1) LCT1 was first isolated and its low expression level is detected in roots and leaves of wheat (Horie 2004). In some systems, LCT1 has shown the similar role as that of NSCCs by having the ability to transport the Rb, K, Cd2 and also confers the Na hypersensitivity (Celmens 1998). Evidence suggests that LCT1 may not be directly involved in the transport of Na, but it has role in enhancing the activity of other transporters (Ammtman et al. 2001).

Apoplast The apoplast is another possible pathway of Na transport into root and into shoot. In this pathway, Na can diffuse into cell wall but by escaping the plasma membrane. Na cycling within plants done by apoplast pathways will occur when the external concentration of Na is higher, which is recently demonstrated in *Arabidopsis* and barley (Coskun et al. 2016). The process of Na transportation in apoplastic manners also known as apoplastic bypass flow due to the direct flow of Na from cell wall to xylem (Kronzucker 2011). Many plant species, such as rice, wheat, maize, barley, and beans, exhibit the apoplastic ways of Na transport, but the degree of transport varies between species (Horie et al. 2004).

Reactive Oxygen Species Sometimes, GIPCs can also stimulate the Ca^{+2} signals indirectly. Microdomain in lipid membrane and GIPCs in these microdomain help in the regulation of signaling in plants. Salt stress also stimulates the generation of reactive oxygen species (ROS). Due to generation of ROS, Ca^{+2} signal will be activated. Salt stress also affects the formation of microdomains in plasma membrane, which further affects the activity of enzyme termed as NADPH oxidases. NADPH oxidases act in the generation of ROS signals (Kudla et al. 2019). Annexin 1 in *Arabidopsis thaliana* mediate against exterior to salt generate ROS-activated Ca^{+2} influx through the plasma membrane (Maathuis et al. 2019).

Protein Phosphorylation and Dephosphorylation There are wide evidence for phosphorylation and dephosphorylation of proteins in response to salt stress. There is a possibility that membrane-bound histidine kinases could act as a Na sensor. Increase in kinase activity may involve in direct or indirect Na pumping activity. Overexpression of Dbf2 kinase in *Arabidopsis* cell also produces tolerance to salinity. There are many other phosphatases and kinases in plants that are likely to response salt condition or may be to promote tolerance to salt. Further genomic and biochemical studies are required in determining how they interact and produce responses against Na (Davenport 2003).

Glycosyl Inositol Phosphorylceramide (GIPC) Jiang and his colleagues identified a plant that had mutation in the gene that codes for protein IPUT1. IPUT1 catalyzes the central step in the formation of lipid glycosyl inositol phosphorhylceramide. GIPCs are the major component of lipid bilayer plasma membrane of plant. By performing further biochemical test, it is revealed that GIPCs can bind to the Na ion and other ions that have single positive charge. Direct binding of sodium Na with GPICs is an important step in sodium sensing by plant and also triggers the Ca^{+2} signals that would result in salt tolerance. GIPCs interact with Ca^{+2} channels directly because Na binding to GIPCs will modulate channels which in turn leads to production of Ca^{+2} signals in cell (Fig. 13.6).

When there is increase of salts in the exterior environment of cell, unknown Ca^{+2} channels will be activated and Ca^{+2} ions present outside cell will enter into cell. Negatively charged GIPC binds directly to Na ions present in the external environment. The subsequent influx of Ca^{+2} ions will generate high salt level response in which Ca^{+2}-binding protein SOS3 will produce, SOS3 will further recruit SOS2,

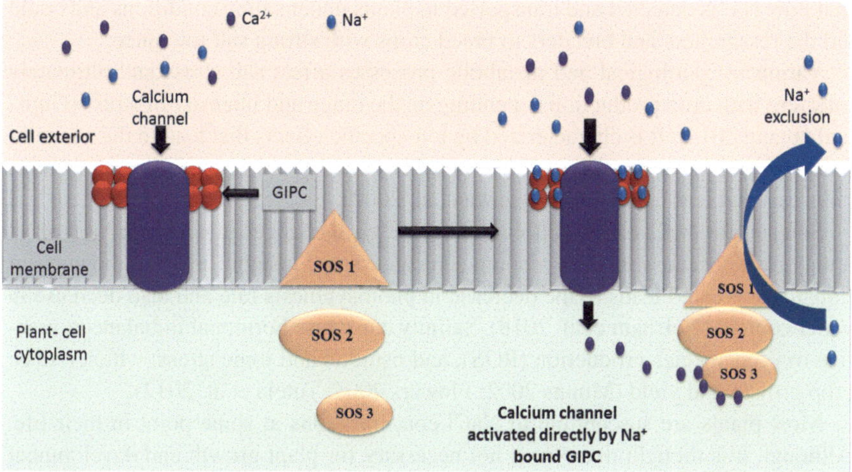

Fig. 13.6 How plants sense Na with help of GIPC and Ca^{+2} channels (a) Low salt level (**b**) Increase in exterior salt

which in turn activates the SOS1 protein. SOS1 will pump out the Na out of the cell (Kudla 2019).

SOS Pathway In *Arabidopsis*, three overly salt-sensitive genes (SOS1, SOS2, SOS3) were identified contributing to the salt tolerance (Quan-Sheng et al. 2002). One of these SOS1s encode for the Na/H antiporter that involves in efflux of Na from the cytosol (Maathuis 2013). SOS evolutionary conserved pathway plays a central role during the salinity stress. In this pathway protein, SOS3 by binding with Ca^{+2} ions decodes the Ca^{+2} signals. SOS3 also activates another protein SOS2 having the kinase activity. The SOS2 enzyme will then activate one more protein present in the cell wall known as SOS1. SOS1 is an antiporter protein that can transport Na^+ out of the cell (Kudla 2019) and has some concerns with Na transport in salinity stress.

13.8 Na^+ as a Nutrient in Plant

Na^+ is not an essential element, but its usage in minute quantities such as micronutrient helps in metabolism and synthesis of chlorophyll. However, it is used in the partial replacement for K^+ and also helps in the opening and closing of stomata, which helps in maintaining internal water balance.

High Na^+ accumulation within the cytosol can not only cause K^+ deficiency but can also disturb various enzymatic processes. This causes an active burden on the cell due to the need of organic solute synthesis to catch up on the export of Na^+ for osmotic adjustment (Munns and Tester 2008). This leads to a better understanding that how Na^+ is detected and transported in plants under saline conditions and could aid the researchers and breeders to breed crops with strong salt tolerance.

Various physiological and metabolic processes affect salt stress and ultimately interfere with crop production depending on the range and intensity of stress (Gupta and Huang 2014). It is characterized by ion-specific effects that lead to the accumulation of toxic concentrations of Na^+ and Cl^- in cells, which appear in the form of chlorosis and leaf necrosis (Hasegawa et al. 2000).

Salt such as NaCl creates different physiological changes in which Na^+ ion plays a major role because its high amount can cause some instabilities, and the nutrient imbalance created leads to the decrease in photosynthesis rate and also decrease in water content (Keisham et al. 2018). Salinity can cause hormonal imbalances, reactive oxygen species production (ROS), and osmotic and ionic stress, which reduce crop growth and yield (Munns 2002; Flowers 2004; Tuteja et al. 2012).

Most plants are susceptible to NaCl concentrations at some point in their life. Although it is thought that Na^+ is not necessary for plant growth and development (except for the C4 subgroup of plants), it is favorable at low concentrations, especially when the concentration of K^+ is low (Maathuis 2014). In several C4 plants and all of CAM plants, Na^+ plays an important role in chloroplast functioning (Subbarao et al. 2003).

There is always limit to the uptake of sodium in plants as it is used in micro quantity, that is why its larger accumulation causes different stresses in plants. As Na^+ is one of the most soluble minerals, which is effortlessly handy to plants and helps to increase osmotic potential, it absorbs water and turgor pressure, and for these reasons Na^+ uptake is desirable (Pardo and Quintero 2002). Na^+ concentration is usually observed to be benign or also useful and may even stimulate growth of many plant species (Schulze et al. 2012). This impact occurs due to the substitution of Na^+ and K^+ in the vacuole of plant cells, making available extra K^+ to the cytosol. For example, under 200 mMol-1 NaCl stimulated the growth shown by halophytes *salicomra europe* as compared to non-saline conditions (Maser et al. 2008; Lv et al. 2012). Researchers have adhered to the mechanism utilized by halophytes to apply Na^+ as a nutrient (Lv et al. 2012; Flowers and Colmer 2008). Na^+ causes ionic stress effect which is not directly the result of the availability of Na^+, but majorly this is because of downward effect of K^+ homeostatic disruption (Tester and Davenport 2003; Kronzucker and Britto 2011; Kronzucker et al. 2013). For example, specific ionic stress of Na^+ results in a reduction in protein synthesis due to high levels of K^+ that are required to bind tRNA to ribosomes (Tester and Davenport 2003; Britto and Kronzucker 2008; Szczerba et al. 2009). The various strategies used by halophytes and glycophytes in response to salt stress are compared in Fig. 13.7.

Sodium as a nutrient helps in metabolism, but it has some specific concentration in plants. Above that concentration, the accumulation of Na^+ in the cytoplasm has detrimental effects on the cell metabolism in plants such as on the photosynthetic activities of plants. Due to these reasons Na^+ is transported outside the cytoplasm. Hence, the accumulation of Na^+ occurs at a high concentration in vacuoles where it is used as osmoticum and helps in different transport mechanisms in halophyte (Nieves-Cordones et al. 2016).

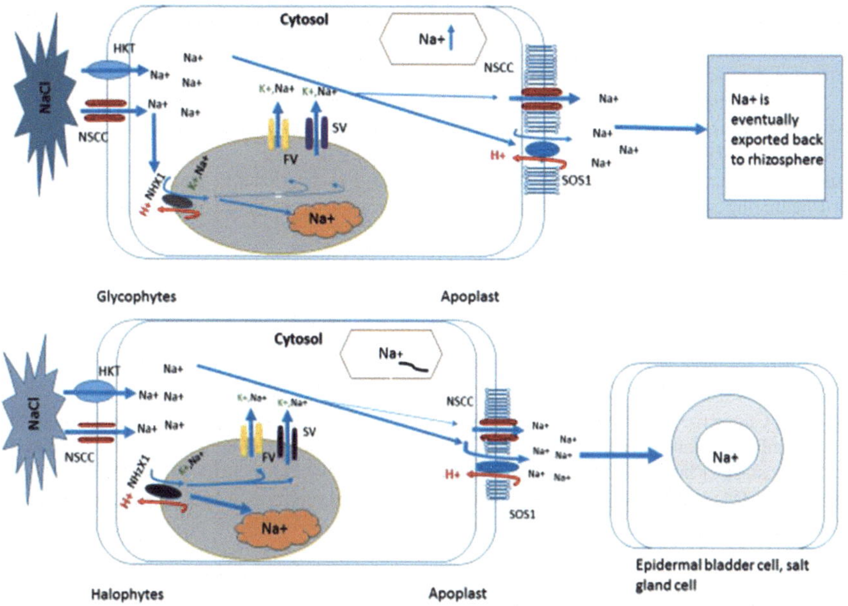

Fig. 13.7 Halophytes and glycophytes show different strategies in response to salt stress

References

Adams E, Shin R (2014) Transport, signalling and homeostais of potassium and sodium in salt stress, J Int Plant Bio 3(56):231–249. https://doi.org/10.1111/jipb.12159

Ahmed Hamam Doctor of Philosophy (2020). On the understanding of sodium transport and its role in salinity stress in plants. Cell & Systems Biology. University of Toronto, Toronto.

Ali A, Maggio A, Bressan RA, Yun DJ (2019) Role and functional differences of HKT1-type transporters in plants under salt stress. Int J Med Sci 20:1059

Amtmann A, Sanders D (1998) Mechanisms of Na uptake by Plants cells. Adv Bot Res 29:75–112. https://doi.org/10.1016/S0065-2296(08)60310-9

An D et al (2017) AtHKT1 drives adaptation of Arabidopsis thaliana to salinity by reducing floral sodium content. PLoS Genet 13(10):e1007086

Apse MP, Aharon GS, Snedden WA, Blumwald E (1999) Salt tolerance conferred by over expression of a vacuole of Na/H antiport in Arabidopsis, Science, 285(5431):1256-8. https://doi.org/10.1126/science.285.5431.1256

Berthomieu P, Conejero G, Nublat A, Brackenbury WJ, Lambert C, Savio C, Uozumi N, Oiki S, Yamada K, Cellier F et al (2003) Functional analysis of AtHKT1 in Arabidopsis shows that Na^+ recirculation by the phloem is crucial for salt tolerance. EMBO J 22:2004–2014

Blumwald E (2000) Sodium transport and salt tolerance in plants. Curr Opin Plant Biol 12:431–434

Blumwald E, Poole RJ (1985) Na^+ /H+ antiport in isolated tonoplast vesicles from storage tissue of Beta vulgaris. Plant Physiol 78:163–167

Bose J, Munns R, Shabala S, Gilliham M, Pogson B, Tyerman SD (2017) Chloroplast function and ion regulation in plants growing on saline soils: lessons from halophytes. J Exp Bot 68(12):3129–3143

Britto DT, Kronzucker HJ (2006) Futile cycling at the plasma membrane: a hallmark of lowaffinity nutrient transport. Trends Plant Sci 11:529–534

Britto DT, Kronzucker HJ (2008) Cellular mechanism of potassium transport in plants. Physiologia Plantarum, 133(4):637–650

Busoms S, Paajanen P, Marburger S, Bray S, Huang XY, Poschenrieder C, Yant L, Salt DE (2018) Fluctuating selection on migrant adaptive sodium transporter alleles in coastal Arabidopsis thaliana. Proc Natl Acad Sci U S A 115:E12443–E12452

Byrt, CS et al. (2007) "HKT1;5-like cation transporters linked to Na^+ exclusion loci in wheat, Nax2 and Kna1." Plant physiology 143(4): 1918–28. https://doi.org/10.1104/pp.106.093476

Byrt CS, Zhao M, Kourghi M, Bose J, Henderson SW, Qiu J, Gilliham M, Schultz C, Schwarz M, Ramesh SA, Yool A, Tyerman S (2017) Non-selective cation channel activity of aquaporin AtPIP2;1 regulated by Ca^{2+} and pH. Plant Cell Environ 40(6):802–815. https://doi.org/10.1111/pce.12832. Epub 2016 Oct 18. PMID: 27620834

Cellier F, Conejero G, Licaud L, Lepettit M, Gosti F, Casse F (2004) Characterization of AtCHX, a member of a cation/H exchanger, CHX family, from Arabidopsis thaliana suggests a role in K homeostasis. Plant Journal 39(6):834–46. https://doi.org/10.1111/j.1365-313X.2004.02177.x

Cheeseman JM (1982) Pump-leak sodium fluxes in low salt corn roots. J Membr Biol 70:157–164

Cheeseman JM (2013) The integration of activity in saline environment: problems and perspectives. Functional Plant Biology 40(9):759–774. https://doi.org/10.1071/FP12285

Chen ZC et al (2016) A cation-chloride cotransporter gene is required for cell elongation and osmoregulation in rice. Plant Physiol 171(1):494–507

Colmenero-Flore JM et al (2007) Identification and functional characterization of cation–chloride cotransporters in plants. Plant J 50(2):278–292

Coskun D, Britto DT, Hunyh QW, Kronzucker HJ (2016) The role of silicon in higher plants under salinity and drought stress. Frontier Plant Science. https://doi.org/10.3389/fpls.2016.01072

Cramer GR, Epstein E, Lauchli A (1991) Effects of sodium, potassium and calcium on salt stress barley. Physiologia Plantarum, 81(2):197–202. https://doi.org/10.1111/j.1399-3054.1991.tb02129.x

Davenport R (1998) Mechanisms of toxic sodium influx into wheat roots. Cambridge University, Cambridge

Davenport RJ, Tester M (2000) A weakly voltage-dependent, nonselective cation channel mediates toxic sodium influx in wheat. Plant physiology 122(3):823–834, https://doi.org/10.1104/pp.122.3.823

Davenport RJ et al (2007) The Na^+ transporter AtHKT1; 1 controls retrieval of Na+ from the xylem in Arabidopsis. Plant Cell Environ 30(4):497–507

Demidchik V, Essah PA, Tester M. (2004) Glutamate activates cation currents in the plasma membrane of Arabidopsis root cells. Planta 219(1):167–75. https://doi.org/10.1007/s00425-004-1207-8. Epub 2004 Feb 7. PMID: 14767768

Demidchik V, Cuin TA, Svistunenko D, Smith SJ, Miller AJ, Shabala S, Sokolik A, Yurin V (2010) Arabidopsis root K^+-efflux conductance activated by hydroxyl radicals: single-channel properties, genetic basis and involvement in stress-induced cell death. J Cell Sci 123:1468–79. https://doi.org/10.1242/jcs.064352. Epub 2010 Apr 7. PMID: 20375061

Demidchik V, Maathuis FJM (2007) Physiological roles of non-selective cations channels in plants: from salt stress to signalling developments. New Phytologist 175(3):387–404. https://doi.org/10.1111/j.1469-8137.2007.02128.x

Demidchik V, Tester M (2002) Sodium fluxes through non-selective cations channels in the plasma membrane of protoplasts from Arabidopsis roots. Plant Physiology 128(2):379–87. https://doi.org/10.1104/pp.010524

Deinlein U, Stephan AB, Horie T, Luo W, Schroeder JI (2014) Plant salt-tolerance mechanisms. Trends in Plant Sci 19(6):371–379. https://doi.org/10.1016/j.tplants.2014.02.001
Dobermann A, Fairhurst T (2000) Rice: nutrient disorders and nutrient management. Potash & Phosphate Institute (PPI), Potash & Phosphate Institute of Canada (PPIC) and International Rice Research Institute (IRRI).
Ed Bloodnick October 5, 2018. Role of sodium and chloride in plant culture. *PRO-MIX®* is a registered trademark.
Essah PA, Davenport R, Tester M (2003) Sodium influx and accumulation in Arabidopsis. Plant Physiol 133:307–318
Fliegel L (2019) Structural and functional changes in the Na^+ /H^+ exchanger isoform 1, induced by Erk1/2 phosphorylation. Int J Med Sci 20:2378
Flowers TJ (2004) Improving crop salt tolerance. J Exp Bot 55:307–319
Flowers TJ, Colmer TD (2008) Salinity tolerance in halophytes. New Phytol 179:945–963
Flowers TJ, Colmer TD (2015) Plant salt tolerance: adaptations in halophytes. Ann Bot 115:327–331
Flowers TJ, Galal HK, Bromham L (2010) Evolution of halo-phytes: multiple origins of salt tolerance in land plants. Funct Plant Biol 37:604–612
Gao J, Chao D, Lin H (2007) Understanding abiotic stress tolerance mechanisms: recent studies on stress response in rice. J Integr Plant Biol 49:742–750
Garciadeblas B, Senn M, Banuelos MA, Navarro AR (2003) Sodium transport and HKT transporters: the rice model. The Plant J 6(34):788–801. https://doi.org/10.1046/j.1365-313X.2003.01764.x
Gobert A, Park G, Amtamann A, Sanders D, Mathuis FJM (2006) Arabidopsis thaliana cyclic nucleotide gated channels 3 transporters involved in germination and cations transport. J Exp Bot 4(57):791–800. https://doi.org/10.1093/jxb/erj064
Golldack D, Quigley F, Michalows CB, Bohnert HJ (2003) Salinity stress tolerant and sensitive rice regulate AKT1 type potassium channels transcripts differently. Plant Molecular Biology, 51(1):71–81. https://doi.org/10.1023/a:1020763218045
Gong Z, Xiong L, Shi H, Yang S, Herrera-Estrella LR, Xu G, Chao D-Y, Li J (2004) Plant abiotic stress response and nutrient use efficiency. Sci China Life Sci 63(5):635–674
Grattan SR, Grieve CM (1998) Salinity-mineral nutrient relations in horticultural crops. Sci Hortic 78:127–157
Guo Y, Halfter U, Ishitani M, Zhu JK (2001) Molecular characterization of functional domains in the protein kinase SOS2 that is required for plant salt tolerance. Plant Cell 13:1383–1400
Gupta B, Huang B (2014) Mechanism of salinity tolerance in plants: physiological, biochemical, and molecular characterization. Int J Genom 701596
Hall D et al (2006) Functional analysis of CHX21: a putative sodium transporter in Arabidopsis. J Exp Bot 57(5):1201–1210
Hamamoto S, Horie T, Hauser F, Deinlein U, Schroeder JI, Uozumi N (2015) HKT transporters mediate salt stress resistance in plants: from structure and function to the field. Curr Opin Biotechnol 32:113–120
Hasegawa PM, Bressan RA, Zhu JK, Bohnert HJ (2000) Plant cellular and molecular responses to high salinity. Annu Rev Plant Physiol Plant Mol Biol 51:463–499
Horie T, Schroeder JI (2004) Sodium transporters in plants. Diverse genes and physiological functions. Plant Physiol 136:2457–2462. https://doi.org/10.1104/pp.104.046664
Horie T, Costa A, Kim TH, Han MJ, Horie R, Leung HY, Miyao A, Hirochika H, An G, Schroeder JI (2007) Rice OsHKT2;1 transporter mediates large Na^+ influx component into K+-starved roots for growth. The EMBO journal, 26(12):3003–3014. https://doi.org/10.1038/sj.emboj.7601732
Horie ST, Motoda J, Kubo M, Yang H, Yoda K, Horie R, Chan WY, Leung HY, Hattori K, Konomi M, Osumi M, Yamagami M, Schroeder JI, Uozumi N (2004) Enhanced salt tolerance mediated by AtHKT1 transporter-induced Na unloading from xylem vessels to xylem parenchyma cells. Plant J 44:928–938. https://doi.org/10.1111/j.1365-313X.2005.02595.x

Jha D, Shirley N, Tester M, Roy SJ (2010) Variation in salinity tolerance and shoot sodium accumulation in Arabidopsis ecotypes linked to differences in the natural expression levels of transporters involved in sodium transport. Plant Cell Environ 33:793–804

Katiyar-Agarwal S, Zhu J, Kim K, Agarwal M, Fu X, Huang A, Zhu JK (2006) The plasma membrane Na^+/H^+ antiporter SOS1 interacts with RCD1 and functions in oxidative stress tolerance in Arabidopsis. Proc Natl Acad Sci U S A 103:18816–18821

Keisham M, Mukherjee S, Bhatla SC (2018) Mechanisms of sodium transport in plants—progresses and challenges. Int J Mol Sci 19(3):647

Ketehouli T et al (2019) Adaptation of plants to salt stress: characterization of Na^+ and K^+ transporters and role of CBL gene family in regulating salt stress response. Agronomy 9(11):687

Kim WY, Ali Z, Park HJ, Park SJ, Cha JY, Perez-Hormaeche J, Quintero FJ, Shin G, Kim MR, Qiang Z et al (2013) Release of SOS2 kinase from sequestration with GIGANTEA determines salt tolerance in Arabidopsis. Nat Commun 4:1352

Kobayashi NI et al (2017) OsHKT1; 5 mediates Na^+ exclusion in the vasculature to protect leaf blades and reproductive tissues from salt toxicity in rice. Plant J 91(4):657–670

Kochian LV, Xin-Zhi J, Lucas WJ (1985) Potassium transport in corn roots : IV. Characterization of the linear component. Plant Physiol 79:771–776

Kragelund BB, Jensen MK, Skriver K (2012) Order by disorder in plant signaling. Trends Plant Sci 17:625–632

Krishnamurthy P, Ranathunge K, Franke R, Prakash HS, Schreiber L, Mathew MK (2009) The role of root apoplastic transport barriers in salt tolerance of rice (Oryza sativa L.). Planta 230(1):119–34. https://doi.org/10.1007/s00425-009-0930-6. Epub 2009 Apr 11. PMID: 19363620

Kronzucker HJ, Britto DT (2011) Sodium transport in plants: a critical review. New Phytol 189:54–81. https://doi.org/10.1111/j.1469-8137.2010.03540.x

Kronzucker HJ, Coskun D, Schulze LM, Wong JR, Britto DT (2013) Sodium as nutrient and toxicant. Plant Soil 369:1–23

Legué V, Blancaflor E, Wymer C, Perbal G, Fantin D, Gilroy S. (1997) Cytoplasmic free Ca^{2+} in Arabidopsis roots changes in response to touch but not gravity. Plant Physiol 114(3):789–800. https://doi.org/10.1104/pp.114.3.789. PMID: 9232870; PMCID: PMC158365

Li Y, Gu M, Zhang X, Zhang J, Fan H, Li P, Li Z, Xu G (2014) Engineering a sensitive visual-tracking reporter system for real-time monitoring phosphorus deficiency in tobacco. Plant Biotechnol J 12:674–684

Lin H, Yang Y, Quan R, Mendoza I, Wu Y, Du W, Zhao S, Schumaker KS, Pardo JM, Guo Y (2009) Phosphorylation of SOS3-LIKE CALCIUM BINDING PROTEIN8 by SOS2 protein kinase stabilizes their protein complex and regulates salt tolerance in Arabidopsis. Plant Cell 21:1607–1619

Liu J, Zhu JK (1998) A calcium signal homolog required for salt tolerance. Science, 5371(280):1943–1945

Liu J, Ishitani M, Halfter U, Kim CS, Zhu JK (2000) The Arabidopsis thaliana SOS2 gene encodes a protein kinase that is required for salt tolerance. Proc Natl Acad Sci U S A 97:3730–3734

Lv S, Nie L, Fan P, Wang X, Jiang D, Chen X, Li Y (2012) Sodium plays a more important role than potassium and chloride in growth of Salicornia europaea. Acta Physiol Plant 34(2):503–513

Maathuis FJM (2006) The role of monovalent cation transporters in plant responses to salinity, J Exp Bot 57(5):1137–1147, https://doi.org/10.1093/jxb/erj001

Maathuis FJM (2013) Sodium in plants: perception, signalling, and regulation of sodium 65(3):849–858. https://doi.org/10.1093/jxb/ert326

Maathuis FJM (2014) Sodium in plants: perception, signalling, and regulation of sodium fluxes. J Exp Bot 65:849–858. mediating plant salt-stress responses. *New Phyto*l 217

Maathuis FJM, Sanders D. (2001) Sodium uptake in Arabidopsis thaliana root is regulated by cyclic nucleotides. Plant physiologia plantarum, 127(4):1617–1625

Malagoli P, Britto DT, Schulze LM, Kronzucker HJ (2008) Futile Na+ cycling at the root plasma membrane in rice (Oryza sativa L.): kinetics, energetics, and relationship to salinity tolerance. J Exp Bot 59:4109–4117

Mäser P, Eckelman B, Vaidyanathan R, Horie T, Fairbairn DJ, Kubo M, Yamagami M, Yamaguchi K, Nishimura M, Uozumi N (2002) Altered shoot/root Na^+ distribution and bifurcating salt sensitivity in Arabidopsis by genetic disruption of the Na^+ transporter AtHKT1. FEBS Lett 531:157–161

Mian A, Oomen RJFJ, Isayenkov S, Sentenac H, Maathuis FJM, Very AA (2011) Overexpression of Na and K permeable HKT transporter in barley improves salt tolerance. Plant Journal, 68(3):468–79. https://doi.org/10.1111/j.1365-313X.2011.04701.x

Miller G, Suzuki N, Ciftci-Yilmaz S, Mittler R (2010) Reactive oxygen species homeostasis and signalling during drought and salinity stresses. Plant Cell Environ 33:453–467

Møller IS, Gilliham M, Jha D, Mayo GM, Roy SJ, Coates JC, Haseloff J, Tester M (2009) Shoot Na+ exclusion and increased salinity tolerance engineered by cell type-specific alteration of Na+ transport in Arabidopsis. Plant Cell 21:2163–2178

Munns R (2002) Comparative physiology of salt and water stress. Plant Cell Environ 25:239–250

Munns R, Tester M (2008) Mechanisms of salinity tolerance. Annu Rev Plant Biol 59:651–681. https://doi.org/10.1146/annurev.arplant.59.032607.092911

Munns R, James RA, Läuchli A (2006) Approaches to increasing the salt tolerance of wheat and other cereals. J Exp Bot 57:1025–1043

Negrão S, Courtois B, Ahmadi N, Abreu I, Saibo N, Oliveira MM (2011) Recent updates on salinity stress in rice: from physiological to molecular responses. CRC Crit Rev Plant Sci 114(30):329–377

Nieves-Cordones M, Al Shiblawi FR, Sentenac H (2016) Roles and transport of sodium and potassium in plants. In: The alkali metal ions: their role for life. Springer, Cham, pp 291–324

Pardo JM, Quintero FJ (2002) Plants and sodium ions: keeping company with the enemy. Genome Biol 3:(Re-views 1017.1–1017.4)

Passioura JB, Munns R (2000) Rapid environmental changes that affect leaf water status induce 115 transient surges or pauses in leaf expansion rate. Funct Plant Biol 27:941

Plett DC, Moller IS (2010). Na transport in glycophytic plants: what we know and would like to know. Plant Cell Environment 33(4):612–626. https://doi.org/10.1111/j.1365-3040.2009.02086.x

Qadir M, Quillérou E, Nangia V, Murtaza G, Singh M, Thomas RJ, Drechsel P, Noble AD (2014) Economics of salt-induced land degradation and restoration. Nat Resour Forum 38:282–295

Qiu QS et al (2003) Na^+/H^+ exchange activity in the plasma membrane of Arabidopsis. Plant Physiol 132(2):1041–1052

Quan R, Lin H, Mendoza I, Zhang Y, Cao W, Yang Y, Shang M, Chen S, Pardo JM, Guo Y (2007) SCABP8/CBL10, a putative calcium sensor, interacts with the protein kinase SOS2 to protect Arabidopsis shoots from salt stress. Plant Cell 19:1415–1431

Quintero FJ, Ohta M, Shi H, Zhu JK, Pardo JM (2002) Reconstitution in yeast of the Arabidopsis SOS signaling pathway for Na^+ homeostasis. Proc Natl Acad Sci U S A 99:9061–9066

Quintero FJ, Atienza JM, Villalta I, Jiang X, Kim WY (2011) Activation of plasma membrane Na/H antiporter Salt overly sensitive 1 (SOS1) by phosphorylation of an auto-inhibitory C-terminal domain.pnas.org, 108(6):2611–2616

Ranathunge K, Schreiber L (2011) Water and solute permeabilities of Arabidopsis roots in relation to the amount and composition of aliphatic suberin. J Exp Bot 62(6):1961–1974. https://doi.org/10.1093/jxb/erq389

Ren XL et al (2013) Calcineurin B-like protein CBL 10 directly interacts with AKT 1 and modulates K^+ homeostasis in Arabidopsis. Plant J 74(2):258–266

Rodríguez-Rosales MP, Gálvez FJ, Huertas R, Aranda MN, Baghour M, Cagnac O, Venema K (2009) Plant NHX cation/proton antiporters. Plant Signal Behav 4:265–276

Sato H, Suzuki T, Takahashi F, Shinozaki K, Yamaguchi-Shinozaki K (2019) NF-YB2 and NF-YB3 have functionally diverged and differentially induce drought and heat stress-specific genes. Plant Physiol 180:1677–1690

Schulze LM, Britto DT, Li M, Kronzucker HJ (2012) A pharma-cological analysis of high-affinity sodium transport in barley (*Hordeum vulgare L.*): a $^{24}Na^+/^{42}K^+$ study. J Exp Bot 63:2479–2489

Shabala S (2013) Learning from halophytes: physiological basis and strategies to improve abiotic stress tolerance in crops. Ann Bot 112:1209–1221

Sentenac H, Bonneud N, Minet F, Lacroute F, Salmon JM, Gaymard F, Grignon C (1992) Cloning and expression of yeast of a plant potassium ions transports system. Science, 256(5057):663–665. https://doi.org/10.1126/science.1585180

Shi H, Ishitani M, Kim C, Zhu JK (2000) The Arabidopsis thaliana salt tolerance gene SOS1 encodes a putative Na^+/H^+ antiporter. Proc Natl Acad Sci U S A 97:6896–6901

Shi H, Quintero FJ, Pardo JM, Zhu JK (2002) The putative plasma membrane Na^+/H^+ antiporter SOS1 controls long-distance Na^+ transport in plants. Plant Cell 14:465–477

Shi H, Lee B, Wu SJ, Zhu JK (2003) Overexpression of a plasma membrane Na^+/H^+ antiporter gene improves salt tolerance in Arabidopsis thaliana. Nat Biotechnol 21:81–85

Shi Y, Wang Y, Flowers TJ, Gong H (2013) Silicon decreases chloride transport in rice (Oryza sativa L.) in saline conditions. J Plant Physiol 170(9):847–53. https://doi.org/10.1016/j.jplph.2013.01.018. Epub 2013 Mar 22. PMID: 23523465

Subbarao G, Ito O, Berry WL, Wheeler RM (2003) Sodium—A functional plant nutrient. CRC Crit Rev Plant Sci 22:391–416

Świeżawska B, Duszyn M, Jaworski K, Szmidt-Jaworska A (2018) Downstream targets of cyclic nucleotides in plants. Front Plant Sci 9:1428

Szczerba MW, Britto DT, Kronzucker HJ (2009) K^+ transport in plants: physiology and molecular biology. J Plant Physiol 166:447–466

Tavakkoli E, Rengasamy P, McDonald GK (2010) High concentrations of Na^+ and Cl^- ions in soil solution have simultaneous detrimental effects on growth of faba bean under salinity stress. J Exp Bot 61:4449–4459

Tester M, Davenport R (2003) Na^+ tolerance and Na^+ transport in higher plants. Ann Bot 91:503–527

Tuteja N, Gill SS, Tiburcio AF (2012) Improving crop resistance to abiotic stress. Wiley-VCH Verlag GmbH & Co. KGaA, Weinheim

Tyermann SD, Skerrett M, Garril A, Findlay GP Leigh RA (1997) Pathways for the permeation of Na and Cl into protoplasts derived from the cortex of wheat roots. J Exp Bot 48:459–480

Wang S-M, Zhang J-L, Flowers TJ (2007) Low-Affinity Na^+ Uptake in the Halophyte Suaeda maritima, Plant Physiology 145:559–571. https://doi.org/10.1104/pp.107.104315

Wang CM, Zhang JL, Li Z, Wu GQ, Cai JY, Flowers TJ, Wang SM (2009) Puccinellia tenuiflora maintains a low Na level under salinity by limiting unidirectional Na influx resulting in a high

Wegner LH, Boer AH (1997) Properties of two outward rectifying channels in root xylem parenchyma cells suggest role in K homestasis and long distance signalling. Plant Physiology 115(4):1707–1719. https://doi.org/10.1104/pp.115.4.1707

Wu H (2018) Plant salt tolerance and Na^+ sensing and transport. Crop J 6(3):215–225

Yang Y, Guo Y (2018a) Elucidating the molecular mechanisms. Unraveling salt stress signaling in plants. J Integr Plant Biol 60(9):796–804.523–539

Yang Y, Guo Y (2018b) Elucidating the molecular mechanisms mediating plant salt-stress responses. New Phytol 217:523–539

Yang Z, Wang C, Xue Y, Liu X, Chen S, Song CP, Yang Y, Guo Y (2019) Calcium-activated 14-3-3 proteins as a molecular switch in salt stress tolerance. Nat Commun 10:1199

Yokoi S, Quintero FJ, Cubero B, Ruiz TM, Bresson RA, Pardo JM (2002) Differential expression and function of Arabidopsis thaliana NHX/H antiporter in salt stress response. The Plant Journal for cell and molecular biology 35(5):529–539. https://doi.org/10.1046/j.1365-313x.2002.01309.x

Zhang JS, Xie C, Chen SY (1999) Expression of plasma membrane H-ATPase gene in response to salt stress in a rice salt tolerant mutant and its original variety. Theoretical and Applied Genetics 99:1006–1011

Zhu JK (2001) Cell signaling under salt, water and cold stresses. Curr Opin Plant Biol 4:401–406
Zhu JK (2016) Abiotic stress signaling and responses in plants. Cell 167:313–324
Zhu K et al (2016) Evolution of an intron-poor cluster of the CIPK gene family and expression in response to drought stress in soybean. Sci Rep 6(1):1–12
Zhu M et al (2017) Physiological and molecular mechanisms mediating xylem Na^+ loading in barley in the context of salinity stress tolerance. Plant Cell Environ 40(7):1009–1020

Chapter 14
Role of Quorum Sensing in Nutrient Acquisition and Synergistic Plant-Microbe Association

Syeda Shaima Meryem, Arshid Pervez, and Abdol Ghaffar Ebadi

14.1 Introduction

Sustainable agricultural practices are intimately tied with soil fertility and health. Rapid growth of human population and built environment, consequences of industrialization, and other anthropogenic activities have resulted in correlative upsurge with regard to marginalized crop lands (Pathak et al. 2018). Agricultural soil is a mixture of organic and inorganic material with different morphology and mineral contents. Location of soil will define the texture, chemical nature, organic content, nutrient availability, and its microbial diversity. Inhabitation of plants, microbes (fungi and bacteria), insects, and worms in soil formulates soil fertility and viability. Most of the barren soil inhabited with few living organisms is considered as "bulk soil." Plant roots and their secretions make "the rhizoplane," whereas counting 1–5 mm of their surrounding area is rhizosphere with elevated microbial abundance and diversity (Zhu et al. 2018; Hassan et al. 2019). Mature roots where metaxylem begins to conduct elements are usually covered with an amalgam of soil particles, plant roots, and microbial cell and their secretions (Helliwell et al. 2019).

Rhizosphere, its biotic and abiotic components, and their progressions have been an interest of study to soil and allied scientists, as they are related to plant and soil

S. S. Meryem (✉)
Department of Environmental Sciences, COMSATS University Islamabad, Abbottabad, Pakistan

Department of Environmental Sciences, The Women University Multan, Multan, Pakistan

A. Pervez
Department of Environmental Sciences, COMSATS University Islamabad, Abbottabad, Pakistan

A. G. Ebadi
Department of Agriculture, Jouybar Branch, Islamic Azad University, Jouybar, Iran

Q. Mahmood (ed.), *Sustainable Plant Nutrition under Contaminated Environments*, Sustainable Plant Nutrition in a Changing World,
https://doi.org/10.1007/978-3-030-91499-8_14

health. Soil and clay particles, micro- and macronutrients, plant root secretions, microorganisms, and other environmental parameters collectively describe the essence of interaction of biota with abiotic bodies. These interactions formulate some of the unique solid products and describe the essence of microbial community in the pedosphere. Input of microbial and plant root secretions articulates pedogenesis. Creation of a sustainable agriculture system is impossible without comprehensive knowledge of processes operating in soil rhizosphere. Hiltner in 1904 calculated the exact soil volume of the narrow zone, where plant roots and microbes interact, and he coined the term "rhizosphere." Conventionally, rhizosphere was considered as the plant root surroundings, in which the soil microbes are influenced by root secretions: the exudates. Major active interactions in rhizosphere are between microbes-plant, soil-plant, soil-microbe, and microbe-microbe (Qu et al. 2020). Later on, intensive studies on rhizosphere revealed root surroundings as many-sided interface and a center of biogeochemical cycle of carbon, nitrogen, phosphorus, and many other chemical elements. The three main components of rhizosphere: soil, plant roots, and microbes have no functional boundaries, where all the biota compete or sometime collaborate with each other for available nutrients. All the components of rhizosphere (roots, microbes, mobile nutrients, soil particles, chelates, moisture, and gaseous components) confined in a zone with clear fractions and distances, depending on the type of ecosystem. Extracellular products of microbial metabolism restrict all other components to stick together along the root surface and make a sheet (solid soil phase) called the rhizosheets (Pang et al. 2017). These solid soil phases protect the plant root surfaces from biotic and abiotic stresses; lead to efficient nutrient cycling, decaying, and deposition of dying cells as organic matter; intensify survival of micronutrients; and ultimately define soil characteristics and plant health. The transport of nutrients through soil solution and solid phases of rhizosphere is supported by mycorrhiza (root-associated fungus), bacterial biofilms, soil moisture diffusion, and mass transfer (Basirat et al. 2019). The complex nature of rhizosphere is due to its myriad microbial biodiversity, with profoundly dynamic interactions (positive/negative) between plant roots and soil-borne microbes. In natural ecosystem, these interactions have found to pose huge impact on the plant vigor and biomass, nutritional uptake, response to environmental, stress and gene expressions (Marasco et al. 2018).

14.2 Rhizosphere Soil

Rhizosphere is the zone, surrounding plant root, influenced by microbes and plant secretions. Layers of top and subsoil host plant roots, and microbiota either present free or keep associated with plant roots, make this precinct a live soil. Rhizosphere soil poses vital progression factors for living roots, e.g., nutrients, moisture, and pH. Plant roots secrete carbon-based metabolites as root exudates, which are added in nutrient availability for soil microbiota (Zhang et al. 2017; Hassan et al. 2019). As reaction to availability and suitability of growth factors, microorganisms associate with plant roots, utilize its exudates, and in return shape up the growth of plant in a variety of ways (synergistic, antagonistic). Response of plants toward these interactions could be seen in their growth and development, nutrient dynamics, and vulnerability to biotic and abiotic stress (Chen et al. 2019).

The soil surrounding the plant roots is divided into distinct layers based on the distance to plant roots. Ectorhizosphere is the zone of soil adjacent to outer surface of plant root almost 2–3 cm, whereas the layer which includes the cortical layer and outermost endodermis is known as endorhizosphere, rhizoplane is the area of endodermis that remains covered with layer of mucilage of polysaccharides secreted by root-associated bacteria (Ur Rehman et al. 2020). The layer of mixture of extrapolymeric substances (EPS), root exudates, soil particles, root-associated microbes makes an amalgam and adheres to root surface in rhizoplane, this envelop around root surface is known as rhizosheath. This layer provides protection to plant roots against biotic and biotic stress in different ways. Plant roots and associated microbes that colonize the adjacent soil influence the rhizoplane and not the bulk soil: the non-rhizosphere soil. These natural phenomena are observed in in vivo hydroponic cultures where nutrient solution adjacent to root showed different chemistry as compared to bulk solution (Clairmont et al. 2019; Yates et al. 2021). To confirm this effect, there is a need to experiment more and overcome the reported methodological hurdles.

Increment in microbial population around rhizosphere is driven by the carbon loss from the root hairs and tips. The organic compounds released by the plants vary in different plant, crop, or their cultivars and consequently modulate the quantity and quality of microbial population around plant roots. Conversely, the microbial population adhering to plant roots or present freely in rhizosphere, affects the plant growth and chemistry of live soil (Deng et al. 2018). Microbial influence on plant growth is reported in different studies, i.e., nutrition uptake, susceptibility to disease, degradation of xenobiotic compounds, bioaccumulation of heavy metals, and resistance to biotic and abiotic stresses. These natural interactions can be exploited to achieve coveted outcomes (Hassan et al. 2019).

14.2.1 A Living Soil

Sustainability of soil is dependent on continuity of soil ecosystem in a resilient manner. Understanding of soil bionetwork would decrease dependency of artificial soil nutrients and aid in upgrading of agricultural practices and plant conservation. Soil aptitude to resist and cope with any stress is infinite, only the screening of live soil's resistance and flexibility toward controlling and perturbation is a matter of concern. Agriculture output to an extent to fulfill requirement of exponentially growing population is a scorching issue (Manter et al. 2017).

Soil pampers diverse micro- and macro living communities, as well indirectly impacts insects and larger organisms. Owing to life-supportive nature of soil, it is considered as black box. Microorganisms hosted by soil are envisaged to resist any environmental pressure and continue to reprocess nutrients (Lehmann et al. 2020). Contamination of soil with organic or inorganic pollutants and global warming shake the ecology of soil; however, some microorganisms have been observed to continue to survive and help soil and plants to cope with stress. Intra- and inter-species interactions however investigated to modulate microbial functional characteristics, nutrient cycling, plant development, and so agriculture productivity. Restricted agriculture productivity could affect human population growth (Srivastava et al. 2017; Wołejko et al. 2020).

14.3 What Does a Plant Need from Below the Ground?

Plant receives O_2, CO_2, and light in phyllosphere, whereas rest of the bio essential sustenance is directly delivered by the rhizosphere. Hence, plants must possess strong root system to support the above ground half of the plant. Physical parameters of soil that need to be appropriate are optimum pH, temperature, moisture, and aeration; moreover, soil should not host plant pathogens and toxic substances. Soil is considered fertile when it maintains the balance of all essential macro- and micronutrients before and after accommodating crop (Laliberté 2017). Life-essential elements, provided by the soil, required to plants in high amount are N, S, P, K, Ca, and Mg, and trace elements required are B, Cu, Cl, Mn, Fe, Zn, Mo, Ni. Plant root cells have special gates to allow those abovementioned elements to enter inside the cell, and elemental soil pollutants may pass those gates by mimicking essential micro element. Ionic toxic elements limit the availability of essential nutrients and stunted the plant growth in all possible ways. In this scenario, plant-microbe interactions come up with solution and protect plants and induce resistance against biotic and abiotic stress. Predominantly bacteria within rhizosphere prevent accumulation of toxic substances in plants and help plant root to acquire biologically essential micronutrients only (Paz et al. 2020).

Inorganic phosphate (Pi) is a limiting factor for plant growth owing to its ability to regulate photosynthesis, energy-driven mechanisms, root maturity, cell division, etc. (Kalayu 2019). Any fluctuation in optimum soil pH (6–7.5) sponsors chelation of phosphorus with soil's parent material such as iron, aluminum, and calcium. Natural form of phosphate, i.e., $H_2PO{-}4H_2PO_4$ in rhizosphere becomes unavailable to plant due to following reasons, 1. if sorbed by soil clay and oxides of Al and Fe in B horizon of live soil, 2. bound via precipitation with soil secondary minerals (Ca, Fe, Al), 3. run off as a consequent of soil erosion (Zheng et al. 2017). To cope with low availability of soluble phosphate or other nutrients, plants have evolved strategies, i.e., increment in carbohydrate secretion from root, alteration in root morphology, spurting of lateral root, mounting root height, and its spread to explore maximum soil, efficacy of Pi carrying proteins. In addition, plant roots excrete phosphatase and RNase to free-bound and immobilized phosphate in soil (Kafle et al. 2019). However, studies showed that besides fighting against Pi deficiency plants fail to cope, whereas efficient root colonization of mycorrhizae, fungi, and bacteria assist plants in immobilization of phosphate. Rhizosphere offers a series of microbe for plant root association and consequently solubilization and availability of all minerals required by plants.

14.4 Underground Plant-Microbial Associations

Root exudates attract a variety of microbial population in the rhizosphere; hence, they show bulk microbial biomass and activity. Microbes firmly attach to plant roots, nodules, and insert and reside inside root tissues, with variable functions and consequences. To explore diversity of rhizosphere bacteria is a challenge because many species are unable to grow in vivo, thus difficult to scrutinize (de Vries and Wallenstein 2017). Young roots host large number of bacterial communities known as r-strategies, whereas upon maturity of roots a shift is observed from dominance of bacterial communities to enhanced affectivity along with deliberate growth rate. Bacteria colonize mature roots firmly and degrade complex root exudates more effectively are known as k-strategies. In general, quantification of bacteria is easy; the hurdle is faced when it is time to study community structure and communal function (Ho et al. 2017).

Advanced molecular genetic studies using polymerase chain reaction and DGGE analysis reveal that the structure of rhizosphere bacterial communities changes as per change in time, distance with plant root, root zones, quality and quantity of root exudates, and plant species. Root colonizing bacteria exhibited less communal diversity as compared to rhizosphere and bulk soil; moreover, soil type also seems to control diversity. Same plant species exhibited different bacterial communities in rhizosphere, e.g., field crop; this variation can be seen both spatially and temporally. Healthy roots show less diversity than pathogen-infected plant root which make it more prone to bacterial outbreak (Badri et al. 2009).

Root exudates vary according to molecular weight, low molecular weight secretions of root cells include amino acids (proline, glutamic acid, valine, lysine etc.), organic acids (benzoic acid, p-hydroxybenzoic acid, and coumaric acid), phenolic compounds (Vives-Peris et al. 2020) and some secondary metabolites, and high molecular weight secretions include proteins and polysaccharides. Plants release 20% of their photosynthetic product, e.g., polysaccharides and proteins, into the soil via root cells, which help to establish plant-microbe interactions. Rhizosphere supports a wide variety of bacteria that can stimulate development of plant. These bacteria are known as PGPR (plant growth promoting rhizobacteria), possess plant growth supporting traits including phosphate solubilization, fixation of atmospheric nitrogen, and production of phytohormones and antimicrobial agents. These considerations established the importance of studying rhizospheric microbial community structure influenced by root exudates. A variety of microbes including fungus, virus, and bacteria reside in rhizosphere and formulate the fate of plant development (Hayat et al. 2017).

Root cell metabolites of low carbon number, amino acids, mucilage of polysaccharides, cell lysates, and secreted enzymes deposited on the root surface are collectively known as rhizodeposition. Total carbon assimilation by a healthy plant is observed to be 10%, whereas nutrient's stressed roots assimilate 44% of net carbon (Vives-Peris et al. 2020). Soil carbon turnover in rhizosphere is directed to rhizodeposition, which becomes a strong interface exchange route between soil and roots. Trio interaction of plant, bacteria, and soil is bridged over rhizodeposition of carbon. Microbes colonize the root and utilize accessible rich carbon source. Bacteria sense hospitality of roots and establish protective symbiotic relationship with plants. Bacteria secrete antibiotics as well induce resistance in plants against pathogenic associations, thereby secure their habitat (Kenawy et al. 2019).

In response to quantity and quality of root exudates, bacteria alter and articulate their community structure and produce various plant's valuable substances. Rhizobacterial population, among other organisms, is given much importance owing to their role in maintenance of root health, trace element uptake, tolerance of biotic abiotic stress, and crop responses (Korenblum et al. 2020). Previous researches reported that plant-bacteria interactions help plant roots to maintain their vigor, to sequester soil carbon, and to regulate element retrieval by degrading parent compounds. PGPR promote plant development by providing essential minerals (P, K, Fe, N), secrete phytohormones, inhibit or decrease deadly effects of pathogens, thereby named as biocontrol agents, biofertilizers, root colonizers, biofilm formers, and environmental protector (Basu et al. 2021).

PGPR may produce plant beneficial substance extracellular (ePGPR) or intracellular (iPGPR) depending on genetic makeup of bacterial species. The iPGPR localize in root nodules of leguminous plants, whereas ePGPR colonize outer root surface facing rhizosphere on the rhizoplane, or partially invade furrows between root cortex cells (Ur Rehman et al. 2020). The iPGPR reside inside the root cell and are termed as endophytes; moreover, in combination with root nodules they fix atmospheric nitrogen. *Rhizobium* genus from the family of Rhizobiaceae is an endophyte, once invaded it forms the root nodules (Korenblum et al. 2020). The genera of *Azorhizobium*, *Frankia*, *Mesorhizobium*, and *Allorhizobium* are likewise endophytes in nature. Root colonizing soil bacteria include genera of *Bacillus*, *Erwinia*, *Burkholderia*, *Azotobacter*, Choromobacterium, and Serratia with the ability to protect root surface as well secrete plant beneficial substances outside the cell (Yu et al. 2019).

Plants and bacteria in rhizosphere communicate with each other via chemicals; sense provision, absence of any nutrient or fluctuation in any environmental factors; and respond accordingly. The delivering of chemicals or enzymes in plant-microbe interaction (PMI) depends on the purpose of interaction (pathogenic or synergistic) (Stringlis et al. 2019). The primary method of chemical release is via diffusion through a transitional medium (amorphous or aqueous) as soluble compounds that may release internally or in an extracellular way. Intracellular chemical secretion of PGPR will encounter the elemental pollutants or organics after engulfing them via vesicle formation or membranous channels (Mohanram and Kumar 2019). PGPR's delivery of chemicals outside the cell may be volatile (2, 3 butanediol, dimethyl disulfide), inorganic, or organic in nature. Traditionally, chemical nature, solubility, and role of extracellular secretions of singular species were investigated; however, bulk chemical's secretions of bacteria as community either dual or multiple species were largely overlooked. However, if studies on PMI will continue to focus on relationship between PGPR secretions and physiology of plants by employing analytical tools, we expect to get more insights in the role of bacterial assemblies and their communal secretions (Ho et al. 2017).

PGPR are also observed to induce resistance in vulnerable plants against biological or environmental stress. Many soils display their ability to suppress the pathogen's attack, which was firmly found to relate with the manifestation of PGPR. The phenomenon of induction of systematic resistance (ISR) observed by different investigator by PGPR includes exudation of antibiotics, direction competition with disease-causing agents, plant growth promotion in a way that plant itself could be strong enough to fight with disease (Verma et al. 2018; Kannojia et al. 2019; Kenawy et al. 2019). Thereby PGPR produce phytohormones, antioxidants; solubilize soil minerals; produce scavengers of overburden; and produce toxic plant secretions. Collectively these activities backup plants to achieve resistance against pathogen. Relative contribution of each of these mechanisms is resolutely based on the stress type, host plant, and bacterial species, which could be best in any one of these mechanisms. Furthermore, genetic makeup of bacterial species will describe the way and extent of assistance to plants. Some PGPR are good antimicrobial agents, whereas others might be good in inducing resistance, nutrient cycling, or production of growth stimulators (Ho et al. 2017; Parewa et al. 2018).

To understand rhizosphere environment and PMI, it is important to dissect chemical mode of action of singular PGPR, and their multicomponent activity. To overcome issues of pest management, nutrient deficiency, global warming affects, and prerequisite of high biomass production, there is a need to identify, understand, and develop environmental-friendly compounds to assist plants in combating these issues. In the past few decades, researchers approached adaptation of chemical profiling of bacterial metabolites and their bioactive agents using advanced molecular genetic techniques. Plant growth promoting traits are necessary to observe and exploit for sustainable agriculture production in future and are discussed hereby (Akram et al. 2017; Ho et al. 2017; Ur Rehman et al. 2020).

14.5 Metabolites of Rhizobacteria

Plant-bacteria interaction for decades was investigated with a hope to manipulate the interaction in future for biologically controlling the phytopathogens, primarily bacteria and fungi. Thereby many casual rhizobacterial species were isolated, identified, and characterized for plant growth promoting traits. Studies were focused on one perspective at one time, to see which trait is more responsible for synergistic relationship and benefitting more (Massalha et al. 2017). Various types of interactions between plants and rhizobacteria have been presented in Fig. 14.1.

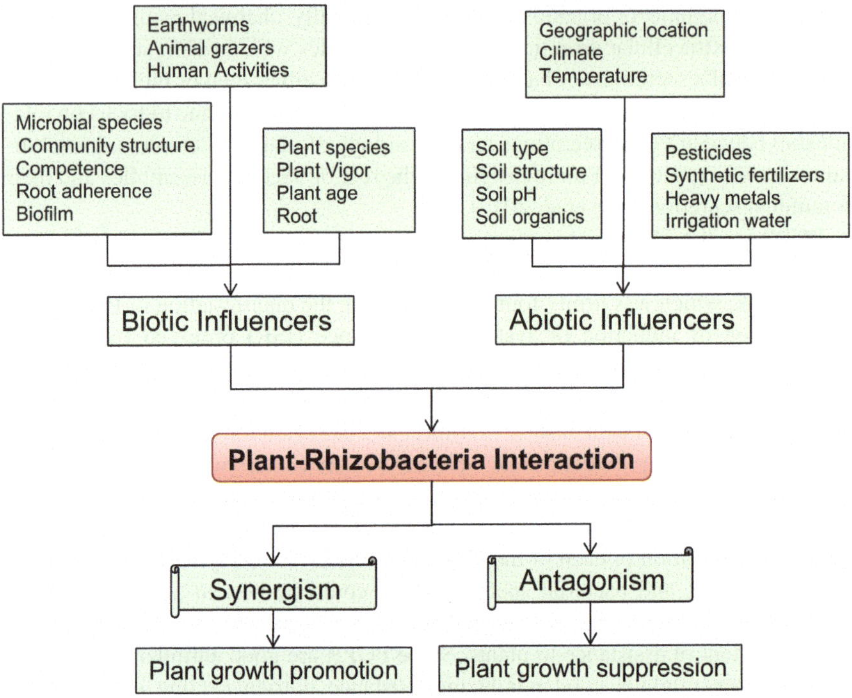

Fig. 14.1 Plant-rhizobacterial interaction

Antimicrobial assays were carried out in vitro to see the inhibition ability of PGPR against pathogens, PGPR grown near to pathogen (dual culture) in solid growth media petri dishes, and the growth inhibition zone by PGPR measured and assessed for resistance extent. These bacteria were later on announced as biocontrol agents that kill pathogens without getting in physical contact, just with the help of antibiotic-natured metabolites (Kenawy et al. 2019). Fractionation of these metabolites to find out their active agents revealed an array of antimicrobial compounds, e.g., chitinase enzymes which degrade outer cell covering, nuclease enzymes which degrade nucleic acid molecules, or some metabolites dehydrate their target organism to an extent which can lead to their death. To identify the antibiotics, chromatographic techniques such as sodium dodecyl sulphate-polyacrylamide gel electrophoresis (SDS-PAGE) mass spectroscopy and high performance column chromatography (HPLC) have been used (Jiao et al. 2021). Rhizobacteria when grown in liquid cultures, release hundreds of soluble metabolites, which need to be differentiated and identified. Thereby new advanced techniques are in research market for better chemical fabrication of metabolites, i.e., electrospray ionization coupled with mass spectroscopy (ESI-MS), and matrix-assisted laser desorption ionization time-of-flight tandem mass spectrometry (MALDITOF-MS/MS). Bands in SDS-PAGE and peaks in spectrophotometric analysis further matched with commercially available standard chemicals (Singh et al. 2021; Mhlongo et al. 2018, 2021; Jayaprakashvel et al. 2019).

The detrimental relationship of bacteria with other pathogenic microbes is known as antibiosis interaction, which along with ISR role has been focused while studying soluble metabolites of rhizobacteria. There are many hurdles faced while pointing out exactly the singular species, its mechanism, and metabolite responsible for killing pathogen and inducing resistance in plants. On the one hand, only 1.5% of soil microbe is culturable in vitro; the rest are still undiscovered, which surely interfere with PGPR and play their role in rhizosphere. Metagenomic studies of rhizobacterial community revealed more than 33 K species in disease-suppressive rhizosphere surpass the diversity found in disease-prone/conducive rhizosphere. Gigantic diversity means more diverse metabolites repertoire and varying magnitude of their working and affectivity (Ijaz et al. 2019; Valverde et al. 2016; Singh et al. 2021). Bacterial genera showed disease-suppressive ability against phytopathogens (fungi and bacteria) including Xanthomonadales, Actinobacteria, Burkholderiaceae, and Firmicutes. Published literature on identified rhizobacterial diversity is considered as the tip of an iceberg, thereby a lot more way to go to study rhizobacterial diversity, community structure, and communal behavior as antibiosis. Antibiosis and ISR are the direct ways to support plant in combating pathogens, below we will discuss indirect ways that PGPR communities harbor to assist plants (Fernández-González et al. 2019; Yadav et al. 2018; Priya et al. 2021).

14.5.1 Soluble PGPR Metabolites Induce ISR

Induced systemic resistance (ISR) is the priming of defense mechanisms in plants by means of rhizosheaths (rhizobacterial colonization). PGPR's colonies secrete ISR activating extracellular polymeric compounds with distinct biochemical nature (secondary metabolites, carbohydrate, proteins, and long and short chain lipids) as revealed by molecular and genetic analysis. The cell wall of rhizobacteria has carbohydrate and protein in their compositions, which are not found to involve in secreting biochemical signals, and that is why more focus is on lipopeptides's role in ISR (Kannojia et al. 2019; Meena et al. 2020). Surfactins and fengycins secreted extracellular by rhizobacteria were found to induce ISR in different plant varieties, e.g., *Vigna*, *Triticum*. Fengycins produced by *Bacillus* species were studied for its antibiosis and were positively correlated with ISR activation. Seed treatment with *Bacillus subtillus* revealed ISR in host plants by activating their oxylipin genes, whereas surfactin-defective species were compromised in ISR (Paraszkiewicz et al. 2017; Al-Ali et al. 2018; Ali et al. 2020). Iturin secreted by rhizobacteria is an extracellular polymeric substance (EPS) that activates resistance genes. Expression of defense-related enzymes (chitinase, glucanase, lipoxygenase), antioxidants (oxidase, peroxidase, superoxide dismutase), and proteins also owned to rhizobacterial EPSs. These abovementioned secretions are good against abiotic stress, whereas abiotic stress, e.g., presence of heavy metals and organic pollutants still need to be studied (Saha et al. 2020; Mohammed 2018; Etesami and Maheshwari 2018).

14.5.2 PGPR Traits

Many researchers have studied plant-microbe interaction and revealed several mechanisms posed by PGPR to help host plant in coping biotic and abiotic stress in combination. PGPR traits observed were mineral nutrition availability (Fe, Cu, Zn, Mn), ISR, production and release of phytohormones (IAA, abscisic acid, ethylene), enzymes (nitrogenase, ACC deaminase, phosphatase, dehydrogenase), carbohydrates, lipids, and metabolites (hydrogen cyanide, fungycin, siderophore). Owing to disease suppression, improved nutrient availability, and plant growth stimulation activities, PGPR may be used as biocontrol agents, biofertilizers, and biostimulants, respectively (Oleńska et al. 2020; Patel and Saraf 2017; Fahad et al. 2015). Active compounds of root exudates formulate the contents and extent of primary and secondary metabolites of PGPR. Rhizobacteria (adhere to root surface) and endophytes (colonize interior of root tissues) may be exploited as biofertilizers by inoculating to seeds, applied to plant surface, or added to soil. Rhizobacteria interact with harsh environment directly and use combination of PGPR traits thereby it is more imperative to scrutinize all PGPR traits. Various PGPR found so far belong to genera of *Azotobacter*, *Bacillus*, *Pseudomonas*, *Enterobacter*, *Azospirillum*, *Serratia*, and *Micrococcus*. Besides the fact many researchers studied PGPR traits, there is a lack of scrutiny on mix culture response on host plant and interaction of single and variant species with each other and with host plant (Purwaningsih et al. 2021; Maheshwari et al. 2019; Soussi et al. 2016; Verma et al. 2019). Figure 14.2 shows various plant growth promoting traits of rhizobacteria.

14.5.3 Mineral Acquisition

In agriculture, macronutrients N2, P, and K are the limiting factors for crop production. Rhizobacteria facilitate provision of plant nutrients and trace minerals fixed in soil chelates and organic molecules. PGPR degrade organic compounds and solubilize the complex inorganic chelates thereby freeing the bound nutrients (Vamerali et al. 2021; Weber et al. 2018; Mogal et al. 2020).

14.5.4 Phosphate Solubilization

A competent PGPR must solubilize the essential phosphate and is known as phosphate solubilizing bacteria (PSB). Inorganic free phosphate in soil may bind and fix as Fe or Al phosphate and calcium phosphate in acidic and alkaline pH conditions,

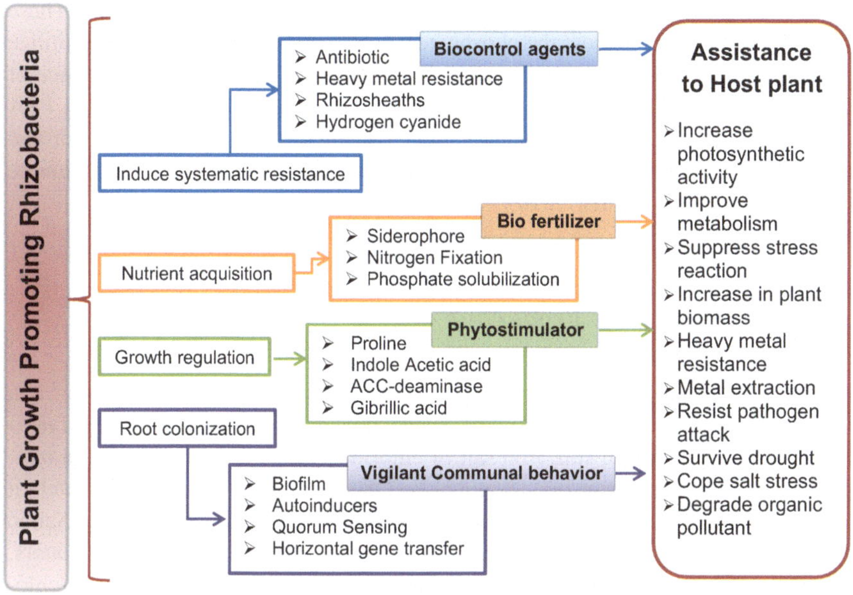

Fig. 14.2 Plant growth promoting traits of rhizobacteria

respectively (Elias et al. 2016). In Pakistan, salinity and soil alkalinity are major issues, thereby lead to deprived agriculture productivity, and coped via addition of inorganic phosphate artificially. Unbridled release of industrial effluent and its use as irrigation water decrease the soil pH, lead to Fe-P and Al-P complex, and deficiency of inorganic free soil phosphate. Inorganic phosphate is component of DNA and required for root development, vigor, strength of shoots, stem and stalk, maturity of plant till flower and seed formation, last of all for resistance to biotic stress (Rafique et al. 2017; Majeed et al. 2015; Manzoor et al. 2017).

Rhizobacteria release short carbon chain organic acids in soil to solubilize inorganic phosphate complex. Gluconic acid and citric acid are known to actively release inorganic pyro-phosphate (pp_i) as compared to citric and ketogluconic acid. Organic phosphate, on the other hand, released by bacterial enzymes, and different enzymes reported so far are phytase, phosphatase, and phosphonoacetate (Wei et al. 2018).

Hydrolysis of phosphoric esters (organic phosphate) in soil is achieved by bacterial enzyme hydrolase and is abundantly produced by rhizobacteria. Population of phosphate solubilizing bacteria comprised up to 50% of total soil bacteria, as well overcomes the fungal counterpart with the same ability. Although fungal species produce highly acidic nature phosphate solubilizing compounds (PSC), transverse extensive rhizosphere area, thereby not suitable for other soil constraints and maintenance of optimum soil pH. On the other hand, PSC produced by bacteria have the ability to solubilize both organic and inorganic phosphate complexes (Billah et al. 2019; Arif et al. 2017). Rhizobacteria species such as *Bacillus*, *Burkholderia*, *Azotobacter*, and *Serratia* are best known to have the maximum PSC producing activity. These species do not solely possess the phosphate solubilizing activity (PSA) rather utilize an array of plant growth promoting and ISR activities. Thereby in the presence of other functional PGPR traits, it is difficult to assess and measure the role of PSA alone. Many in vitro seed inoculation studies resulted in enhanced plant vigor and mineral accumulation as compared to controlled uninoculated set of experiment. (Sharon et al. 2016; Kumar 2016).

14.5.5 Siderophore Production

Iron (Fe^{3+}) is an essential macronutrient required by all living organisms; it is a limiting factor in crop lands; and plants needed Fe^{3+} for metabolic systems and optimum cellular growth. In soil, iron salt of chlorides tends to make chelates in oxygenic states and transforms into insoluble $Fe(OH)_3$ and oxyhydroxides, ultimately it lowers its availability for plants and soil organisms (Pii et al. 2016; Rout and Sahoo 2015). Rhizobacteria produce ligands accompanied siderophores; iron chelators, which bound the soil iron prior to any complex formation or chelation of iron chloride, with soil chemicals. Siderophore is low molecular weight organic compound that helps in solubilization of iron chelates (Khan et al. 2018; Niehus et al. 2017).

Different rhizobacterial species were observed to produce siderophore of different chemical nature, some of the widely found are catecholates, carboxylates, and hydroxamates; all of them were different due to different ligands attached to them. Plants coordinate with siderophore response and assimilate Fe^{3+} via ligand exchange reaction or uptake siderophore- Fe^{3+} complex directly. *P. fluorescent* produce greenish yellow pigmentation in agar medium, identified as pyoverdines which under UV light gives fluorescence; later termed as siderophore (Khan et al. 2018; Niehus et al. 2017). Bacterial genera reported as siderophore positive includes *Pseudomonas*, *Enterobacter*, *Rhizobium*, *Bacillus*, and *Azotobacter*. Application of *P. fluorescens* to *Arabidopsis thaliana* resulted in improved iron acquisition, photosynthetic activity, growth, and vigor. Similar results were observed in in vitro studies of PGPR *Bacillus* treatment to maize, *A. thaliana* cassava plants and *Pseudomonas* treatment to wheat, corn, rice, *Vigna radiata,* and chickpeas. All the abovementioned interactions revealed high Fe^{3+} bioaccumulation in host plants compared to uninoculated controls. Iron accumulation was not restricted to particular plant parts rather aerial parts and roots found same in Fe^{3+} concentrations; thereby, further focus was given on whole plant iron content (Kramer et al. 2020; Ferreira et al. 2019).

Sufficient iron increases plants' height, dry matter accumulation, branching of shoots, and lateral roots and diminishes effects of severe environmental conditions. PGPR also compete with pathogens via lowering the availability of iron, thereby indirectly help host plant to fight against phytopathogens (Pii et al. 2016; Rout and Sahoo 2015). Purified siderophore has been used as antifungal and antibacterial agent in plant growth media. Fungal pathogens that surrender against siderophore are *F. oxysporum*, *A. niger*, *A. flavus*, *A. oryzae,* and *S. rolfsii*. On occasions, siderophores are considered as secondary metabolites when they act as antibiotics. Siderophore-producing *Rhizobium* species often used not only as PGPR but also as biocontrol agent. Siderophore-producing PGPR were found effective against *F. oxysporum*, a deadly phytopathogen and enhanced plant biomass and root nodulation of leguminous plants (Patel et al. 2018). Siderophore positive *R. meliloti* was isolated from medicinal plant's rhizosphere and found to be effective against the attack of *M. phaseolina* (Patel et al. 2017; Brígido et al. 2019).

14.5.6 Nitrogen Fixation

Frankia alni is the filamentous species of actinomycetes, animate in nodules of actinorhizal plants, secrete the enzyme nitrogenase, and convert atmospheric nitrogen into useful plant fertilizer ammonia, in a process of biological nitrogen fixation (BNF). Otherwise, 78% of air nitrogen cannot be directly utilized or converted into beneficial form by any of the plant, so they need bacterial assistance in this regard. Besides *F. alni*, several other actinomycetes play their ecological role by fixing atmospheric nitrogen and are active members of rhizosphere microbial communities (Van Nguyen and Pawlowski 2017). Several species from the genera of *Actinoplanes*, *Micromonospora*, and *Actinomadura* are actinomycetes and stimulate plant growth directly or indirectly. Commercially used root colonizing bacteria as biocontrol agents against fungal pathogens belong to the genera of *Streptomycetes*, *Spirillospora*, *Streptoverticillium*, and *Streptosporangium* (Averill et al. 2019; Lehnert et al. 2018; Mukherjee and Sen 2021). Various *Bacillus* species (*B. brevis*, *B. cereus*, *B. licheniformis*, *B. subtilis*) have the ability to convert inert N_2 into plant-utilizable NH_4, using their nitrogenase catalytic activity. Endophytic bacteria colonize inside the tissues of root nodules and perform BNF activity (Igiehon and Babalola 2018).

14.5.7 Phytostimulators

PGPR could be an alternative to synthetic chemical fertilizers in crop land as they release growth stimulator signals in the form of phytohormones, which regulate every aspect of plant development. Phytohormones also shield the plant against biotic and abiotic stress in soil. Different phytohormones or plant growth regulators (PGRs) have been discovered, isolated, and characterized in different rhizobacteria. Following are the most popular PGRs: A tool of PGPR (Borriss 2020; Ortiz-Castro et al. 2017; Ortiz-Castro and López-Bucio 2019; Soni and Keharia 2021).

14.5.7.1 Auxin

Auxin or indole acetic acid (IAA) is a secondary metabolite of rhizobacteria, and 80% of studied rhizobacteria were found to synthesize and release extracellular IAA in rhizosphere soil. Production of IAA was linked with roots vigor, and inoculum of IAA-producing PGPR helps in the enhancement of root length, root hairs, spread of lateral root, directly tangled with nutrient uptake. Biosynthesis of IAA in rhizobacteria is dependent on tryptophan, an amino acid natured forerunner, which is released by plants in roots exudates (Yousef 2018). IAA biosynthesis pathway studies revealed independency of some bacterial species on tryptophan for IAA production. During plant-microbe interaction, IAA of bacterial origin defines the root-adhering strategies opted by the PGPR. Most often, IAA acts as a chemical signal and regulates phytostimulation and positive outcome of plant immunity (Wagi and Ahmed 2019; Mohite 2013). In literature, bacterial strains commencing *Rhizobium*, *Pseudomonas*, *Azotobacter*, and *Bacillus* genera were found proficient in IAA biosynthesis. Several in vitro PGPR inoculation studies indicated IAA as precursor of improved plant growth (Yousef 2018; Chen et al. 2017; Subair 2015; Bhutani et al. 2018; Wagi and Ahmed 2019).

14.5.7.2 Ethylene and ACC Deaminase

Ethylene ($H_2C{=}CH_2$) is a gaseous phytohormone and plays a vital role in seed germination, tissue differentiation, root elongation, fruit ripening, flowering initiation, opening and senescence, root gravitropism, and absorptive activity of cell membrane (Singh et al. 2015). All seed-bearing plants synthesize ethylene and use ACC synthase as a precursor for the biosynthesis of ethylene. However, fewer bacterial genera have been reported as producers of ethylene or any other plant growth controlling volatiles (Ali and Kim 2018). Some volatile compounds produced by bacteria were dependent on plant growth media, inoculum quantity, and IAA of bacterial origin, which help in the transcription of ACC synthase. In response to environmental stress, plants release excess ethylene known as stress ethylene, which inhibits the seedling growth and restricts IAA signaling and transport (Gamalero and Glick 2015; Ravanbakhsh et al. 2018). Bacteria produce ACC deaminase, which permit IAA stimulation by suppressing the ethylene-mediated IAA repression, ultimately resulting in improved plant growth. ACC deaminase along with IAA boosts up root colonization strategy, circumvention of pathogen attack, and plant growth stimulation (Saleem et al. 2007; Barnawal et al. 2014; Gamalero and Glick 2015).

14.5.8 Root Colonization and Biofilm

Bacteria start colonizing the plant root surface and ultimately establish biofilms to secure a solid stable root surface and in response of abundant nutrient availability around rhizoplane. Colonization activity depends on plant type, bacterial species, and bacterial genetic traits. A multistep process involved in colonization, i.e., sensing nutrient availability, migration toward root surface, adherence and distribution, biofilm construction, and rhizosheath formation (Podile et al. 2014).

Pioneer colonizers are influenced by root exudates and its derived nutrients, i.e., amino acids, vitamin, nitrogenous basis, carbohydrates, and other secondary metabolites. Firm adhesion of rhizobacteria improves symbiotic relation via biofilm formation, phytostimulation, pollutant resistance, etc.; plants show response toward bacterial hosts and resist pathogens via signal communication while they welcome symbionts and allow them to deliver beneficial functions (Kang et al. 2014; Meliani and Bensoltane 2016). This preference generates competence in bacterial species for rhizospheric niches and nutrients. Population size of bacteria and its root attachment at right place and on right time indirectly endorses the biocontrol activity. PGPR exploitation as biofertilizers, biocontrol agents, growth stimulators, nutrient provider, and as bioremediators had restricted beneficial attainment so far. The main reason reported frequently was the completion of bacterial inoculum with numerous other microorganisms in natural environmental conditions (Rafique et al. 2015; Seneviratne et al. 2010).

Most PGPR did not deliver their best in field trials; however, in vitro plant-microbe interaction studies under controlled conditions showed successful results in all aspects. Hence, synergistic relation has been credited to bacterial incompetence to effectively adhere and colonize the plant root surface. In field trials, microbial competition and displacement were the main reasons behind partial colonization, which ultimately lead to inconsistent results and deprived ecological performance (Kang et al. 2014; Meliani and Bensoltane 2016; Ahmad and Husain 2017). Other colonization controlling factors reported under field experiments were chemistry of root exudates, bacterial biofilms and quorum sensing, soil type, seasonal fluctuation of environmental factors, plant genetics, plant tissues, PGP traits of bacteria, adequate bacterial population size, and inter- and intra-bacterial species communication. Bacterial traits that determine the adherence and colonization extents are motility, presence and absence of flagella or pilli, chemotaxis phenomena, structure of outer cell membrane and presence of lipid/carbohydrate moieties and OPrF proteins, and secretion of extra-polymeric substances. Studies aimed to enhance synergism via root colonization focused mostly on bacterial and host plant genetics, biotic and abiotic factors, and mono or multiple species inoculum (Xu et al. 2019; Zhang et al. 2014; Kang et al. 2014).

Rhizospheric bacteria have the potential to invade the plant tissue and reside there and colonize inside tissues as endophytes. Endophytes secure their niches, adapt themselves in intimate association, and apparently cause no harm to their host plant. Exploitation of endophytes in sustainable agriculture production limits its range of application, as they reside inside and leave the host to fight against environmental stress (Azri et al. 2018).

Selection of PGPR starts from screening experiments and tailored according to the soil type and crop permutation. Scientists confound their studies on quantifying the contribution of PGPR traits in plant growth promotion under environmental stress. Environmental factors such as temperature, soil pH, heavy metal pollution,

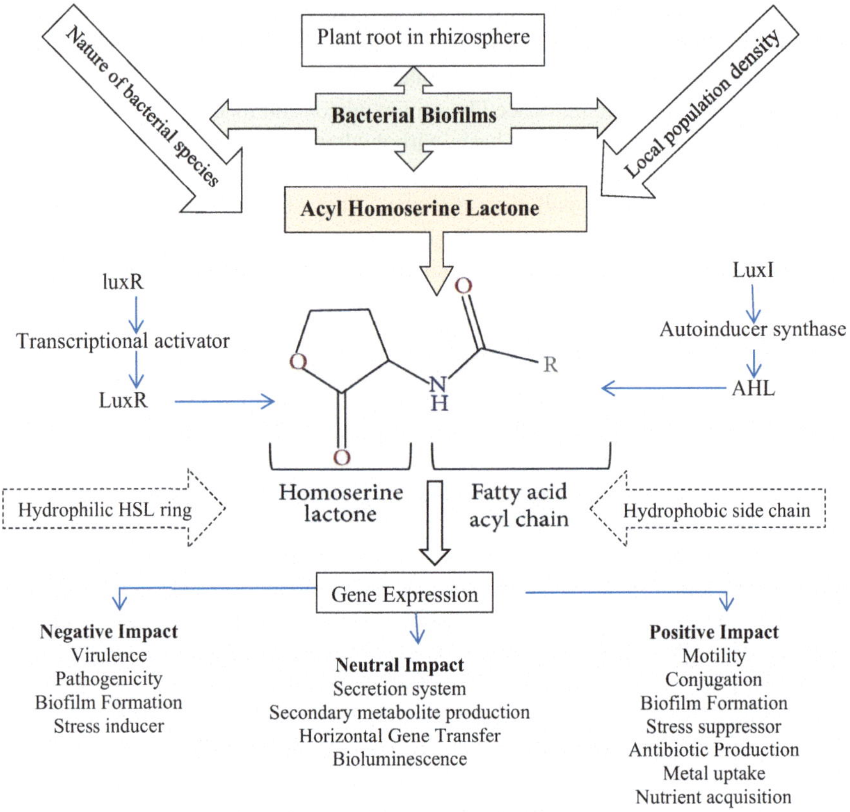

Fig. 14.3 Quorum sensing in bacterial biofilms

added fertilizers, and natural resident microflora are important growth controllers parallel to PGPR traits (Santoyo et al. 2021; Jha et al. 2018).

Rhizobacteria are found to produce signal molecules and contain quorum sensing system through which gene expression is being regulated. The root colonization is regulated by network of quorum sensing system, which depends on small molecules known as autoinducers produced and recognized by the network of rhizobacterial biofilm (Fig. 14.2). Among autoinducers, acyl homoserine lactones are widely studied which is found in gram-negative bacteria. Various researchers worked on chemistry of autoinducers and other signal molecules found in bacterial networks. Quorum sensing in a root-associated biofilm regulates and controls other traits of plant growth regulation and ultimately defines the mineral acquisition of plants from active rhizosphere. Several ecological and interdependent key characters of bacteria, such as antibiotic, siderophore, or enzyme secretion, virulence factors of phytopathogens, as well as plant-microbe communications, are coordinated through quorum sensing (QS). Quorum sensing in bacterial biofilms has been shown in Fig. 14.3.

References

Ahmad I, Husain FM (2017) Biofilms in plant and soil health. Wiley, Hoboken

Akram MS, Shahid M, Tahir M, Mehmood F, Ijaz M (2017) Plant-microbe interactions: current perspectives of mechanisms behind symbiotic and pathogenic associations. In: Plant-microbe interactions in agro-ecological perspectives. Springer, Singapore, pp 97–126

Al-Ali A, Deravel J, Krier F, Béchet M, Ongena M, Jacques P (2018) Biofilm formation is determinant in tomato rhizosphere colonization by Bacillus velezensis FZB42. Environ Sci Pollut Res Int 25:29910–29920

Ali S, Kim W-C (2018) Plant growth promotion under water: decrease of waterlogging-induced ACC and ethylene levels by ACC deaminase-producing bacteria. Front Microbiol 9:1096

Ali S, Hameed S, Shahid M, Iqbal M, Lazarovits G, Imran A (2020) Functional characterization of potential PGPR exhibiting broad-spectrum antifungal activity. Microbiol Res 232:126389

Arif MS, Shahzad SM, Yasmeen T, Riaz M, Ashraf M, Ashraf MA, Mubarik MS, Kausar R (2017) Improving plant phosphorus (P) acquisition by phosphate-solubilizing bacteria. In: Essential plant nutrients. Springer, Cham

Averill C, Bhatnagar JM, Dietze MC, Pearse WD, Kivlin SN (2019) Global imprint of mycorrhizal fungi on whole-plant nutrient economics. Proc Natl Acad Sci 116:23163–23168

Azri MH, Ismail S, Abdullah R (2018) An endophytic Bacillus strain promotes growth of oil palm seedling by fine root biofilm formation. Rhizosphere 5:1–7

Badri DV, Weir TL, Van der Lelie D, Vivanco JM (2009) Rhizosphere chemical dialogues: plant–microbe interactions. Curr Opin Biotechnol 20:642–650

Barnawal D, Bharti N, Maji D, Chanotiya CS, Kalra A (2014) ACC deaminase-containing Arthrobacter protophormiae induces NaCl stress tolerance through reduced ACC oxidase activity and ethylene production resulting in improved nodulation and mycorrhization in Pisum sativum. J Plant Physiol 171:884–894

Basirat M, Mousavi SM, Abbaszadeh S, Ebrahimi M, Zarebanadkouki M (2019) The rhizosheath: a potential root trait helping plants to tolerate drought stress. Plant Soil 445(1):565–575

Basu A, Prasad P, Das SN, Sadaf Kalam RZ, Sayyed MSR, El Enshasy H (2021) Plant growth promoting rhizobacteria (PGPR) as green bioinoculants: recent developments, constraints, and prospects. Sustainability 13:1140

Bhutani N, Maheshwari R, Negi M, Suneja P (2018) Optimization of IAA production by endophytic Bacillus spp. from Vigna radiata for their potential use as plant growth promoters. Isr J Plant Sci 65:83–96

Billah M, Khan M, Bano A, Hassan TU, Munir A, Gurmani AR (2019) Phosphorus and phosphate solubilizing bacteria: keys for sustainable agriculture. Geomicrobiol J 36:904–916

Borriss R (2020) Phytostimulation and biocontrol by the plant-associated bacillus. In: Phytomicrobiome in stress regulation, vol 1. Springer, Singapore

Brígido C, Menéndez E, Paço A, Glick BR, Belo A, Félix MR, Oliveira S, Carvalho M (2019) Mediterranean native leguminous plants: a reservoir of endophytic bacteria with potential to enhance chickpea growth under stress conditions. Microorganisms 7:392

Chen B, Luo S, Wu Y, Ye J, Wang Q, Xu X, Pan F, Khan KY, Feng Y, Yang X (2017) The effects of the endophytic bacterium Pseudomonas fluorescens Sasm05 and IAA on the plant growth and cadmium uptake of Sedum alfredii Hance. Front Microbiol 8:2538

Chen Y, Palta JA, Wu P, Siddique KH (2019) Crop root systems and rhizosphere interactions. Plant Soil 439:1–5

Clairmont LK, Stevens KJ, Slawson RM (2019) Site-specific differences in microbial community structure and function within the rhizosphere and rhizoplane of wetland plants is plant species dependent. Rhizosphere 9:56–68

de Vries FT, Wallenstein MD (2017) Below-ground connections underlying above-ground food production: a framework for optimising ecological connections in the rhizosphere. J Ecol 105:913–920

Deng S, Ke T, Li L, Cai S, Zhou Y, Liu Y, Chen L, Zhang D (2018) Impacts of environmental factors on the whole microbial communities in the rhizosphere of a metal-tolerant plant: Elsholtzia haichowensis Sun. Environ Pollut 237:1088–1097

Elias F, Woyessa D, Muleta D (2016) Phosphate solubilization potential of rhizosphere fungi isolated from plants in Jimma Zone, Southwest Ethiopia. Int J Microbiol 2016(3):1–11

Etesami H, Maheshwari DK (2018) Use of plant growth promoting rhizobacteria (PGPRs) with multiple plant growth promoting traits in stress agriculture: action mechanisms and future prospects. Ecotoxicol Environ Saf 156:225–246

Fahad S, Hussain S, Bano A, Saud S, Hassan S, Shan D, Khan FA, Khan F, Chen Y, Chao W (2015) Potential role of phytohormones and plant growth-promoting rhizobacteria in abiotic stresses: consequences for changing environment. Environ Sci Pollut Res 22:4907–4921

Fernández-González AJ, Villadas PJ, Cabanás CG-L, Valverde-Corredor A, Belaj A, Mercado-Blanco J, Fernández-López M (2019) Defining the root endosphere and rhizosphere microbiomes from the World Olive Germplasm Collection. Sci Rep 9:1–13

Ferreira MJ, Silva H, Cunha A (2019) Siderophore-producing rhizobacteria as a promising tool for empowering plants to cope with iron limitation in saline soils: a review. Pedosphere 29:409–420

Gamalero E, Glick BR (2015) Bacterial modulation of plant ethylene levels. Plant Physiol 169:13–22

Hassan MK, McInroy JA, Kloepper JW (2019) The interactions of rhizodeposits with plant growth-promoting rhizobacteria in the rhizosphere: a review. Agriculture 9(7):142

Hayat S, Faraz A, Faizan M (2017) Root exudates: composition and impact on plant–microbe interaction. In: Biofilms in plant and soil health. Wiley, Chichester, pp 179–193

Helliwell JR, Sturrock CJ, Miller AJ, Whalley WR, Mooney SJ (2019) The role of plant species and soil condition in the structural development of the rhizosphere. Plant Cell Environ 42(6):1974–1986

Ho Y-N, Mathew DC, Huang C-C (2017) Plant-microbe ecology: interactions of plants and symbiotic microbial communities. In: Plant ecology–traditional approaches to recent trends. IntechOpen Limited, London, pp 93–119

Igiehon NO, Babalola OO (2018) Rhizosphere microbiome modulators: contributions of nitrogen fixing bacteria towards sustainable agriculture. Int J Environ Res Public Health 15:574

Ijaz M, Shahzadi R, Rahman M-u, Iqbal M (2019) Molecular approaches to study plant growth-promoting rhizobacteria (PGPRs). In: Plant growth promoting rhizobacteria for agricultural sustainability. Springer, Singapore

Jayaprakashvel M, Chitra C, Mathivanan N (2019) Metabolites of plant growth-promoting rhizobacteria for the management of soilborne pathogenic fungi in crops. In: Secondary metabolites of plant growth promoting rhizomicroorganisms. Springer, Singapore

Jha P, Panwar J, Jha PN (2018) Mechanistic insights on plant root colonization by bacterial endophytes: a symbiotic relationship for sustainable agriculture. Environ Sustain 1:25–38

Jiao X, Takishita Y, Zhou G, Smith DL (2021) Plant associated rhizobacteria for biocontrol and plant growth enhancement. Front Plant Sci 12:420

Kafle A, Cope KR, Raths R, Krishna Yakha J, Subramanian S, Bücking H, Garcia K (2019) Harnessing soil microbes to improve plant phosphate efficiency in cropping systems. Agronomy 9(3):127

Kalayu G (2019) Phosphate solubilizing microorganisms: promising approach as biofertilizers. Int J Agron 2019:7

Kang Y, Shen M, Yang X, Cheng D, Zhao Q (2014) A plant growth-promoting rhizobacteria (PGPR) mixture does not display synergistic effects, likely by biofilm but not growth inhibition. J Microbiol 83:666–673

Kannojia P, Choudhary KK, Srivastava AK, Singh AK (2019) PGPR bioelicitors: induced systemic resistance (ISR) and proteomic perspective on biocontrol. In: PGPR amelioration in sustainable agriculture. Elsevier, Amsterdam

Kenawy A, Dailin DJ, Abo-Zaid GA, Malek RA, Ambehabati KK, Zakaria KHN, Sayyed RZ, El Enshasy HA (2019) Biosynthesis of antibiotics by PGPR and their roles in biocontrol of

plant diseases. In: Plant growth promoting rhizobacteria for sustainable stress management. Springer, Singapore
Khan A, Singh P, Srivastava A (2018) Synthesis, nature and utility of universal iron chelator–Siderophore: a review. Microbiol Res 212:103–111
Korenblum E, Dong Y, Szymanski J, Panda S, Jozwiak A, Massalha H, Meir S, Rogachev I, Aharoni A (2020) Rhizosphere microbiome mediates systemic root metabolite exudation by root-to-root signaling. Proc Natl Acad Sci 117:3874–3883
Kramer J, Özkaya Ö, Kümmerli R (2020) Bacterial siderophores in community and host interactions. Nat Rev Microbiol 18:152–163
Kumar A (2016) Phosphate solubilizing bacteria in agriculture biotechnology: diversity, mechanism and their role in plant growth and crop yield. Int J Adv Res 4:116–124
Laliberté E (2017) Below-ground frontiers in trait-based plant ecology. New Phytol 213(4):1597–1603
Lehmann J, Bossio DA, Kögel-Knabner I, Rillig MC (2020) The concept and future prospects of soil health. Nat Rev Earth Environ 1(10):544–553
Lehnert N, Dong HT, Harland JB, Hunt AP, White CJ (2018) Reversing nitrogen fixation. Nat Rev Chem 2:278–289
Maheshwari DK, Saraf M, Dheeman S (2019) Plant growth-promoting rhizobacteria (PGPR) as protagonists of ever-sustained agriculture: an introduction. In: Field crops: sustainable management by PGPR. Springer, Cham
Majeed A, Kaleem Abbasi M, Hameed S, Imran A, Rahim N (2015) Isolation and characterization of plant growth-promoting rhizobacteria from wheat rhizosphere and their effect on plant growth promotion. Front Microbiol 6:198
Manter DK, Delgado JA, Blackburn HD, Harmel D, de León AAP, Honeycutt CW (2017) Opinion: why we need a national living soil repository. Proc Natl Acad Sci 114(52):13587–13590
Manzoor M, Kaleem Abbasi M, Sultan T (2017) Isolation of phosphate solubilizing bacteria from maize rhizosphere and their potential for rock phosphate solubilization–mineralization and plant growth promotion. Geomicrobiol J 34:81–95
Marasco R, Mosqueira MJ, Fusi M, Ramond JB, Merlino G, Booth JM, Maggs-Kölling G, Cowan DA, Daffonchio D (2018) Rhizosheath microbial community assembly of sympatric desert speargrasses is independent of the plant host. Microbiome 6(1):1–18
Massalha H, Korenblum E, Tholl D, Aharoni A (2017) Small molecules below-ground: the role of specialized metabolites in the rhizosphere. Plant J 90(4):788–807. Wiley Online Library
Meena M, Swapnil P, Divyanshu K, Kumar S, Tripathi YN, Zehra A, Marwal A, Upadhyay RS (2020) PGPR-mediated induction of systemic resistance and physiochemical alterations in plants against the pathogens: current perspectives. J Basic Microbiol 60:828–861
Meliani A, Bensoltane AJ (2016) Biofilm-mediated heavy metals bioremediation in PGPR pseudomonas. J Bioremed Biodegr 7:2
Mhlongo MI, Piater LA, Madala NE, Labuschagne N, Dubery IA (2018) 'The chemistry of plant–microbe interactions in the rhizosphere and the potential for metabolomics to reveal signaling related to defense priming and induced systemic resistance. J Front Plant Sci 9:112
Mhlongo MI, Piater LA, Steenkamp PA, Labuschagne N, Dubery IA (2021) Metabolomic evaluation of tissue-specific defense responses in tomato plants modulated by PGPR-priming against Phytophthora capsici infection. Plants 10:1530
Mogal C, Jha S, Suthar H, Parekh V, Rajkumar BK (2020) Efficiency of Plant Growth Promoting Rhizobacteria (PGPR) consortia for modulation of phytohormone and better nutrient acquisition. Plant Sci 3:20–29
Mohammed AF (2018) Effectiveness of exopolysaccharides and biofilm forming plant growth promoting rhizobacteria on salinity tolerance of faba bean (Vicia faba L.). Afr J Microbiol Res 12:399–404
Mohanram S, Kumar P (2019) Rhizosphere microbiome: revisiting the synergy of plant-microbe interactions. Ann Microbiol 69:307–320

Mohite B (2013) Isolation and characterization of indole acetic acid (IAA) producing bacteria from rhizospheric soil and its effect on plant growth. J Soil Sci Plant Nutr 13:638–649
Mukherjee R, Sen S (2021) Agricultural sustainability through nitrogen fixation: approaches and techniques. Harvest 6:48–55
Niehus R, Picot A, Oliveira NM, Mitri S, Foster KR (2017) The evolution of siderophore production as a competitive trait. Evolution 71:1443–1455
Oleńska E, Małek W, Wójcik M, Swiecicka I, Thijs S, Vangronsveld J (2020) Beneficial features of plant growth-promoting rhizobacteria for improving plant growth and health in challenging conditions: a methodical review. Sci Total Environ 743:140682
Ortiz-Castro R, López-Bucio J (2019) Phytostimulation and root architectural responses to quorum-sensing signals and related molecules from rhizobacteria. Plant Sci 284:135–142
Ortiz-Castro R, López-Bucio JS, López-Bucio J (2017) Physiological and molecular mechanisms of bacterial phytostimulation. In: Singh HB, Sarma BK, Keswani C (eds) Advances in PGPR research. CABI, São Paulo, pp 16–28
Pang J, Ryan MH, Siddique KH, Simpson RJ (2017) Unwrapping the rhizosheath. Plant Soil 418(1):129–139
Paraszkiewicz K, Bernat P, Siewiera P, Moryl M, Paszt LS, Trzciński P, Jałowiecki Ł, Płaza G (2017) Agricultural potential of rhizospheric Bacillus subtilis strains exhibiting varied efficiency of surfactin production. Sci Hortic 225:802–809
Parewa HP, Meena VS, Jain LK, Choudhary A (2018) Sustainable crop production and soil health management through plant growth-promoting Rhizobacteria. In: Role of rhizospheric microbes in soil. Springer, Singapore
Patel T, Saraf M (2017) Biosynthesis of phytohormones from novel rhizobacterial isolates and their in vitro plant growth-promoting efficacy. J Plant Interact 12:480–487
Patel D, Patel S, Thakar P, Saraf M (2017) Siderophore producing Aspergillus spp as bioinoculant for enhanced growth of mung bean. Int J Adv Agric Sci Technol 6:111–120
Patel P, Trivedi G, Saraf M (2018) Iron biofortification in mungbean using siderophore producing plant growth promoting bacteria. Environ Sustain 1:357–365
Pathak J, Maurya PK, Singh SP, Häder DP, Sinha RP (2018) Cyanobacterial farming for environment friendly sustainable agriculture practices: innovations and perspectives. Front Environ Sci 6:7
Paz A, Gagen EJ, Levett A, Zhao Y, Kopittke PM, Southam G (2020) Biogeochemical cycling of iron oxides in the rhizosphere of plants grown on ferruginous duricrust (canga). Sci Total Environ 713:136637
Pii Y, Borruso L, Brusetti L, Crecchio C, Cesco S, Mimmo T (2016) The interaction between iron nutrition, plant species and soil type shapes the rhizosphere microbiome. Plant Physiol Biochem 99:39–48
Podile AR, Vukanti RVNR, Sravani A, Kalam S, Dutta S, Durgeshwar P, Papa Rao V (2014) Root colonization and quorum sensing are the driving forces of plant growth promoting rhizobacteria (PGPR) for growth promotion. Proc Indian Natl Sci Acad 80:407–413
Priya P, Aneesh B, Harikrishnan K (2021) Genomics as a potential tool to unravel the rhizosphere microbiome interactions on plant health. J Microbiol Methods 185:106215
Purwaningsih S, Agustiyani D, Antonius S (2021) Diversity, activity, and effectiveness of Rhizobium bacteria as plant growth promoting rhizobacteria (PGPR) isolated from Dieng, central Java. Iran J Microbiol 13:130
Qu Q, Zhang Z, Peijnenburg WJGM, Liu W, Lu T, Hu B, Chen J, Chen J, Lin Z, Qian H (2020) Rhizosphere microbiome assembly and its impact on plant growth. J Agric Food Chem 68(18):5024–5038
Rafique M, Hayat K, Mukhtar T, Khan AA, Afridi MS, Hussain T, Sultan T, Munis MFH, Imran M, Chaudhary HJ (2015) The battle against microbial pathogens: basic science technological advances and educational programs. In: Méndez-Vilas A (ed) Bacterial biofilm formation and its role against agricultural pathogens. Formatex, Badajoz, pp 373–382

Rafique M, Sultan T, Ortas I, Chaudhary HJ (2017) Enhancement of maize plant growth with inoculation of phosphate-solubilizing bacteria and biochar amendment in soil. Soil Sci Plant Nutr 63:460–469

Ravanbakhsh M, Sasidharan R, Voesenek LACJ, Kowalchuk GA, Jousset A (2018) Microbial modulation of plant ethylene signaling: ecological and evolutionary consequences. Microbiome 6:1–10

Rout GR, Sahoo S (2015) Role of iron in plant growth and metabolism. Rev Agric Sci 3:1–24

Saha I, Datta S, Biswas D (2020) Exploring the role of bacterial extracellular polymeric substances for sustainable development in agriculture. Curr Microbiol 77:1–16

Saleem M, Arshad M, Hussain S, Bhatti AS (2007) Perspective of plant growth promoting rhizobacteria (PGPR) containing ACC deaminase in stress agriculture. J Ind Microbiol Biotechnol 34:635–648

Santoyo G, Urtis-Flores CA, Loeza-Lara PD, Orozco-Mosqueda M, Glick BR (2021) Rhizosphere colonization determinants by Plant Growth-Promoting Rhizobacteria (PGPR). Biology 10:475

Seneviratne G, Weerasekara MLMAW, Seneviratne KACN, Zavahir JS, Kecskés ML, Kennedy IR (2010) Importance of biofilm formation in plant growth promoting rhizobacterial action. In: Plant growth and health promoting bacteria. Springer, Berlin

Sharon JA, Hathwaik LT, Glenn GM, Imam SH, Lee CC (2016) Isolation of efficient phosphate solubilizing bacteria capable of enhancing tomato plant growth. J Soil Sci Plant Nutr 16:525–536

Singh RP, Shelke GM, Kumar A, Jha PN (2015) Biochemistry and genetics of ACC deaminase: a weapon to "stress ethylene" produced in plants. Front Microbiol 6:937

Singh AD, Kour J, Dhiman S, Khanna K, Kumar P, Kaur R, Kapoor N, Bhardwaj R (2021) Prospects of PGPR-mediated antioxidants and S and P metabolism in plants under drought stress. In: Antioxidants in plant-microbe interaction. Springer, Singapore

Soni R, Keharia H (2021) Phytostimulation and biocontrol potential of Gram-positive endospore-forming Bacilli. Planta 254:1–24

Soussi A, Ferjani R, Marasco R, Guesmi A, Cherif H, Rolli E, Mapelli F, Ouzari HI, Daffonchio D, Cherif A (2016) Plant-associated microbiomes in arid lands: diversity, ecology and biotechnological potential. Plant Soil 405:357–370

Srivastava V, Sarkar A, Singh S, Singh P, de Araujo AS, Singh RP (2017) Agroecological responses of heavy metal pollution with special emphasis on soil health and plant performances. Front Environ Sci 5:64

Stringlis IA, De Jonge R, Pieterse CMJ (2019) The age of coumarins in plant–microbe interactions. Plant Cell Physiol 60:1405–1419

Subair H (2015) Isolation and Screening Bacterial Exopolysaccharide (EPS) from potato rhizosphere in highland and the potential as a producer Indole Acetic Acid (IAA). Proc Food Sci 3:74–81

Ur Rehman F, Kalsoom M, Nasir TA, Adnan M, Anwar S, Zahra A (2020) Chemistry of plant–microbe interactions in rhizosphere and rhizoplane. Indian J Pure App Biosci 8(5):11–19

Valverde JR, Gullón S, Mellado RP (2016) Looking for rhizobacterial ecological indicators in agricultural soils using 16S rRNA metagenomic amplicon data. PLoS One 11:e0165204

Vamerali T, Panozzo A, Visioli G, Dal Cortivo C (2021) Root–microbe interactions influencing water and nutrient acquisition efficiency. In: The root systems in sustainable agricultural intensification. Wiley, Hoboken, pp 159–192

Van Nguyen T, Pawlowski K (2017) Frankia and actinorhizal plants: symbiotic nitrogen fixation. In: Rhizotrophs: plant growth promotion to bioremediation. Springer, Singapore

Verma RK, Sachan M, Vishwakarma K, Upadhyay N, Mishra RK, Tripathi DK, Sharma S (2018) Role of PGPR in sustainable agriculture: molecular approach toward disease suppression and growth promotion. In: Role of rhizospheric microbes in soil. Springer, Singapore

Verma M, Mishra J, Arora NK (2019) Plant growth-promoting rhizobacteria: diversity and applications. In: Environmental biotechnology: for sustainable future. Springer, Singapore

Vives-Peris V, de Ollas C, Gómez-Cadenas A, Pérez-Clemente RM (2020) Root exudates: from plant to rhizosphere and beyond. Plant Cell Rep 39:3–17

Wagi S, Ahmed A (2019) Bacillus spp.: potent microfactories of bacterial IAA. Peer J 7:e7258
Weber NF, Herrmann I, Hochholdinger F, Ludewig U, Neumann G (2018) PGPR-induced growth stimulation and nutrient acquisition in maize: do root hairs matter. Sci Agric Bohem 49:164–172
Wei Y, Zhao Y, Shi M, Cao Z, Qian L, Yang T, Fan Y, Wei Z (2018) Effect of organic acids production and bacterial community on the possible mechanism of phosphorus solubilization during composting with enriched phosphate-solubilizing bacteria inoculation. Bioresour Technol 247:190–199
Wołejko E, Jabłońska-Trypuć A, Wydro U, Butarewicz A, Łozowicka B (2020) Soil biological activity as an indicator of soil pollution with pesticides–a review. Appl Soil Ecol 147:103356
Xu Z, Zhang H, Sun X, Liu Y, Yan W, Xun W, Shen Q, Zhang R (2019) Bacillus velezensis wall teichoic acids are required for biofilm formation and root colonization. Appl Environ Microbiol 85:e02116–e02118
Yadav AN, Verma P, Kumar S, Kumar V, Kumar M, Sugitha TCK, Singh BP, Saxena AK, Dhaliwal HS (2018) Actinobacteria from rhizosphere: molecular diversity, distributions, and potential biotechnological applications. In: New and future developments in microbial biotechnology and bioengineering. Elsevier, Amsterdam
Yates CF, Guo J, Bell TH, Fleishman SM, Bock HW, Trexler RV, Eissenstat DM, Centinari M (2021) Tree-induced alterations to soil properties and rhizoplane-associated bacteria following 23 years in a common garden. Plant Soil 461(1):591–602
Yousef NMH (2018) Capability of plant growth-promoting rhizobacteria (PGPR) for producing indole acetic acid (IAA) under extreme conditions. Eur J Biol Res 8:174–182
Yu K, Pieterse CMJ, Bakker PAHM, Berendsen RL (2019) Beneficial microbes going underground of root immunity. Plant Cell Environ 42:2860–2870
Zhang N, Wang D, Liu Y, Li S, Shen Q, Zhang R (2014) Effects of different plant root exudates and their organic acid components on chemotaxis, biofilm formation and colonization by beneficial rhizosphere-associated bacterial strains. Plant Soil 374:689–700
Zhang R, Vivanco JM, Shen Q (2017) The unseen rhizosphere root–soil–microbe interactions for crop production. Curr Opin Microbiol 37:8–14
Zheng BX, Hao XL, Ding K, Zhou GW, Chen QL, Zhang JB, Zhu YG (2017) Long-term nitrogen fertilization decreased the abundance of inorganic phosphate solubilizing bacteria in an alkaline soil. Sci Rep 7(1):1–10
Zhu D, Wei Y, Zhao Y, Wang Q, Han J (2018) Heavy metal pollution and ecological risk assessment of the agriculture soil in Xunyang mining area, Shaanxi Province, Northwestern China. Bull Environ Contam Toxicol 101(2):178–184

Index

Q. Mahmood (ed.), *Sustainable Plant Nutrition under Contaminated Environments*, Sustainable Plant Nutrition in a Changing World,
https://doi.org/10.1007/978-3-030-91499-8

D

E

N

O

P

Q

R

S

T

The manufacturer's authorised representative in the EU is Springer Nature Customer Service Centre GmbH, Europaplatz 3, 69115 Heidelberg, Germany. If you have any concerns regarding our products, please contact ProductSafety@springernature.com

Printed and bound by CPI Group (UK) Ltd, Croydon, CR0 4YY
30/09/2026
02225496-0010